AF619269

Carbon–Carbon Composites

Carbon–Carbon Composites

G. Savage

SPRINGER-SCIENCE+BUSINESS MEDIA, B.V.

First edition 1993

© 1993 G. Savage
Originally published by Chapman & Hall
Softcover reprint of the hardcover 1st edition 1993

Typeset in 10/12pt Times by Graphicraft Typesetters Ltd, Hong Kong

ISBN 978-94-010-4690-9

Apart from any fair dealing for the purposes of research or private study, or criticism or review, as permitted under the UK Copyright Designs and Patents Act, 1988, this publication may not be reproduced, stored, or transmitted, in any form or by any means, without the prior permission in writing of the publishers, or in the case of reprographic reproduction only in accordance with the terms of the licences issued by the Copyright Licensing Agency in the UK, or in accordance with the terms of licences issued by the appropriate Reproduction Rights Organization outside the UK. Enquiries concerning reproduction outside the terms stated here should be sent to the publishers at the London address printed on this page.

This publisher makes no representation, express or implied, with regard to the accuracy of the information contained in this book and cannot accept and legal responsibility or liability for any errors or omissions that may be made.

A catalogue record for this book is available from the British Library

Library of Congress Cataloging-in-Publication data enclosed

Savage, G. (Gary)
Carbon–carbon composites / G. Savage. — 1st ed.
p. cm.
Includes index.
ISBN 978-94-010-4690-9 ISBN 978-94-011-1586-5 (eBook)
DOI 10.1007/978-94-011-1586-5

1. Carbon composites. 2. Carbon fibers. 3. Fibrous composites.
I. Title.
TA418.9.C6S249 1992
620.1′93—dc20 92-27294
CIP

Printed on permanent acid-free text paper, manufactured in accordance with the proposed
ANSI/NISO Z 39.48–199X and ANSI Z 39.48–1984

Contents

For Grandad

Preface

Carbon fibre reinforced carbon composites form a very specialized group of materials. They may be considered as a development of the family of carbon fibre reinforced polymer composites which are becoming ever more prevalent in modern engineering. Since the early 1960s a large number of so-called 'advanced materials' have appeared on the scene. Carbon–carbon is arguably the most successful of all these products finding many and varied applications. In the field of Formula 1 motor racing for example, the present levels of performance simply could not be achieved without the use of carbon–carbon brakes and clutches. Despite the materials' obvious assets, they have not, and will not, reach their full potential until their inherent problems of excessive production costs and oxidation resistance have been addressed properly. In this respect the 'carbon–carbon story', of much potential but only limited success, serves as a lesson to all those involved in materials research, development and application.

In writing this book I have tried to set up a logical progression of what the materials are, how they are made, what their assets and deficiencies are, what they are used for and to what extent they are commercially exploited. Each specialized chapter may be considered in isolation or as part of a sequence, whereas the final chapter provides a summary of the principal concepts as well as a basic review of the economic situation past, present and, hopefully, future.

Carbon itself is a unique material around which a whole branch of science has developed. The first chapter introduces carbon–carbon composites with respect to their relationship to other carbon materials, from whence they derive many of their properties. This chapter should be of particular interest to those not familiar with carbon science in that it introduces many of the principles and much of the terminology expanded later in the text. Like any other fibre reinforced composite, the engineering and many of the physical properties of carbon–carbon are dominated by those of the reinforcing fibres. The second chapter therefore consists of an introduction to carbon fibres, covering their production and properties as well as the techniques used to process them into intermediate products such as fabrics and preforms. It is hoped that this review will be equally apposite to those involved in the 'conventional' composites industry. Each of the major production methods is covered in a separate chapter highlighting not only the attractive aspects of the technique but also the problems which have resulted in high costs and inefficiency. Aside from high cost the other major drawback of carbon–carbon is its susceptibility to oxidation. The problem is so acute that it merits a complete chapter. The oxidative degradation is covered in detail along with remedies which have been and are currently being developed.

One of the primary aims of this book is to aid the reader who wishes to engage in the research and development of carbon–carbon materials. In most texts on composites carbon–carbon (if mentioned at all) it is usually only given a few lines at the end of the book. As a redress, Chapter 7 consists of a detailed description of the techniques used to fabricate, test and analyse the materials on a laboratory scale. The remaining chapters cover the properties of the various forms of carbon–carbon, its uses and applications and finally a brief overview of the technology and the market and economic potential. Throughout the book, a theme is developed of carbon–carbon not being a single material but rather a whole family whose properties are capable of being tailored to meet specific applications. To this end, the chapter on materials' properties consists not of a list of the properties of commercial products, but a discussion of the principles governing the way properties are developed and may be altered.

In preparing this book a great deal of help in terms of information and photographs has been provided by a number of people and their organizations whom I would now like to thank very much, in no particular order: Gavin Gray of ICI, Shelley Gildersleve of Capricorn–Fulton, Pergamon Press, Elsevier Science publishers, Rob Simmonds of AP racing, Graham Goldsmith of Gaybo International Kayaks Ltd, Aoki-San from the Tonen Corporation, Chris Mellings of Ciba-Geigy, Perry Bruno and Neil Hansen of Hercules/RTAC, Dr Jim Williamson, Dr Rees Rawlings, Ian Davies and Margaret Campbell from Imperial College, Brian O'Rourke of Williams Grand Prix Engineering, Dr Richard Alexander of Mitsubishi Kasei Corporation, Steve Beebe, Peter Stulgaitis and Graham Rogers of Instron.

Many thanks are due to my friend Dr John Runnacles of Fiberite for his help in proof reading the manuscript and in provision of information expecially on CVD, to my 'oppo' Paul Cox for his painstaking draughting of many of the figures and Dr Jim Williamson of Imperial College for refereeing the text and helping to make it more 'readable'. Finally I would like to thank my father Malcolm Savage for arranging the typing and providing encouragement, along with my wife Liz, when my enthusiasm was flagging. Once again, thank you all very much.

Gary Savage
1992

Introduction

1

1.1 CARBON

Carbon has an atomic weight of 12.011 and is the sixth element in the periodic table. Three isotopes are known to exist, these being C^{12}, C^{13} and C^{14}, the first two of which are stable. C^{12} accounts for around 99% of the naturally occurring carbon and is used as the reference definition of atomic mass. It is defined as having a 'relative atomic mass' of 12[1]. C^{13} has a magnetic moment (spin = $\frac{1}{2}$) which results in its being used as a probe in nuclear magnetic resonance (NMR) studies, although its low abundance induces lengthy acquisition times. The radioactive isotope C^{14} is generated in the earth's upper atmosphere by the interaction of neutrons with nitrogen:

$$N^{14} + n \rightarrow C^{14} + H^{1}. \tag{1.1}$$

C^{14} has a very long half-life of 5730 years and is used extensively in the dating of archaeological artefacts and as a 'label' in the study of organic reaction mechanisms.

The properties of carbon-based materials depend upon its electronic configuration. Despite the electronic ground state of carbon being $1s^2$, $2s^2$, $2p^2$, energetic advantage is gained from involving all four outer orbital electrons in bonding between other atoms or carbon atoms themselves. Carbon displays 'catenation' (bonding to itself) to such a degree that the number of resulting chains, rings and networks are almost limitless.

1.2 BONDING IN CARBON MATERIALS

1.2.1 Quantum mechanics

Quantum or wave mechanics is based on the fundamental principle that electrons show the properties not only of particles, but also of waves (they may, for example, be diffracted), and as a result, therefore, a wave equation can be written for them, in the same sense that light and sound waves,

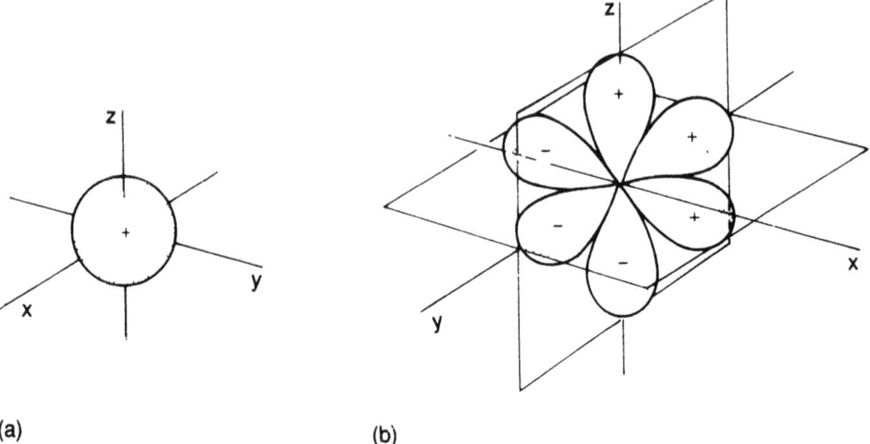

Fig. 1.1 (a) the 1s orbital, (b) the three 2p orbitals.

etc. may be described by wave equations. In 1926 Erwin Schrödinger, at the University of Zürich, derived a mathematical expression to describe the motion of an electron in terms of its energy. The so-called Schrödinger equation for a one-electron system is written

$$\frac{\partial^2\psi}{\partial x^2} + \frac{\partial^2\psi}{\partial y^2} + \frac{\partial^2\psi}{\partial z^2} + \frac{8\pi^2 m}{h^2}(E - V) = 0, \tag{1.2}$$

where m is the mass of the electron, E its total energy, V its potential energy and h Planck's constant.

A wave equation has a series of solutions called wave functions, each corresponding to a different energy level for the electron. In physical terms, the wave function ψ expresses the square root of the probability of finding the electron at any position defined by the coordinates x, y and z where the origin is at the nucleus.

The Schrödinger equation is a differential equation, solutions to which are themselves equations. Such solutions are, however, not differential equations, but simple equations for which graphical solutions can be drawn. The graphs are three-dimensional pictures of the electron density known as orbitals or electron clouds. Figure 1.1 shows the familiar shapes of the 1s and 2p atomic orbitals.

1.2.2 Atomic orbitals

A wave equation is not able to divulge the exact position of an electron at any time nor how fast it is moving. Similarly, it does not allow the plotting of a precise orbit about the nucleus. Rather it tells us the probability of finding the electron at any particular location. As previously stated, the region in space where an electron is likely to be found is called an orbital. There are many different types of orbital each with different sizes and shapes which are arranged about the nucleus in specific ways. The characteristics of the orbital occupied by an electron depend very much upon the energy of that electron. It is the sizes and shapes of these orbitals and their disposition with respect to one another which determine the arrangement in space of the atoms of a molecule and its chemical behaviour.

The most convenient way of picturing an orbital is that it is smeared out to form a cloud rather like a blurred photograph of the rapidly moving electron. The cloud is densest in those regions where the probability of finding the electron is highest, i.e. in those regions where the average negative charge, or electron density, is greatest. An orbital has no definite boundary since there is a probability, albeit a very small one, of finding the electron separated from the atom. That probability, however, decreases very rapidly beyond a certain distance from the nucleus so that the distribution of charge is fairly well represented by the electron clouds depicted in Fig. 1.1.

Each p orbital has a node, i.e. a region in space where the probability of finding the electron is extremely small. In Fig. 1.1 some lobes of the orbitals are labelled + and others –. The signs do not refer to positive or negative charges since both lobes of an electron cloud must, by definition, be negatively charged. Rather, they are the signs of the wave function. When two parts of any orbital are separated by a node they must always have opposite signs on the two sides of the node. There are a number of 'rules' that determine the way in which the electrons of an atom may be distributed. That is to say, there are certain conditions which govern the electronic configuration of an atom. The most fundamental rule is the Pauli exclusion principle: no more than two electrons may be present in any orbital and they must have opposite spins.

If it were possible to solve the Schrödinger equation for molecules containing two or more electrons, we would be able to acquire a precise picture of the shape of the orbitals available to each electron and the energy for each orbital. Unfortunately, exact solutions to the equation can only be obtained for one-electron systems such as the hydrogen atom. Since perfect solutions are not possible, drastic approximations must be made. There are two important methods of approximation of the electronic configuration of large atoms and molecules: the molecular-orbital method and the valence-bond method.

1.2.3 Covalent bonding

The molecular orbital method considers bonding to arise from the overlap of atomic orbitals. When any number of atomic orbitals overlap, they are

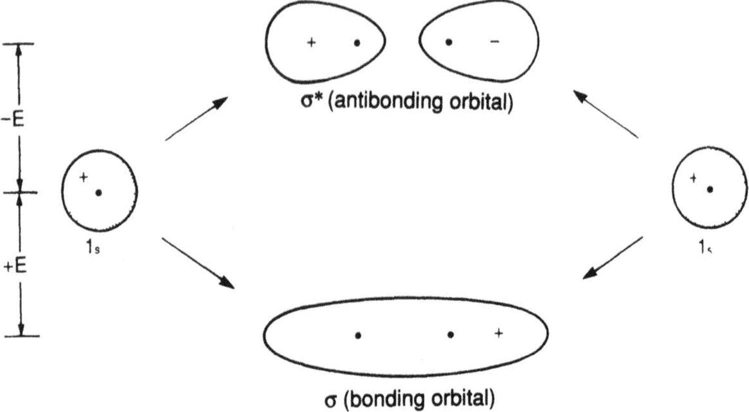

Fig. 1.2 The overlap of two 1s orbitals gives rise to a σ and a σ* orbital in the hydrogen molecule.

replaced by an equal number of new orbitals known as molecular orbitals. Molecular orbitals differ from atomic orbitals in that they are clouds which surround the nuclei of two or more atoms rather than just one as is the case with atomic orbitals. In a covalent bond, two atomic orbitals, each containing one electron, overlap so that two molecular orbitals are generated. One of these, known as a bonding orbital, has a lower energy than the original atomic orbitals, so that a bond may form. The other, known as an antibonding orbital, has a higher energy. Orbitals of lower energy fill up first. Since the two original atomic orbitals each held one electron, and any orbital can hold two electrons, both of these electrons can go into the new molecular bonding orbital. In the ground state the antibonding orbital remains empty. The greater the degree of overlap, the stronger will be the bond. Total overlap is prevented, however, by repulsion between the nuclei. Figure 1.2 illustrates the bonding and antibonding orbitals that arise by the overlap of two 1s electrons to form a hydrogen molecule. The antibonding orbital has a node between the nuclei. In that area there is, therefore, only negligible electron density such that bond formation is very unlikely. Molecular orbitals formed by the overlap of two atomic orbitals when the centres of electron density are on the axis common to the two nuclei are called sigma (σ) orbitals. Similarly, the bonds so formed are referred to as σ bonds. The corresponding antibonding orbitals are denoted σ*. σ bonds may be formed by the overlap of any of the different kinds of atomic orbital (s, p, d or f) whether the same or different. The two lobes that overlap must have the same sign. A positive s orbital can only form a bond by overlapping with another positive s orbital or with the positive lobe of a p, d or f orbital. Any σ orbital may be represented approximately as an ellipse irrespective of the kind of atomic orbitals from which it has arisen.

A univalent atom such as hydrogen has only one orbital available for

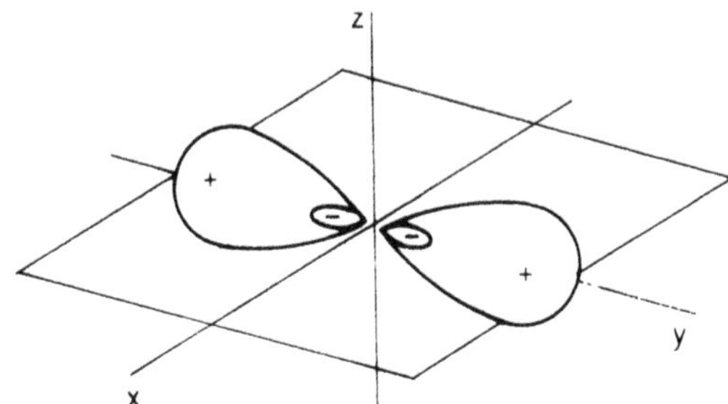

Fig. 1.3 The two sp hybrid orbitals formed by beryllium.

bonding. Atoms with a valence of two or more are required to form bonds by using at least two orbitals. An oxygen atom has two half-filled orbitals resulting in a valence of two. Single bonds are formed by the overlap of these orbitals with the orbitals of two other atoms. Similarly, nitrogen has three half-filled p orbitals and, therefore, forms three single bonds.

1.2.4 Hybridization

The electronic structure of beryllium $1s^2$, $2s^2$ has no unpaired electrons. Although there are no half-filled orbitals, beryllium has a valence of two and forms two covalent bonds. It is possible to account for this by imagining that one of the 2s electrons is 'promoted' to a vacant 2p orbital, producing a $1s^2$, $2s^1$, $2p^1$ configuration. There are now two half-filled orbitals but they are not equivalent. If bonding were to occur, beryllium would have a valence of two but the overlap of these orbitals with the orbitals of external atoms would not be identical. The bond formed from the 2p orbital would be more stable than that formed from the 2s orbital since a greater degree of overlap is possible with the former. In reality, beryllium forms two equivalent bonds as, for example, in beryllium chloride $BeCl_2$. This, more stable, situation, arises from the combination of the 2s and 2p orbitals to form two new orbitals that are equivalent as shown in Fig. 1.3.

The two new orbitals are called hybrid orbitals since they are a mixture of the two original orbitals. Each orbital is formed from the merger of an s and a p orbital and is, therefore, referred to as an sp orbital. The sp orbitals consist of a large lobe and a very small one. They are atomic orbitals but they only arise in the bonding process and in no way represent a possible structure for the free atom. A beryllium atom forms its two bonds by overlapping each of the large lobes shown in Fig. 1.3 with an orbital from an external atom. The external orbital may be any of the atomic orbitals previously considered or another hybrid orbital, provided they are of the same sign. The molecular orbital formed fits our previous definition and is described as a σ bond.

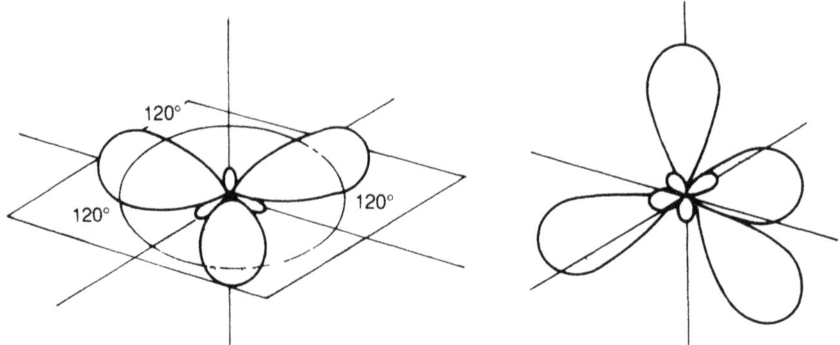

Fig. 1.4 The four sp^3 and three sp^2 bonding orbitals formed by carbon.

Carbon ($1s^2$, $2s^2$, $2p^2$) has an unpaired electron in each of the two p orbitals. One would, therefore, expect carbon to have a valence of two, forming compounds such as CH_2. Carbon does form CH_2 but it is a highly reactive molecule, whose properties centre about the need to provide carbon with two more bonds to achieve stability. The tendency is to form as many bonds as possible: in this case, to combine with four hydrogen atoms to form the compound methane, CH_4.

In order to provide four unpaired electrons it is necessary to promote one of the 2s electrons into the emply p orbital:

$$1s^2 2s^2 p_x^1 2p_y^1 \xrightarrow{\text{promotion}} 1s^2 2s^1 2p_x^1 2p_y^1 2p_z^1. \tag{1.3}$$

The most strongly directed orbitals are hybrid orbitals, sp^3, formed from the mixing of one s and three p orbitals:

$$1s^2 2s^1 2p_x^1 2p_y^1 2p_z^1 \xrightarrow{\text{hybridization}} 1s^2(sp^3)^4. \tag{1.4}$$

The four equivalent sp^3 orbitals point to the corners of a regular tetrahedron as shown in Fig. 1.4.

Consider now the ethene molecule (C_2H_4) in terms of the molecular orbital concepts. The carbon atom forms σ bonds with the three other atoms to which it is connected (one carbon and two hydrogen) using sp^2 orbitals (Fig. 1.4). The sp^2 orbitals arise from hybridization of the $2s^1$, $2p_x^1$ and $2p_y^1$ electrons of the promoted state. Each carbon also has another electron in the $2p_z$ orbital, which by the principle of maximum repulsion, lies perpendicular to the plane of the sp^2 orbitals. The two parallel $2p_z$ orbitals can overlap sideways to generate two new orbitals, a bonding and an antibonding orbital (Fig. 1.5). In the ground state, both electrons will go into the bonding orbital, the antibonding orbital remaining vacant. The molecular orbitals formed by the overlap of atomic orbitals whose axes are parallel are called π orbitals if they are bonding and $\pi*$ if antibonding.

In our model of ethene, the two orbitals that combine to form the double

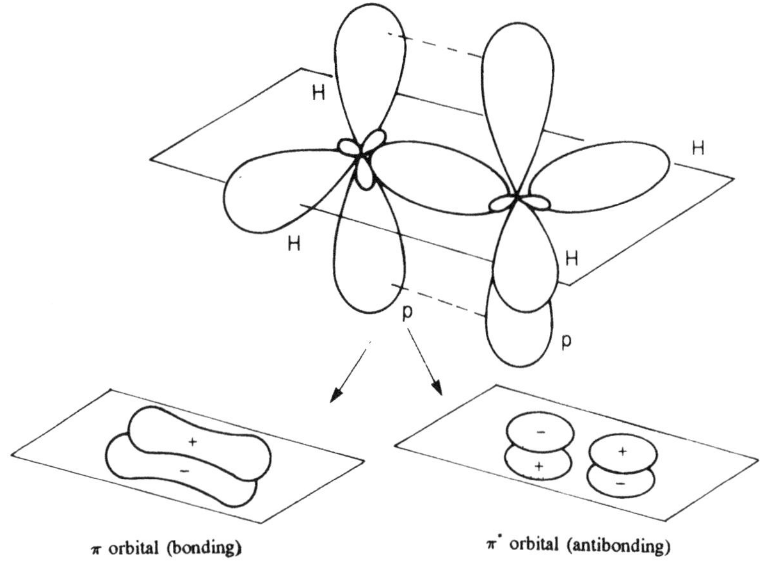

Fig. 1.5 Overlapping p orbitals form a π and a π* orbital.

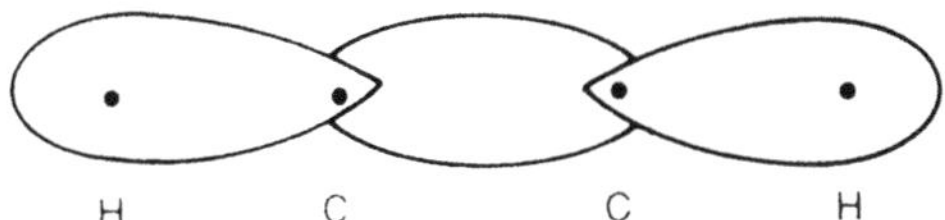

Fig. 1.6 The σ orbitals of the ethyne molecule.

bond are not equivalent. The σ orbital is ellipsoidal and symmetrical about the *c–c* axis. The π orbital has the shape of two ellipsoids, one above and the other below the σ orbital plane. The plane itself represents a node for the π orbital. In order to maintain maximum overlap of the p orbitals, they must be parallel. As a result, free rotation about the double bond is not possible, otherwise the two p orbitals would have to reduce their overlap to allow one H–C–H plane to rotate with respect to the other. The six atoms of a double bond are thus in a plane with angles of around 120°. Since maximum stability is obtained when the p orbitals overlap as much as possible, double bonds are shorter than the corresponding single bond. The double bonds formed between carbon and oxygen or nitrogen may be similarly represented, consisting of one σ and one π orbital.

In triple bond compounds such as ethyne (C_2H_2) carbon is connected to only two other atoms and, therefore, uses sp hybridization to form σ bonds. The four atoms lie in a straight line as shown in Fig. 1.6. Each carbon has two p orbitals remaining, with one electron in each, which are perpendicular to each other and to the *c–c* axis. They overlap as shown in Fig. 1.7 to form two π orbitals. A triple bond is thus composed of one σ and two π

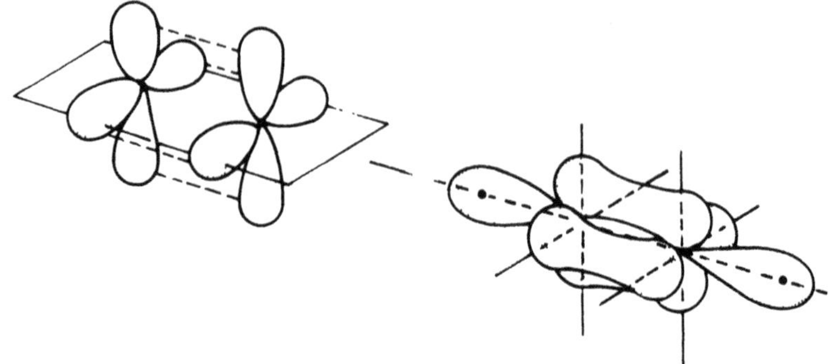

Fig. 1.7 Overlap of p orbitals to form a triple bond.

Table 1.1 Bond energies of common carbon bonds

Bond	*Bond energies* ($kJ\ mol^{-1}$)
C–C	3.47
C=C	6.11
C≡C	8.37
C–H	4.14
C–O	3.60
C=O	7.36
C–S	2.72
C–N	3.05
C–F	4.89
C–Cl	3.26
C–Br	2.72
C–I	2.38

orbitals. The triple bonds formed between carbon and nitrogen may be similarly represented.

1.2.5 The stability of carbon bonds

The formation of σ and π bonds between carbon and other atoms results in the complex and extensive range of structures to which a whole branch of chemistry has been devoted. The principal feature of organic chemistry is the stability of carbon bonds, in particular, the multiple bonding available via π orbitals. Table 1.1 illustrates the stability of the most common carbon bonds.

Bonding in carbon compounds is dominated by two principal regimes as described below.

1. *σ bonds only* – diamond or aliphatic type. This results in chains of carbon atoms such as polyolefines, or three-dimensional structures which are rigid and isotropic.

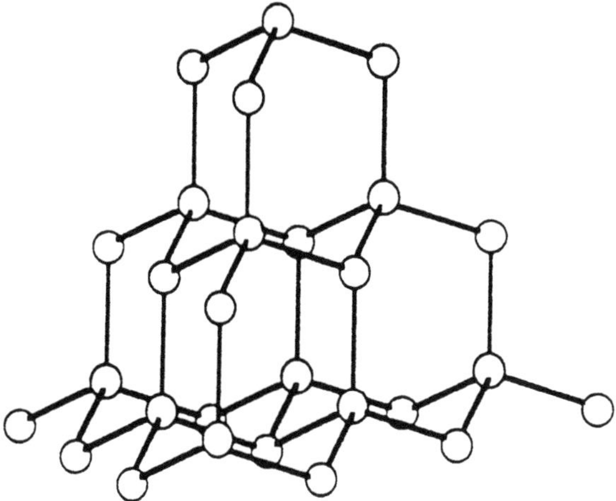

Fig. 1.8 The diamond crystal structure.

2. *A mixture of σ and π bonds* – graphite or aromatic type. This results in predominantly layered structures with a high degree of anisotropy.

The majority of carbonaceous materials contain examples of both bonding regimes with an immense range of complexity.

1.2.6 Crystal structures of carbon

Carbon exists in two regularly ordered crystalline forms, diamond and graphite. At ambient conditions, graphite is the most thermodynamically stable allotrope:

$$C_{(\text{diamond})} \rightarrow C_{(\text{graphite})} \; \Delta H = -2.1 \text{ KJ mol}^{-1}. \tag{1.5}$$

From a kinetic point of view, however, the change is extremely slow at normal temperatures due to the very large number of bonds which require to be broken in the process. The diamond to graphite transition is, though, rapid above 1600 °C.

Diamond

The diamond structure consists of a regular three-dimensional network of sp^3 σ bonds providing for a very rigid, stable tetrahedral structure (Fig. 1.8). As a result, diamond is the hardest material known. Bonding electrons within the diamond lattice are fixed between atoms such that electrical conductivity is very low, tending towards insulation. Diamond, because of its higher density (3.51 g cm^{-3} compared with 2.25 g cm^{-3} for graphite), is the most stable allotrope at high pressures (>600 GPa at 20 °C).

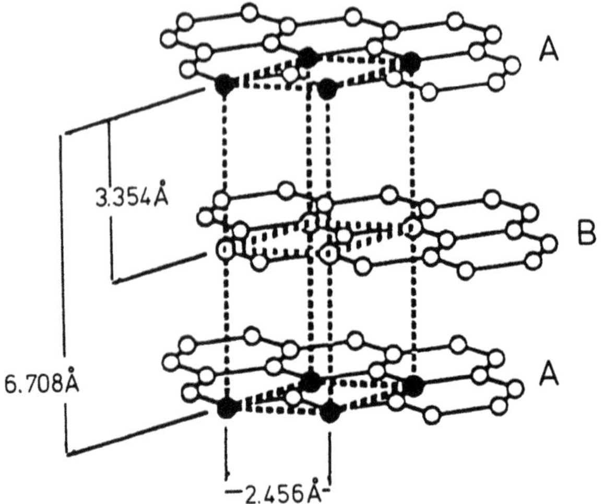

Fig. 1.9 The graphite crystal lattice.

Graphite

In the graphite structure the atoms are held together in two-dimensional hexagonal networks by a mixture of both sp^2 σ and π bonding. The layers are held together very loosely by weak van der Waals' forces. The layers are hexagonally stacked in the AB AB sequence (Fig. 1.9). A small amount of material is stacked according to ABC ABC. This material in known as the rhombohedral form and accounts for less than 10% of the graphite. Graphite is a very soft material with good lubricating properties, since the energy required to slide the layers over one another is very low. The properties of graphite tend to be very anisotropic as a result of its crystal structure. This anisotropic behaviour can be illustrated using the electrical conductivity of graphite. The conjugated π bonding within the layered configuration results in the delocalization of electrons throughout the structure. A means of electrical conductivity similar to the conduction band in metals is thus provided. In direct contrast, there is no electron movement across the layers such that conduction in that direction is at a minimum.

1.3 ORDER AND DISORDER IN CARBON MATERIALS

The overwhelming majority of carbon materials are a mixture of well-ordered material, often of short range, surrounded by disordered material. The proportions of, and relationship between, the ordered and disordered regions contribute greatly to the properties of the material.

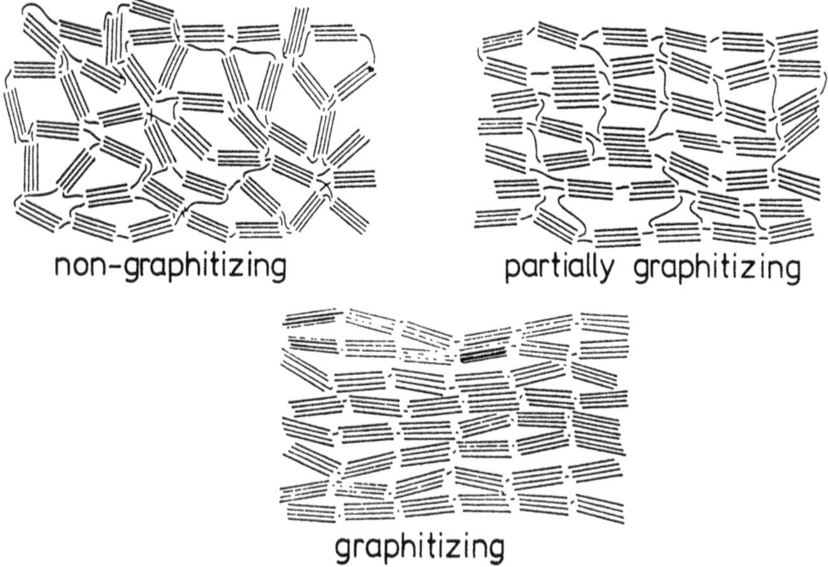

Fig. 1.10 The Franklin models of carbon structure [2].

1.3.1 Regular structures

Regular carbon structures possess what is generally considered to be a graphite lattice. Small volumes are often observed to exhibit an almost perfect graphite crystal structure. Increasing the size of that volume results in a corresponding increase in the presence of defects, distortions and heteroatoms which destroy the regularity and produce a very disordered material. It can be demonstrated that very little energy is needed to slide graphitic layers over one another. Similarly, twisting the layers so that they are no longer aligned is another possibility. The twisting leads to structures which have roughly parallel and equidistant layers but with random orientation. This arises from an irregular disposition of carbon atoms within a layered type of over-structure. A great deal of study has been carried out in order to categorize the degree of ordering within graphitic materials. Franklin, in the 1950s, derived an equation relating the d-spacing between the layers measured by X-ray diffraction to the degree of randomness in the alignment of the layers, denoted p [2].

$$d_{(002)} = 3.440 - 0.086\ (1 - p^2). \tag{1.6}$$

The work has been very closely scrutinized over the years but has stood the test of time. The models she proposed for carbon structure, illustrated in Fig. 1.10, are not dissimilar to those more elaborate descriptions which have been attempted since. The diamond-like parts of most carbon materials tend to have even shorter range than the graphitic regions, although a large proportion of the disordered parts are aliphatic in nature. A whole host of defects and irregularities are again present in any long-range structures.

1.3.2 Irregular structures

For the most part, carbon materials may be regarded as analogous to metal alloys in that they are made up of constituents which can be separately identified within the overall structure. The principal phases may be described as:

1. ordered or graphitic, in which case there is a high degree of anisotropy; or
2. disordered material which, under most characterization techniques, exhibits an isotropic structure.

The processes whereby disordered material is converted into an ordered state and vice versa are one of the primary concerns in the fabrication and properties of products made by the carbon industry in general and carbon–carbon composites in particular. Those processes, whether by heat treatment, ageing or some other method of energy input, are again analogous to metals production. They will be discussed in detail, where appropriate to carbon–carbon, later in this book. As a prelude, it is interesting to consider the gradual changes which occur during the carbonization of a disordered carbon material, such as a pitch coke, towards an ordered graphite structure. The process, illustrated in Fig. 1.11, illustrates the range of irregular structures which may be encountered in carbon science [3].

1.3.3 Range of order

The ordered material within carbons may be considered as crystallites of graphite. The range of order can be estimated by measuring the size and distribution of the crystallites. A consideration of the orientation of the crystallites is often necessary in the calculation, since the actual range of order may be several orders of magnitude higher than the size of the individual crystallites. The type and range of ordering considered within the structures will depend very much on the analysis technique employed. It is appropriate, therefore, to review the methods used to investigate carbon structures and to discuss the implications of their observations. When assessing the properties, performance and applications of carbon materials, the degree of isotropy and anisotropy present is of great significance. Although the macroscopic properties of many carbons are isotropic, their structure on a microscopic level exhibits a high degree of anisotropy. Further, the individual components of the structure may appear anisotropic to one examination technique but isotropic to another. This observation can be exemplified by considering the optical microstructure of a carbon

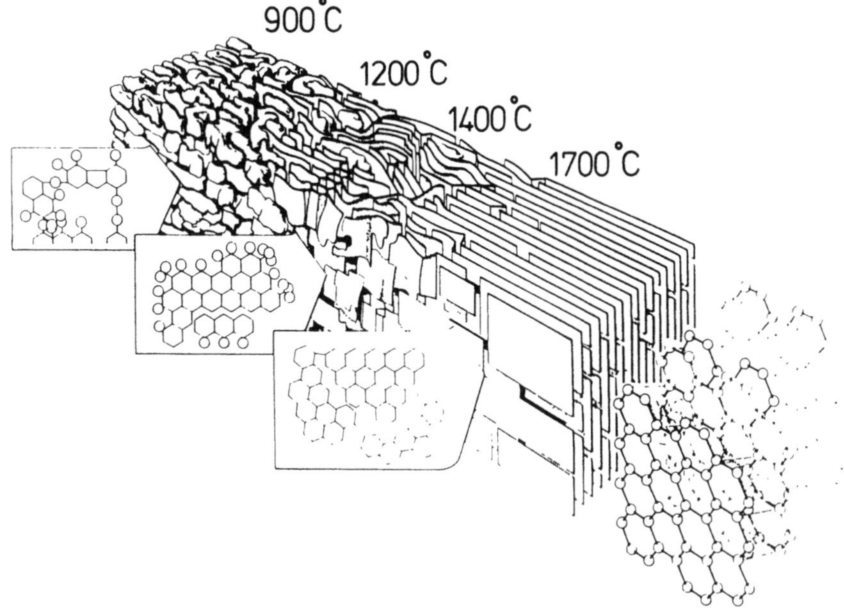

Fig. 1.11 A schematic model of the changes in the lamellar structure of a graphitizing carbon with increasing heat treatment temperature (HTT) (according to Griffiths and Marsh [3]).

when compared with an electron microscope image. Optical microscopes are limited to a resolving power of around 1 μm. Some structures may thus be discerned as isotropic using optical microscopy, whereas they may be shown to be highly anisotropic at the nanometre level observed in the electron microscope.

1.4 TECHNIQUES FOR CHARACTERIZING THE STRUCTURE OF CARBONS

1.4.1 Optical microscopy

Optical microscopy is used to obtain information by light transmission through or reflected from the surface of a material. Reflected light microscopy, as used in metallography, is a widely employed technique to study the interaction of compounds in multiphase carbon materials [4]. Magnified images up to 1000× may be produced, permitting examination of structures too small for unaided visual observation, to a limit of resolution of the order of 1 μm [5].

Specimens are prepared by mounting in a cold setting resin such as an epoxy or polyester. The surface is polished to optical flatness using SiC

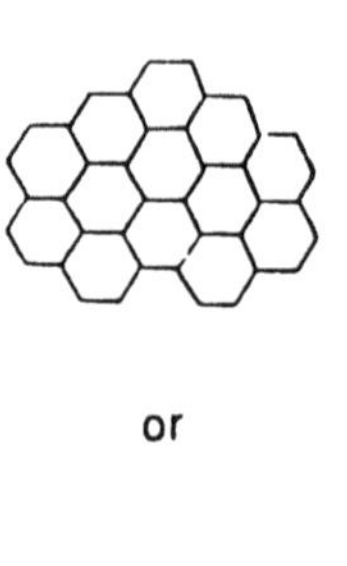

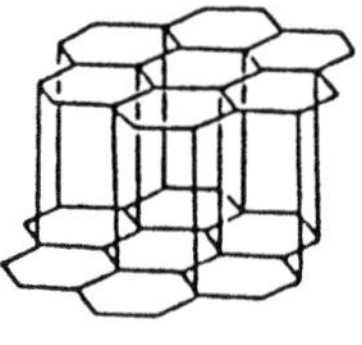

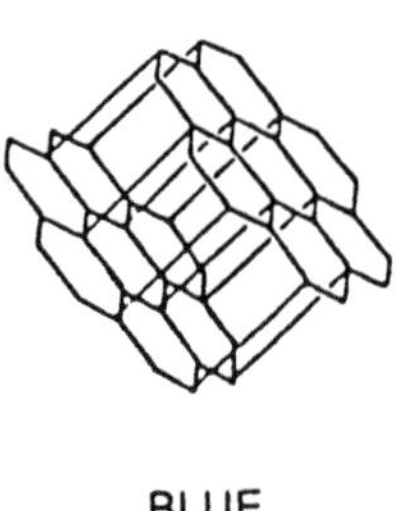

Fig. 1.12 Orientation of carbon lamellae layers to produce colours.

papers and alumina or diamond paste. Polarized light is often used to observe the interference colours generated due to the orientation of the graphitic lamellae at the surface (Fig. 1.12). Observations are made using crossed polars, the general arrangement of which is shown in Fig. 1.13. Examined in this way, carbons can be shown to possess a unique series of microstructures dependent upon the range of shared orientations and the presentation of the constituent aromatic/graphitic lamellae to the polished surface.

The interference colours, usually yellows, blues and purples, can be used to characterize the carbon in terms of size of coloured (isochromatic) areas and the actual colour. The overall appearance of the surface is referred to as its 'optical texture'. The size and shape of the isochromatic areas can be estimated and may vary from around 1 μm to several hundreds of microns. Table 1.2 lists some of the nomenclature which has been

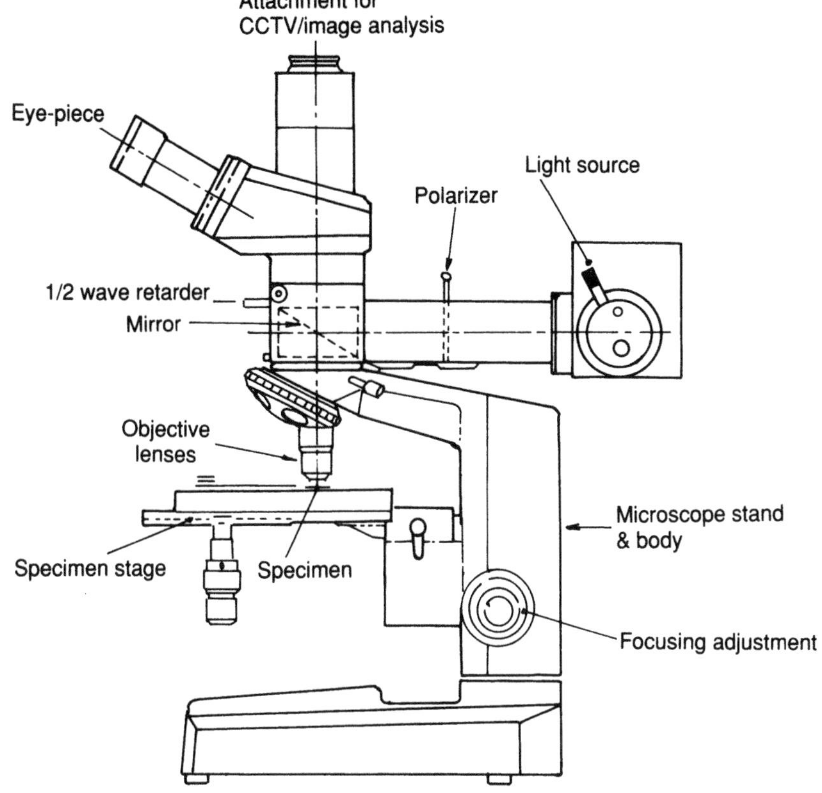

Fig. 1.13 Polarized light optical microscope.

Table 1.2 Nomenclature used to describe optical texture

Type of structure	*Size* (μm)	*OTI factor*
Isotropic (Is and Ip)	No optical activity	0
Fine mosaics (F)	<0.8 in diameter	1
Medium mosaics (M)	>0.8 <2.0 in diameter	3
Coarse mosaics (C)	>2.0 <10.0 in diameter	7
Granular flow (GF)	>2 in length >1 in width	7
Coarse flow (CF)	>10 in length >2 in width	20
Lamellar (L)	>20 in length >10 in width	30

adopted to describe the features observed in carbon microstructures, together with a definition of the optical texture index (OTI).

The individual components of the image can be identified and catalogued using an image analyser. The OTI of the sample is calculated by multiplying the fraction of the image by its corresponding OTI factor and summing the values. This number gives a measure of the overall anisotropy of the carbon. The yellow, blue and purple colours are also used to determine the orientation of the constituent lamellar planes of the carbon at the polished surface. Yellows and blues are indicative of prismatic edges exposed in the polished surface. Yellows change to blue and vice versa on rotation of the specimen stage by 90° or by reversal of the compensator plate. The purples are indicative of basal planes. That is to say, the surfaces of constituent lamellar planes of the carbon lying parallel or perpendicular to the plane of the polished surface. The purple colour represents an essentially isotropic surface and remains unchanged during rotation of the specimen stage of the microscope. Non-graphitic, glassy carbon [6] or graphitic carbon in which the size of the isochromatic regions is below the limit of resolution of the microscope, will also exhibit a purple colour. The purple colour of anisotropic carbon is generally darker in tone than that from glassy carbons and the two can be distinguished, but great care must be taken in the interpretation of the observations made.

Optical microscopy can thus be used to assess shape, size and degree of porosity, the presence or absence of anisotropy and, via the OTI, estimate the shape and size of the anisotropic constituent structures. It must be stressed, however, that, even when optical micrographs are coupled with automated image analysis techniques [7,8], this is only a comparative technique and that material can only be characterized to the level of resolution of the equipment used.

1.4.2 Scanning electron microscopy/electron probe microanalysis

Scanning electron microscopy (SEM) is used primarily for the study of surface topography of solids. It provides a depth of focus far greater than optical or transmission electron microscopy [9]. The resolving power of the SEM is around 30 Å which is approximately 300 times greater than the optical microscope and one order of magnitude less than the transmission electron microscope.

The SEM operates by focusing an electron beam passing through an evacuated column on to the specimen surface using electromagnetic lenses. The beam is raster-scanned over the surface of the specimen in synchronism with the beam of a cathode ray tube (CRT) display screen. Inelastically scattered secondary electrons emitted from the sample surface are collected by a scintillator-counter and the signal from this is used to modulate the brightness of the image on the CRT. Differences in secondary emission result from changes in surface topography. If elastically (backscattered) electrons are collected, an image can be formed from the contrast resulting from compositional differences across the surface of the specimen.

Information may be obtained by examination of both the natural surface of the sample and that exposed by either fracture or sectioning [10]. Rough topographic features and void content are easily revealed.

Changes in topography following processing such as oxidation and heat treatment can be monitored accordingly. Carbon surfaces which have been polished for optical microscopic examination will generally exhibit very few features under SEM as a result of their limited topography. Etching the polished surface either chemically, using chromic acid, or by ion bombardment can reveal a wealth of detail which can often be related to the optical texture of the sample. Forrest [11] and Markovic *et al.* [12] have developed a 'same-area' optical microscopy → etching → SEM sequence to reveal the structure of carbon–carbons; the technique takes a specific region of polished surface which has been identified and characterized optically and is re-examined by SEM following etching.

When materials are bombarded by a high-energy (10–50 keV) electron beam (as in an electron microscope), characteristic X-ray fluorescence radiation is produced. It is possible to obtain X-ray spectra directly on the area as seen by the electron beam by incorporating either energy or wavelength dispersive spectrometers directly into the instrument. Both qualitative and quantitative elemental data can thus be acquired from an area with a diameter of the order of 3 μm for the elements boron to uranium. Data can be obtained from an isolated region of the sample (spot mode) along a pre-selected linear trace (line profiling) or from an area (X-ray distribution mapping) [13]. The technique is known as electron probe microanalysis (EPMA) and can be used to reveal the presence and distribution of inclusions, impurities and fillers in a section through a sample. Figure 1.14 shows a schematic diagram of an SEM/EPMA unit.

1.4.3 Conventional and scanning transmission electron microscopy

Transmission electron microscopy (TEM) is used to study the structure and morphology of materials by examining the diffracted and transmitted electron intensities. A beam of high-energy electrons (100–400 keV) is collimated by electromagnetic lenses and passed through the specimen. The resulting diffraction pattern can be imaged on a fluorescent screen below the specimen. It is possible to obtain the lattice spacings of the structure under consideration from the diffraction pattern [14]. Diffraction information may be extracted from areas less than 0.1 μm in size. Alternatively, the transmitted beam or one of the diffracted beams can be used to form a magnified image of the sample on the viewing screen. These are referred to respectively as the bright field and dark field imaging modes

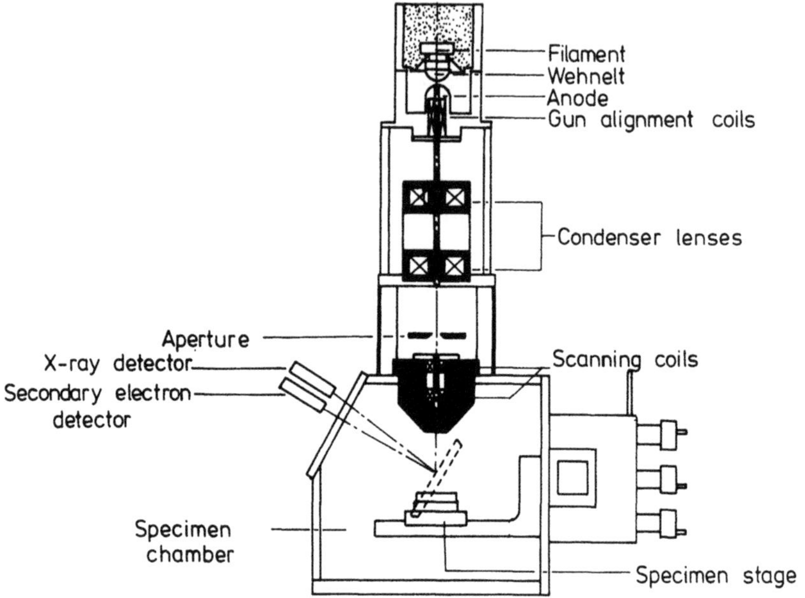

Fig. 1.14 Schematic diagram of a combined scanning electron microscope–microprobe analyser.

and provide information about the size and shape of the microstructure of the material down to a resolution of around 0.2 nm.

By allowing the incident beam to raster over a predetermined area of the specimen and detecting said beam using a scintillator rather than a photographic plate, a scanning transmission electron microscopy (STEM) image is obtained on a television screen. The signal from the CRT may be further interfaced with a micro- or minicomputer allowing the performing of a number of advanced functions such as digital imaging, image enhancement and image analysis. Since the incident electron beam interacts with the specimen, characteristic X-rays will be emitted from the sample, which can be detected and analysed. The elemental composition for regions down to 0.05 μm in size can be studied (Fig. 1.15).

To date, very few authors have published work describing TEM of carbon–carbon composites. This suggests that a serious gap exists in the microstructural knowledge of scientists working on the material; however, TEM provides access to a number of analytical techniques with extremely high spatial resolution and cannot be ignored if a complete understanding of the microstructural evolution of any material during processing is to be obtained. In the case of brittle fibre/brittle matrix composites, such as carbon–carbon, TEM analysis should prove a particularly important aid to the understanding of the crucial fibre/matrix interface [15,16].

The major obstacle to effective TEM of carbon–carbon is simply one of specimen preparation. In order to employ TEM and associated techniques,

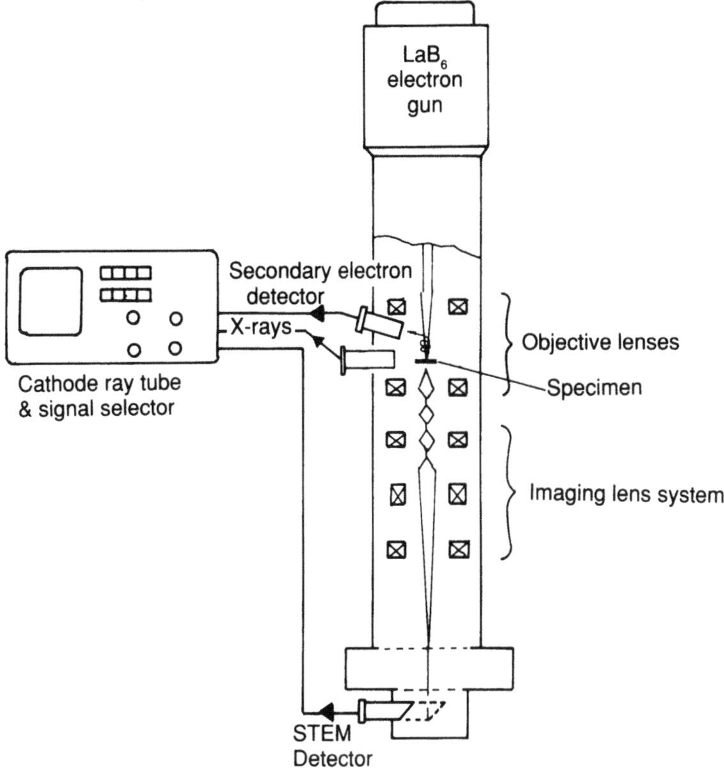

Fig. 1.15 Schematic diagram of scanning transmission electron microscope with detection instrumentation.

specimens must be less than 100 nm thick, and preferably even thinner. Electron diffraction, high-resolution TEM (HREM) and analytical techniques such as energy dispersive X-ray analysis (EDX) and electron energy loss spectroscopy (EELS) require thin samples which are transparent to the electron beam, otherwise the large degree of scattering of the electrons which is possible in solid materials will impair the resolution by beam spreading. The interpretation of results obtained by the aforementioned techniques becomes increasingly difficult when transmitted electrons suffer multiple scattering events due to excessive specimen thickness. Books and papers covering the various techniques and information which they can give have been produced by a number of authors. In-depth and informative volumes have been published by Williams [14], Goodhew [17] and Loretto [18].

Carbonaceous materials are generally prepared by grinding or microtoming. Grinding a material has the obvious limitation that it is very difficult to relate the observed structure to the spatial arrangement of the original material by the time particles are small enough to be electron transparent. Microtoming, the cutting of very thin slices, also results in mechanical disruption of regions of particular interest, such as the fibre/matrix interface. A number of problems, such as lattice distortion, splitting, folding, microcracking and mechanical thinning, have been identified with these preparation techniques [19–22]. Microtomy has been successfully applied to the TEM of carbon fibres [19,23], which are generally embedded in an epoxy resin prior to slicing, thus considerably reducing the volume of brittle material to be cut. The skill and experience of the researcher can usually overcome the difference in mechanical response between resin and fibre. Unfortunately, the same cannot be said for carbon–carbon, where the fibre content may be very high (up to 70%), and the surrounding medium is very brittle.

One technique which can produce good-quality thin sections of carbon–carbon is ion beam thinning or atom milling. The process involves low-energy (5 keV) inert gas ions or atoms impinging at high velocity on the sample from both sides at low angles (between 10 and 45° to the plane of the sample). The process involves the slow removal of material from this surface. The rate of removal is of the order of a few microns of material per hour depending on material type and milling conditions. It is usually used, therefore, on samples which have first been mechanically thinned to around 50 μm. Ion or atom milling is a relatively clean and gentle method. When examined, a material perforated in this way is a truer representation of the actual structure than that likely to be affected by mechanical thinning. Aside from the lack of speed, the main problems associated with milling are: implantation of the inert gas, heating of the specimen and the redeposition of sputtered material. The problems can be overcome or at least minimized by using a very low milling angle towards the completion of thinning and liquid nitrogen cooling of the specimen stage. The book by Goodhew [17] gives detailed descriptions of the techniques used to prepare TEM specimens. Kowbel and Don published a paper comparing microtomed and atom-milled bulk carbon produced from mesophase pitch [24]. They were able to show that a greater degree of damage resulted from microtomy and that this could lead to a misinterpretation of the microstructure. Numerous studies based on electron diffraction measurements of graphite crystallite size and orientation as a function of temperature have demonstrated the importance of TEM in the study of pyrolytic carbons, the results being summarized by Kowbel *et al.* [23].

Papers published with successfully thinned carbon samples highlight their complexity and demonstrate the importance of the interfaces between the various types of carbon present. These different types of carbon include the fibres (whether from rayon, polyacrylonitrile (PAN) or pitch precursors), the initial pyrolysis matrix and any carbon introduced for densification purposes (either by liquid impregnation or chemical vapour infiltration) [23–25]. The application of TEM to carbon–carbon systems will eventually be extended to the analysis of oxidation protection systems. Three basic methods of oxidation protection exist (see Chapter 6):

1. a surface coating of refractory material (often a multi-layer) protection;
2. glass-forming particulates included in the matrix to fill cracks which may form in the surface coating as a result of thermal/mechanical cycling;
3. fibre coatings to prevent oxidation occurring preferentially at any exposed fibre/matrix interfaces.

In each case, the efficient operation of the protective system can be checked with the aid of TEM. At high temperatures the refractory additives are limited by reaction with the surrounding carbon. The extent of this carbothermal reduction will require TEM-associated and complementary techniques for a complete analysis of its microstructural effects.

1.4.4 X-ray diffraction

X-ray powder diffraction is used to obtain information about the average bulk structure of carbon materials. The technique provides a measure of the amount of ordered material present and can be used to give an indication of the size of the crystallites which make up the ordered structure. Samples are prepared as powders either in capillaries or spread on a flat sample holder. The minimum amount of material required is a few milligrams. Greater accuracy is achieved, however, if up to 1 g of the sample is available [26].

When a beam of monochromatic X-radiation is directed at a crystalline material, diffraction of the X-rays is observed at various angles with respect to the primary beam (Fig. 1.16). The relationship between the wavelength of the X-ray beam, λ, the angle of diffraction, 2θ, and the distance between each set of atomic planes of the crystal lattice, d, is given by the Bragg equation

$$n\lambda = 2d \sin \theta, \tag{1.7}$$

where n is the order of diffraction. Using this equation, the interplanar distances of the crystalline material under study can be calculated. The interplanar spacings depend solely on the arrangement of atoms in the crystal unit cell while the intensities of the diffracted rays are a function of both the diffracting power and the placement of the atoms within the unit cell [27].

The diffraction pattern obtained represents the amount of scattering over a range of scattering angles, 2θ. The pattern can be considered as a 'fingerprint', each crystalline structure having, within limits, a unique diffraction pattern which can be analysed in terms of diffraction peaks, their positions and their widths. For the most accurate work, a standard

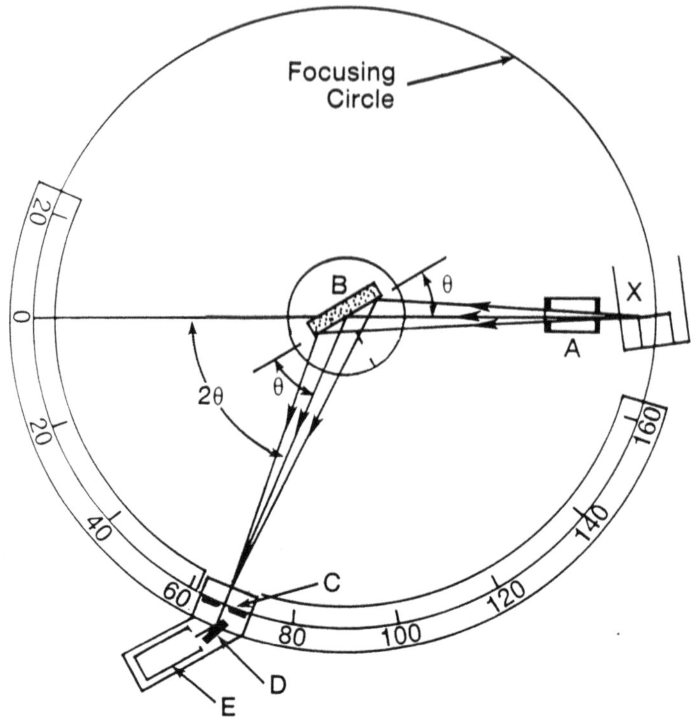

Fig. 1.16 Schematic diagram of an X-ray diffractometer. A, collimation assembly; B, sample; C. slit; D, exit beam monochromator; E, detector; X, source of X-rays.

compound, usually a highly crystalline salt, is added to the powder sample to afford internal calibration of the peak positions and widths, thus permitting any instrumental factors to be accounted for.

An indication of the degree of disorder within the specimen may be obtained since, although the principal scattering is due to the ordered material, the amount of background scatter may be analysed. Furthermore, the broadening of the diffraction peaks allows an estimation of the mean particle size to be made. The approximate crystallite size, t, can be calculated from the amount of broadening, β, using the Scherrer equation

$$t = c\lambda/\beta \cos 2\theta, \tag{1.8}$$

where c is the cell dimension, λ the X-ray wavelength and 2θ the scattering angle. β is the amount of broadening due to the sample. The observed broadening β requires to be corrected for the instrumental broadening b, generally using a relationship such as (Fig. 1.17)

$$\beta^2 = B^2 - b^2. \tag{1.9}$$

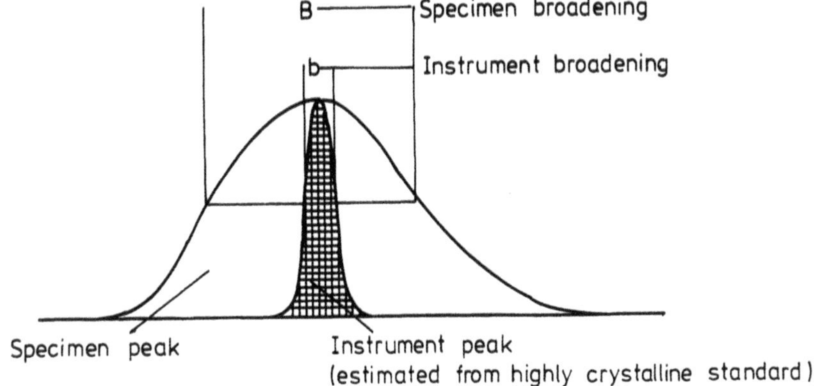

Fig. 1.17 Line broadening in X-ray diffraction experiments.

The parameters most often quoted from X-ray diffraction experiments on carbon materials are:

$d_{(002)}$	Interlayer spacing of [002] planes
L_c	Stack height
L_a	Stack width

Small-angle scattering of X-rays and neutrons (SAXS and SANS) can be used to study the electron-density fluctuations and contrasts which exist over a range of 10–1000 Å. It is, therefore, a useful technique with which to study the pore structure of carbons [28].

1.4.5 Surface analysis

A number of surface-sensitive techniques have been applied to carbon materials. The success of such techniques has unfortunately been somewhat limited because the extreme heterogeneity of the surfaces makes a full analysis as difficult as that of the bulk structures. Furthermore, all of the techniques require to be applied under high vacuum, making the *in situ* observation of surface changes, due to oxidation for example, impossible.

(a) X-ray photoelectron spectroscopy

X-ray photoelectron spectroscopy (XPS) or electron spectroscopy for chemical analysis (ESCA) [29] is used to provide information about the composition and structure of the outermost surface layers of a solid. When a solid is exposed to a flux of X-ray photons of known energy, photoelectrons are emitted from the solid. The photoelectrons originate from discrete electronic energy levels associated with those atoms in the analysis volume. The energy of the emitted photoelectrons is given by

$$E_K = h\nu - E_B - \phi, \tag{1.10}$$

where $h\nu$ is the characteristic photon energy of the excitation source, E_K and E_B are the measured photoelectron energy and binding energy respectively of a specific core or valence-level electron and ϕ is an experimental parameter depending on the spectrometer and sample being analysed. Ionization may occur, with varying probability, in any shell for a particular atom. The spectrum of that element is, therefore, usually comprised of a series of peaks corresponding to electron emission from the different energy levels. The energy separation and relative intensities of the peaks for a given element are well known, thus allowing unambiguous elemental identification. In addition, ESCA can be used to probe the electronic state of atoms so that the bonding configuration in the case of surface species can be determined which can, for example, reveal the difference between oxygen bound as atoms to the surface of the carbon in carbonyl or hydroxyl groups.

(b) Low-energy electron diffraction

Low-energy electron diffraction (LEED) patterns of single-crystal graphites have been obtained and used to study changes in the surface following treatments such as gasification. The technique is unfortunately very limited in application because only regular crystal surfaces can be observed with any acceptable degree of understanding.

(c) Auger electron spectroscopy

In Auger electron spectroscopy (AES) the excitation source is a finely focused electron beam which impinges on the sample surface. The two interactions of interest in AES and scanning Auger microscopy (SAM) are the generation of Auger electrons via the Auger process and the production of secondary electrons providing topographic information. The Auger process is initiated by electrons in the primary beam causing the ejection of core-level electrons from the atoms of the specimen. Once a core hole has been created, an electron from a higher energy level can fall into the core vacancy. Figure 1.18 illustrates the Auger process schematically in the form of an energy-level diagram. A transition of this type results in the release of energy by (a) the emission of characteristic X-rays or (b) the ejection of a second (Auger) electron.

The process in which a K shell ionization occurs followed by an L shell electronic transition which gives rise to an Auger electron ejected from an L shell is denoted a KLL transition. Similarly, transitions such as KLM and LMM may occur as shown in Fig. 1.18. Auger electrons are collected by the instrument's detection system which provides a display of the number

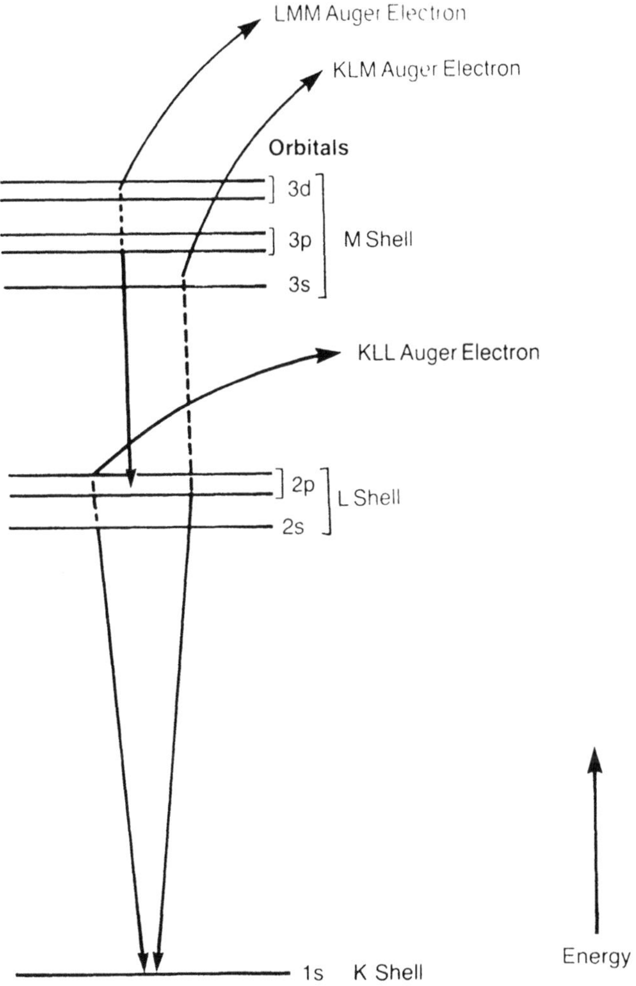

Fig. 1.18 The Auger process.

of electrons $N(E)$ vs their kinetic energy (KE) [30]. Auger spectra are generally displayed in the derivative mode d$N(E)$ vs KE because of the relatively high secondary electron background count [31]. AES also provides information about the elemental and chemical composition of the sample. The major advantage of AES is in the small spot size of the electron beam probe which allows analysis of features of the sample beyond the spatial resolution of other surface techniques. Auger electron microscopy has been used to study carbon deposits on both single-crystal and polycrystalline metal surfaces. The inherently shallow sampling depth and high spatial resolution of AES allow its combination with ion beam milling for the analysis of a sample as a function of depth below the surface. This method may well prove important in assessing the performance of multi-component oxidation protection systems for carbon–carbon materials.

1.5 DEFINITIONS OF CARBON FORMS AND PROCESSES

Carbon science is concerned with solid carbon materials, of which the overwhelming majority, with the exception of diamond, possess the basic structural arrangements of carbon atoms in a hexagonal planar array network. Carbon science and the carbon industry, in common with metallurgy, have their origins in antiquity. As a result, much of the nomenclature applied to the more 'modern' aspects of carbon science is derived from earlier work. It is appropriate, therefore, to define that terminology as used today since its use applies to all aspects of the subject. Solid carbons are generally derived from organic precursors by a pyrolysis process known as carbonization, and exist in graphitic and non-graphitic forms.

1.5.1 Carbonization

Carbonization is the process of formation of a material with increasing carbon content from an organic material, usually by pyrolysis, and ending with an almost pure carbon residue at temperatures over 1200 °C.

1.5.2 Graphitization

Graphitization is the solid state transformation of metastable non-graphitic carbon into a graphite structure by thermal activation. The degree of graphitization depends upon the heat treatment temperature, the time allowed for rearrangement of the atoms and the applied pressure. Most graphitizable carbons pass through a fluid stage during the carbonization process.

1.5.3 Graphitic carbons

Graphite is the allotropic form of carbon consisting of layers of hexagonally arranged carbon atoms in a planar condensed ring system. Chemical bonds within the layers, which are stacked parallel to one another, are covalent with sp^2 hybridization. Bonding between the layers is relatively weak and of the van der Waals' type. Graphitic carbons are all varieties of material consisting of the element carbon in the allotropic form of graphite, irrespective of the presence of structural defects. Natural graphite is a mineral consisting of graphitic carbon regardless of its crystalline perfection. Some natural graphites show a high degree of perfection but most are mined in the form of flake graphites containing other mineral matter. Synthetic graphite is defined as a material consisting mainly of graphitic carbon which has been obtained by means of a graphitization heat treatment of a non-graphitic carbon or by chemical vapour deposition (CVD) from hydrocarbons at temperatures above 1800 °C to give a deposit with the graphite structure.

1.5.4 Non-graphitic carbons

Non-graphitic carbons are all varieties of substances consisting mainly of the element carbon with two-dimensional long-range order of the carbon atoms in planar hexagonal networks, but without any measurable crystallographic order in the direction perpendicular to the planes (*c*-direction). Many non-graphitic carbons can be converted to graphite by graphitization heat treatment to 2200 °C or above. Non-graphitizable carbons are those which cannot be transformed into graphitic carbon solely by heat treatment at temperatures of 3000 °C or above under atmospheric or lower pressures.

1.5.5 Pitches

Pitches are carbonaceous materials derived from organic precursors by relatively low temperature processes below 400 °C such as distillation. They are oligomers containing a wide range of molecular types and masses. Most pitches melt on heating to yield an isotropic fluid. Continued heating above 375 °C results in the alignment of lamellar molecules leading to nematic discotic liquid crystals. The further development of this liquid crystal or 'mesophase' system provides the basic microstructure of the final carbon product, dictating its optical texture.

Volatile matter is released from the bulk of the pitch material throughout the carbonization process which, together with the complex packing within and between the carbonized particles, results in porosity in the final product. The degree and nature of the porosity depend upon the precursor and the conditions of the carbonization process. Two principal types of pitch are used as matrix precursors and will be covered in detail in Chapter 5; coal-tar pitch is a product of coal distillation, generally containing fused aromatic ring systems with minimal aliphatics. Petroleum pitches are the heavy residues of petroleum processing, consisting mainly of alicyclic rings with some aromatic, methylene and alkyl groups. Petroleum pitch cokes are generally the more graphitizable of the two.

1.5.6 Cokes

A coke is a highly ordered carbonaceous product of the pyrolysis of organic material, at least parts of which have passed through a liquid or liquid–crystalline state during the carbonization process and which consists of non-graphitic carbon. Their structure is a mixture of varying sizes of optical texture, from the optically isotropic to domain and flow anisotropy. At the crystallographic level, only the short-range order associated with non-graphitic carbons exists. Various types of coke can be defined. A **'green' coke** is the primary solid carbonization product obtained from high boiling carbon fractions at temperatures below 600 °C. A **calcined coke** is a petroleum or coal-tar derived pitch coke with a mass fraction of hydrogen less than 0.1%. It is obtained by the heat treatment of a green coke to about 1275 °C. A **petroleum coke** is the carbonization product of a petroleum pitch. Similarly, **a coal-tar pitch coke** is the primary industrial solid carbonization product from coal-tar pitch. **Metallurgical coke** is produced by the carbonization of coals or coal blends at temperatures of up to 1100 °C to produce a microporous carbon material of high strength. Finally, a **needle coke** is a commonly used term for a specialized form of coke with an extremely high graphitizability resulting from a strong preferred orientation of the microcrystalline structure.

1.5.7 Chars

A char is a carbonization product of a natural or synthetic organic material which has not passed through a fluid stage during carbonization.

1.5.8 Coals

Coals result from the 'coalification' of organic materials, mainly of plant origin. They possess a wide range of structures at both the microscopic and molecular levels. Coalification is a geological process of dehydrogenation occurring within the earth's crust by gradual transformation at moderate temperatures (around 200–250 °C) and high pressures. The process is progressive with respect to time. It is possible to define the degree of coalification or coal rank at any stage by the C/H ratio of the material. The structure changes from a peat through lignites to anthracites via 'intermediates' such as sub-bituminous and bituminous coals. The carbon content increases from 50 to over 95%. The variation in precursor plant forms and conditions results in varying degrees of coalification and the coals show a wide variation in properties depending on their source.

1.5.9 Carbon fibres

Carbon fibres are filaments consisting of non-graphitic carbon produced by the carbonization of synthetic or natural organic fibres or of fibres spun from organic precursors such as resins and pitches. In the majority of cases (i.e. except for those fabricated using very high heat treatment temperatures (>2700 °C) the fibres remain as non-graphitic carbon, contrary to the nomenclature often used in the United States! The alignment of the lamellar planes along the fibre axes exploits the anisotropic properties of carbon materials. There are three principal types of carbon fibre: those based on PAN, rayon and mesophase pitch. All will be considered in greater detail in Chapter 2.

1.5.10 Deposited carbons

Chemical vapour deposition of carbon from volatile hydrocarbon compounds on to carbon, metal or ceramic substrates provides a processing route to obtain carbon materials with a homogeneous microstructure. The method is used extensively in the densification of carbon–carbon composites and is reviewed in Chapter 3. When the substrate is an active catalyst for carbon deposition, the growth of whiskers or filaments by a solution/deposition mechanism, with small metal particles at the tip, has been observed. Such filaments are often highly graphitic [32].

1.5.11 Other 'miscellaneous' forms of carbon

Charcoal is a traditional term used for a char obtained from wood and certain related natural organic materials. The form of the parent material is retained, often with a highly developed pore structure. At the microscopic level the basic structure is disordered, producing an isotropic optical texture. The crystallographic-level structure also exhibits very little order with no detectable graphitic properties. **Carbon blacks** are produced by vapour phase growth of particles which are, on the whole, spherical with no regular long-range order. Some carbon blacks are the closest approach to a truly 'amorphous' carbon. Activated carbons are porous carbon materials, usually chars, which have been subjected to reaction with gases either during or after carbonization with the purpose of increasing their microporosity. The degree and type of porosity may be controlled to provide materials with large and, if required, specific adsorption capacity. Such materials are widely used as filters and catalyst supports, etc.

1.6 CARBON COMPOSITES

Composites are best defined as materials in which two or more constituents have been brought together to produce a new material nominally, at least, of more than one component, with resultant properties different from those of the individual constituents. The carbon materials thus far defined, although heterogeneous in nature, are all homogeneous in terms

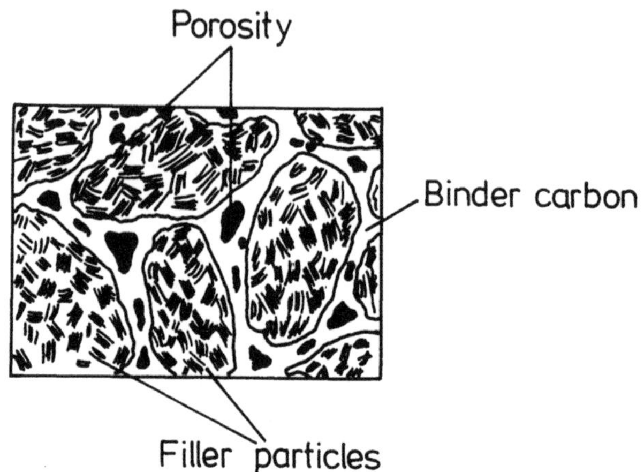

Fig. 1.19 Schematic representation of the microstructure of particulate carbon composites.

of treatment. One of the most critical aspects determining the nature and properties of a composite material is the interfaces between components. The strength of the composite is often governed by the adhesion forces which can be chemical, physical or a combination of the two.

The great majority of the industrial engineering carbons and graphites fall into the category of particulate filled composites. They are manufactured by a process in which a carbon filler is bonded with a liquid organic precursor, shaped, carbonized and, if so desired, graphitized up to 2700 °C. The filler phase is generally some form of graphitizable coke produced by a liquid phase pyrolysis route. The binder is either a resin, such as a phenolic or furan, or a pitch. Both graphitizable and non-graphitizable binders can, therefore, be used although pitch is most often employed. A wide range of particulate carbon composites may thus be produced by the variation in the size distribution of the filler particles, the types of filler and binder, the binder content and the heat treatment temperature (HTT) [33]. Figure 1.19 shows a general schematic microstructure for this family of composites. The materials are porous, although the porosity may be significantly reduced, if required, by reimpregnation with pitch or resin binder followed by recarbonization. The porosity is a result of that in the original filler, that developed due to shrinkage of the binder during carbonization and voidage due to inefficient impregnation by the binder. Carbon composites of this type, whose properties are listed in Table 1.3, are widely used in a number of important applications.

Graphite electrodes may be enormous structures when, for example, used for the production of steel in electric arc furnaces. They are highly graphitic, exhibiting an isotropic optical texture, and are capable of carrying a heavy electrical current at high temperatures under large thermal

Table 1.3 The range of properties of particulate carbon composites

Flexural strength	10–120 MPa
Flexural modulus	up to ≈ 14 GPa
Electrical resistivity	10^{-6}–10^{-4} Ω m
Thermal conductivity	0.05–0.4 W $m^{-1}K^{-1}$
Coefficient of thermal Expansion (CTE)	1–10 × 10^{6} K^{-1} (dependent on structure)

stress. They are made from an elongated grain filler with a preferred orientation from the alignment of the mesophase domains brought about by bubble percolation during a liquid phase pyrolysis treatment. Shaping by extrusion results in a highly anisotropic structure which is paramount in determining thermomechanical and electrical properties. **Nuclear** graphites, used as moderators and for other applications, require extreme chemical purity so as to avoid the adsorption of low-energy neutrons. A high degree of dimensional stability is also demanded, requiring no preferred bulk orientation of the graphitic crystallites. High-purity constituents are used in the manufacturing process and halogen gases are passed through the graphitization furnace in order to remove any inorganic impurities as volatile halides. The resultant optical texture is mostly fine grain with random orientation [34].

The manufacture of **carbon electrodes** is similar to that for graphitic composite electrodes, and involves mixing, shaping, prebaking, densification (when necessary) and heat treatment. Since carbon electrodes do not require to be graphitized, the HTT is kept below 1700 °C. The choice of components is controlled mainly by cost and availability. The electrodes used in aluminium smelting, for example, comprise a calcined coke filler and coal-tar pitch binder. The purpose of the electrodes is to provide a high current to the cell and to act as the reductant in the electrochemical process. They are also required to be resistant to gaseous oxidation. The role of the binder pitch is to provide, on carbonization, coke bridges between the particles of filler coke which hold the structure together and provide electrical contact. The pitch must be of low enough viscosity to enter the pores of the filler during baking but not so fluid as to flow away from around the filler. Good-quality anodes exhibit successful binding obtained by the coke bridges 'keying' into the surface of the filler to provide a strong mechanical bond which allows the passage of electricity.

1.7 CARBON–CARBON COMPOSITES

Carbon and graphite are attractive materials for use at elevated temperatures in inert atmosphere and ablative environments. The use of

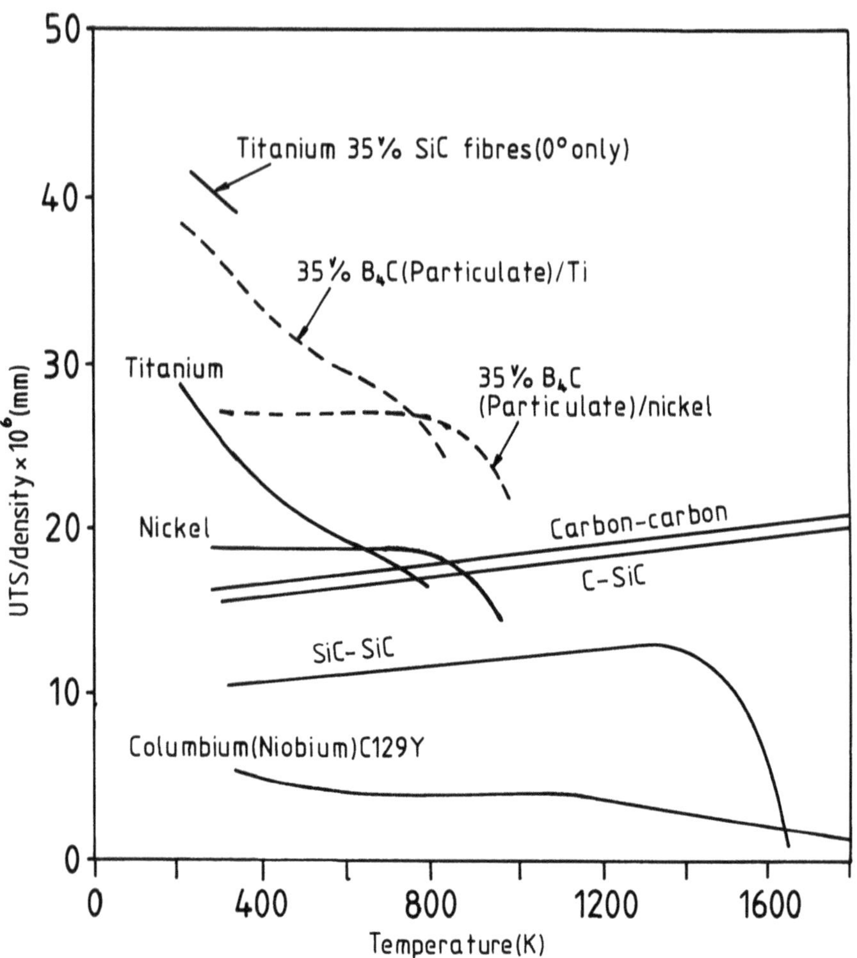

Fig. 1.20 Strength to density ratio for different classes of high-temperature materials with respect to temperature.

monolithic carbon or particulate composites is greatly limited by brittle mechanical behaviour, flaw sensitivity, variability in properties, anisotropy and fabrication difficulties associated with large and complex components and structures [35]. Carbon fibre reinforced carbon matrix composites consist of carbon fibres embedded in a carbonaceous matrix. The aim of these materials is to combine the advantages of fibre-reinforced composites such as high specific strength, stiffness and in-plane toughness with the refractory properties of structural ceramics [36]. A retention of mechanical properties at high temperatures, superior to any other material (Fig. 1.20) has resulted in the exploitation of carbon–carbon composites as structural materials in space vehicle heat shields, rocket nozzles and aircraft brakes. Additionally, properties such as biocompatibility and chemical inertness have led to new applications in medicine and industry. Carbon–carbon composites can perhaps lay claim to represent the ultimate development of carbon science.

Although carbon composites, in the form of polygranular synthetic graphite, were used in the fins of Second World War German rockets [37], it was not until the advent of carbon fibre technology in the late 1950s that the potential for the development of truly structural components was realized.

The advancement of carbon–carbon composite materials technology was initially very slow, but by the late 1960s it had begun to emerge as a major new genre of engineering materials [38,39]. During the 1970s, carbon–carbon structures were under extensive development in the USA and Europe [40–45], mainly for military use. The original carbon–carbon composites, to be used in rocket nozzles and re-entry parts, were produced using reinforcements in the form of woven fabrics of low-modulus rayon precursor carbon fibres. The matrix was derived from pyrolysed high-char yield resins such as phenolic and furan. Fibre-reinforced plastic moulding techniques were used to fabricate precursor composite structures which were subsequently carbonized.

Since those early days, it is now possible to use the whole variety of available types of carbon fibres (see Chapter 2) with their individual characteristics. The fibres may be combined in a wide variety of woven, 'knitted', braided and filament wound forms to provide one-, two- and multi-directional reinforced composites. Additionally, lower-cost composites employ a wide range of felts, fabric and short fibre systems. The reinforcing fibres can be combined with any of the forms of carbon, previously described. The matrix, for example, could be a vapour-deposited carbon, a glassy carbon (resulting from the pyrolysis of a resin) or the coke from the liquid phase pyrolysis of a mesophase pitch. Furthermore, by careful control of the precursor and the HTT, the degree of graphitization of the matrix may be varied considerably, thus imparting a wide range of thermomechanical properties to the composite. Carbon–carbon is not, therefore, a single material, but rather, a family of materials, many combinations of which have yet to be realized and evaluated. To some extent carbon–carbon composites resemble the particulate carbon composites, save that the granular filler is replaced by fibres. The characteristics of the precursors and the processes which take place during manufacture are critical in determining the ultimate properties.

There are two key developments required in order to secure a more widespread deployment of carbon–carbon. Many possible markets are severely limited as a result of the excessive costs of the composites. If carbon–carbon is to be exploited in cost-sensitive applications such as replacement of asbestos in motor vehicle and passenger train brakes, more efficient and economically viable precursors and processes need to be developed. In very 'high technology' usage, where cost is justified by improved performance (such as aircraft engine parts), prevention of oxidation is the paramount concern. Furthermore, a number of less critical, but still significant problems such as poor matrix mechanical properties, for example, need to be addressed.

To address the aforementioned problems, we are required to move away from carbon–carbon in isolation to define and develop what are described as materials systems. We first define a primary function of the material such as retention of strength at high temperatures, which is accomplished by carbon–carbon. The environment in which the material is required to operate will dictate a number of secondary functions/properties before any component can perform successfully. In order to impart attributes such as oxidation stability, fatigue resistance and 'out of plane toughness', it will be necessary to combine the 'basic' composite with a number of other compounds such as oxidation-protective coatings and matrix additives and develop a materials system capable of meeting all of the requirements.

REFERENCES

1. *Int. Bur. Stds* (1961).
2. Franklin, R. E. (1951) *Acta Cryst.* **4**, 253.
3. Griffiths, J. A. and Marsh, H. (1981) *Proc. 15th Biennial Conf. on Carbon*, University of Pennsylvania, Philadelphia, USA, 22–26 June.
4. Cornford, C., Forrest, R. A., Kelly, B. T. and Marsh, H. (1984) in *Chemistry and Physics of Carbon*, **19** (ed. P. A. Thrower), Marcel Dekker, New York, p. 211.
5. Spencer, M. (1982) *Fundamentals of Light Microscopy*, Cambridge University Press, Cambridge, UK.
6. Jenkins, G. M. and Kawamura, K. (1976) *Polymeric Carbons, Carbon Fibre, Glass and Char*, Cambridge University Press, London.
7. Fischmeister, H. (1981) Digital image analysis in quantitative metallography, in *Computers in Materials Technology*, Pergamon Press, New York, pp. 109–29.
8. Van der Voort, G. F. (1986) Image analysis, in *Metals Handbook*, 9th end, Vol. 10, *Materials Characterization*, American Society for Metals, Metals Park, pp. 309–22.
9. Goldstein, J., Yakowitz, H., Newbury, D., Lifshin, E., Colby, J. and Coleman, J. (1975) *Practical Scanning Electron Microscopy*, Plenum Press, New York.
10. Wells, O. C. (1974) *Scanning Electron Microscopy*, McGraw-Hill, New York.
11. Forest, M. A. and Marsh, H. (1981) *Proc. 15th Biennial Conf. on Carbon*, University of Pennsylvania, Philadelphia, USA, 22–6 June.
12. Markovic, V., Ragan, S. and Marsh, H. (1984) *J. Mat. Sci.*, **19**, 3287.
13. Heinrich, K. F. (1981) *Electron Beam X-ray Microanalysis*, Van Nostrand Reinhold Company, New York.

14. Williams, D. B. (1984) *Practical Analytical Electron Microscopy*, Phillips Electronic Instruments Publishing, Mahwah, New Jersey.
15. Peebles, L. H. Meyer, R. A. and Jortner, J. (1988) in *Interfaces in Polymer, Ceramic and Metal Matrix Composites* (ed. H. Ishida), Elsevier, Amsterdam, p. 1.
16. Evans, A. G. and Thonless, M. D. (1988) *Acta Metall.*, **36**, 517.
17. Goodhew, P. J. (1985) *Practical Methods in Electron Microscopy*, Vol. II, Elsevier, Oxford.
18. Loretto, M. H. (1984) *Electron Beam Analysis of Materials*, Chapman and Hall, London.
19. Bennet, S. C. and Johnson, D. J. (1979) *Carbon*, **17**, 25.
20. Tidjani, M. (1986) *Carbon*, **24**, 447.
21. Shiraishi, M. (1978) *J. Mat. Sci.*, **13**, 702.
22. Ehrburger, P. and Lahaye, J. (1981) *Carbon*, **19**, 1.
23. Kowbel, W., Hippo, E. and Murdie, N. (1989) *Carbon*, **27**, 219.
24. Kowbel, W. and Don, J. (1987) *Proc. 18th Biennial Conf. on Carbon*, Worcester, Mass., 19–24 July 1987, p. 286.
25. Weizhou, P., Tianyon, P., Hansin, Z. and Qias, Y. (1988) in *Interfaces in Polymer, Ceramic and Metal Matrix Composites*, (ed. H. Ishida), Elsevier, Amsterdam.
26. Azaroff, L. V. (1968) *Elements of X-ray Crystallography*, McGraw-Hill, New York.
27. Klug H. P. and Alexander, L. E. (1974) *X-ray Diffraction Procedures for Polycrystalline and Amorphous Materials*, Wiley, New York.
28. Glatter, O. and Kratky, O. (1982) *Small-angle X-ray Scattering*, Academic Press, New York.
29. Briggs, D. (ed.) (1972) *Handbook of X-ray and Ultraviolet Photoelectron Spectroscopy*, Heyden Press, London.
30. Carlson, T. A. (1975) *Photoselection and Auger Spectroscopy*, Plenum Press, New York.
31. Spruger, R. W., Haas, T. W., and Grant, J. T. (1978) *Quantitative Surface Analysis of Materials*, ASTM STP 634, American Society for Testing and Materials, Philadelphia.
32. Tibbetts, G. G. (1989) *Carbon*, **27**(5), 745.
33. Rand, B. (1989) *Proc. HIPERMAT 89 Conf.*, City Conference Centre, I. Marine Eng., London, 27–28 Sept. 1989.
34. Kelly, B. T. (1981) *Physics of Graphite*, Applied Science Publishers, London.
35. Fabrication of composites, in *Handbook of Composites*, (1986) Vol. 4 (eds A. Kelly and S. T. Mileiko North-Holland, pp. 111–75.
36. Savage, G. M. (1988) *Metals and Materials*, **4**, 544.
37. Vohler, O. and Fitzer, E. (1961) *Jahrbuch der Wissenschaftlichen Gesellschaft für Luftfahrt e.v. WGL*, pp. 467–73, Munich.
38. Schmidt, D. L. (1972) *SAMPE J.*, **8**, 9.
39. Frye, E. R. and Stoller, H. M. (1969) *Proc. AIAA/AIME 10th Structures, Structural Dynamics and Materials Conf.*, New Orleans, p. 193.
40. McAllister, L. E. and Taverna, A. R. (1976) *Proc. Int. Conf. on Composite Materials*, Vol. 1, Met. Soc. of AIME, New York, p. 307.

41. Fitzer, E. and Burger, A. (1971) *Int. Conf. on Carbon Fibres, their Composites and Applications*, London paper no. 36.
42. Lamieq, P. (1977) *Proc. AIAA/SAE 13th Propulsion Conf.*, Paper no. 77–882, Orlando.
43. Fitzer, E., Geigle, K. H. and Huttner, W. (1978) *Proc. 5th London Int. Carbon and Graphite Conf.*, Vol. 1, p. 493.
44. Girard, H. (1978) *Proc. 5th London Int. Carbon and Graphite Conf.*, Vol. 1 (Soc. Chem. Ind. London), p. 483.
45. Thomas, C. R. and Walker, E. J. (1978). *Proc. 5th London Int. Carbon and Graphite Conf.*, Vol. 1 (Soc. Chem. Ind. London), p. 520.

Carbon Fibres

2

2.1 INTRODUCTION

The measured strengths of materials are several orders of magnitude less than those calculated theoretically. This discrepancy is believed to be due to the presence of inherent flaws within the material [1]. It follows that the strength of a material can therefore be enhanced by eliminating or minimizing such imperfections. Cracks lying perpendicular to the direction of applied loads are the most detrimental to the strength. Fibrous or filamentary materials thus exhibit high strengths and moduli along their lengths because in this direction the large flaws present in the bulk are minimized.

Fibres readily support tensile forces, but alone will offer virtually no resistance and buckle under compression. In order to be directly usable in engineering applications they must be embedded in matrix materials to form fibrous composites. The matrix serves to bind the fibres together, transfer loads to the fibres and protect them against handling damage and environmental attack. Fibre-reinforced composite materials are generally fabricated to improve mechanical properties such as strength and toughness per unit weight, although protective applications such as vehicle armour are gaining in significance [2]. Carbon–carbon composites are usually produced in order to exploit their retention of properties at very high temperatures [3].

Composites can be divided into two classes: those with long fibres (continuous fibre-reinforced composites) and those with short fibres (discontinuous fibre-reinforced composites). In a discontinuous fibre composite, the material properties are affected by the fibre length, whereas in a continuous fibre composite it is assumed that the load is transferred directly to the fibres and that the fibres in the direction of the load are the principal load-bearing constituent.

The most common matrices for fibre-reinforced composites are polymeric materials which may be subdivided into two distinct types: thermosetting and thermoplastic. Thermosetting polymers are resins which crosslink during curing into a glossy, brittle solid (examples are polyesters, vinylesters and epoxies). Thermoplastic polymers are high-molecular-weight, long-chain molecules which can either become entangled (amorphous), for example polycarbonate, or partially crystalline, such as nylon and polyetheretherketone (PEEK), at room temperature to give strength and shape. Crystalline thermoplastic systems offer enhanced environmental resistance, creep resistance and toughness at the expense of increased cost and difficulties in processing.

Designers of weight-sensitive structures such as aircraft and racing cars require materials which combine good mechanical properties with low weight. Aircraft originally employed wood and fabric in their construction, but over the last 50 years aluminium alloys have been the dominant materials. During the last decade advanced composite materials have been increasingly employed for certain aircraft structures. At this point it is appropriate to define what has become a 'buzz-word' in the composites industry, the term 'advanced'; an advanced composite is simply one in which the mechanical properties are dominated by the fibres. The composites used in aircraft applications have been mainly those based on continuous carbon fibres and to a lesser extent glass and aramid. The use of composites in Formula 1 racing cars is far more widespread, such that the whole structure is normally of carbon fibre reinforced composite (Fig. 2.1).

The driving force for the increasing substitution of metals by composite materials is demonstrated in Table 2.1. Although the strengths and stiffnesses are not very different from those of metals, their densities are much lower, hence composites have greatly improved specific properties. The weight savings obtained in practice are not as great as Table 2.1 implies because the fibre composites are anisotropic and this must be accounted for in any design calculations. Even so, savings in weight of around 30% are readily achieved over aluminium [4,5]. An increasing number of applications, such as re-entry components and rocket nozzles, demand higher and higher operating temperatures. Metals creep at such temperatures and polymeric composites decompose, thus rendering them useless. Carbon fibres on the other hand, provided they are protected from oxidation, are shown to retain their properties to temperatures well in excess of 2000 °C. It is logical therefore to develop carbon fibre/carbon matrix composites in order to solve the problems of high-temperature operation. Since carbon fibres form a major part of carbon–carbon composites, and to a great extent control their mechanical properties, it is important to review their production, properties and processing so that their interaction within the composites may be more clearly understood.

In Chapter 1 it was discussed how the element carbon has two low-density allotropes, graphite and diamond. Both forms exhibit strong covalent bonding between the carbon atoms. Graphite has a hexagonal structure in which the strong sp^2 hybrid bonding within the hexagonal-layer planes generates the highest absolute and specific moduli and theoretical tensile

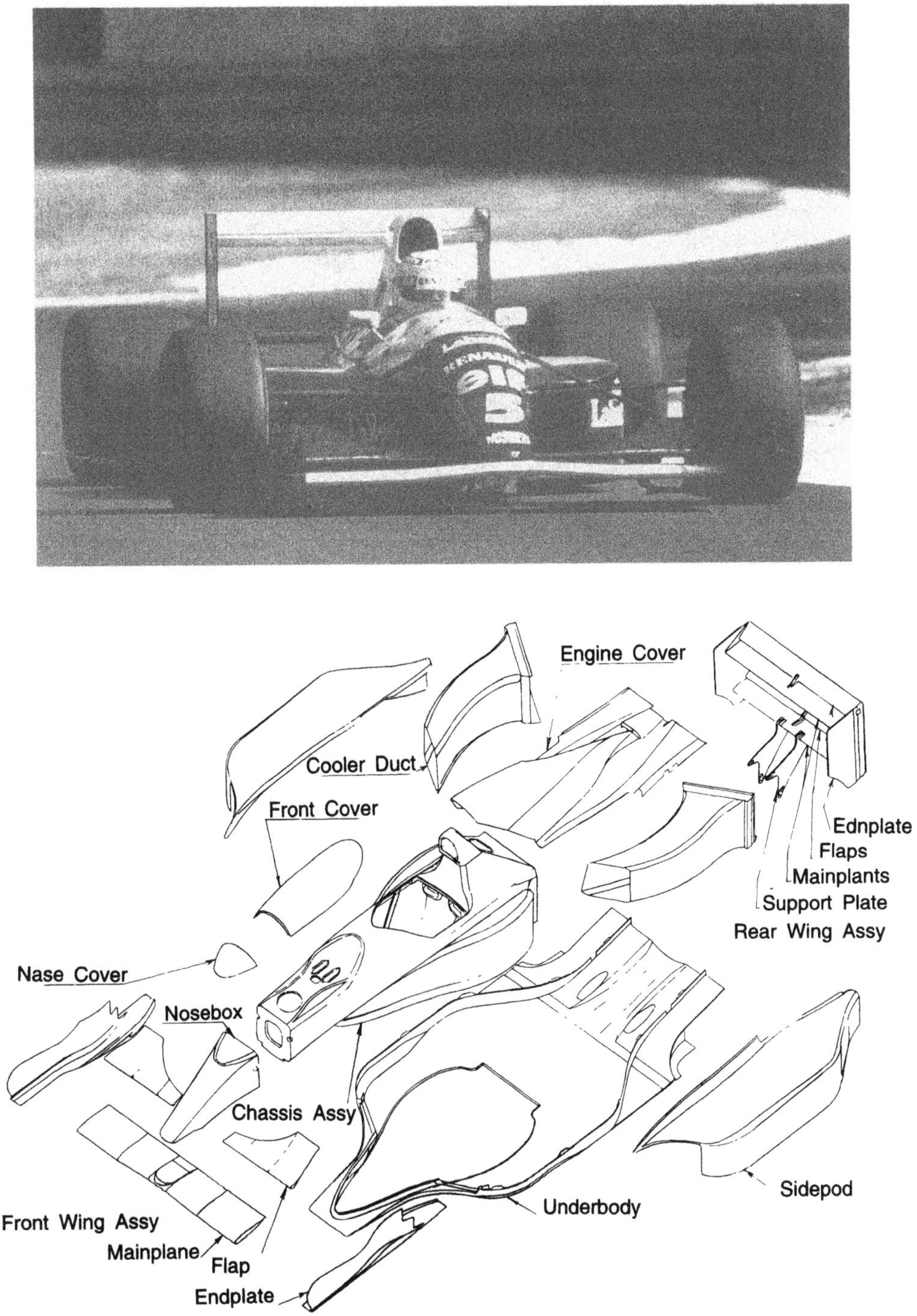

Engine Cover
Cooler Duct
Front Cover
Ednplate
Flaps
Mainplants
Support Plate
Rear Wing Assy
Nase Cover
Nosebox
Chassis Assy
Sidepod
Underbody
Front Wing Assy
Mainplane
Flap
Endplate

Fig. 2.1 The Williams FW14 Formula 1 racing car showing its composite components. (By kind permission of Brian O'Rourke, Williams Grand Prix Engineering Ltd.)

Table 2.1 Comparison of mechanical properties of composites and metals

Material	*Density* ($g\ cm^{-3}$)	*Tensile strength* ($GN\ m^{-2}$)	*Young's modulus* ($GN\ m^{-2}$)	σ/ρ	E/ρ
*Composite**					
E glass	2.1	1.1	45	0.5	20
IM carbon	1.5	2.6	170	1.7	113
HM carbon	1.6	1.6	257	1.0	161
Aramid	1.4	1.4	75	1.0	50
Metals					
Steel	7.8	1.3	200	0.2	26
Aluminium	2.8	0.3	73	0.1	26
Titanium	4.0	0.4	100	0.1	25

* Parallel to basal plane.

Table 2.2 Theoretical or measured properties of graphite and diamond

Allotrope	*Density* ($g\ cm^{-3}$)	*Tensile strength* ($GN\ m^{-2}$)	*Tensile modulus* ($GN\ m^{-2}$)	σ/ρ	E/ρ
Graphite*	2.26	150	1020	66	451
Diamond	3.51	90	20	25	177

* Parallel to basal plane.

strength of all known materials (Table 2.2). Very weak dispersive bonding between the planes, however, results in a low shear modulus which, when translated to a fibrous structure, produces a low cross-plane Young's modulus that is detrimental to fibre properties [6]. Diamond possesses a cubic crystallographic structure based on sp^3 bonding, exhibiting the next highest absolute and specific modulus, and does not suffer from a low shear modulus as does graphite [7]. To date, no one has developed fibres based on a diamond-like structure, although one might postulate such fibres would be very useful, especially in compressively loaded structures.

The carbon fibres with which we are familiar are based on the **graphene** hexagonal layer networks present in natural graphite. In the case when these graphene-layer planes stack with three-dimensional order, the material is defined as graphite [8]. Disorder frequently occurs, however, as a result of the weak bonding between planes. Rotations and/or translations arise such that only the two-dimensional ordering within the layers is generally present. In the early literature, two-dimensionally ordered structures are referred to as **turbostratic** graphite although the term is now falling into disuse.

Another gross misnomer whose use persists, much beloved by the advertising industry, is the improper application of the term 'graphite fibres' to carbon fibres which have only two-dimensional ordering.

Carbon fibres offer the highest modulus and highest strength of all reinforcing fibres. The fibres are not susceptible to stress corrosion or stress rupture failures at room temperature unlike glass and polymeric fibres. Ongoing process development work promises significant improvements in the ratio of performance/cost which would greatly increase the use and applications of these fibres.

2.2 PROCESSING OF CARBON FIBRES

2.2.1 Historical

Carbon fibres have been produced inadvertently from natural cellulosic fibres such as cotton or linen for millennia. The first recorded purposeful transformation of cellulose to carbon fibres was by Thomas Edison in 1878. Edison converted cotton and later, bamboo strips, into carbon for use as filaments in incandescent electric lamps [9]. After 1910, the lamp industry changed to the use of tungsten filaments so that the production of carbon filaments was terminated.

Interest in carbon fibres was renewed in the late 1950s with the advent of jet propulsion which precipitated the development of faster and larger aircraft. Laboratory tests on graphite whiskers uncovered strength values of 20 GPa and a modulus of 1000 GPa. Mechanical properties of this magnitude represented a 30-fold increase in specific tensile strength and a 17-fold improvement in specific modulus over the structural metals such as steel, aluminium and titanium [10]. Under sponsorship from the USAF, the Union Carbide Company fabricated carbon fibres in a similar way to Edison save that they used a rayon precursor. Two types of fibre were produced: the VIB type were treated to carbonization temperatures of around 1000 °C whereas the WYB type were post heat treated to 'graphitization' temperatures in excess of 2200 °C [11]. The fibres retained the fibrous structure of their rayon precursors but were isotropic as defined by their X-ray diffraction pattern. It is possible that the incorrect term 'graphite fibres' was introduced at that time in order to make a distinction between the WYB fibres which were pure carbon, as a result of high temperature heat treatment, and the VIB type which still contained many heteroatoms such as oxygen, hydrogen and ash-forming residues.

During the 1960s Union Carbide were able to demonstrate that hot stretching of the fibres at the highest temperatures (>2200 °C) resulted in higher modulus fibres. A number of contributory factors, such as the plastic deformability of solid carbon, the fibrillar structure of the rayon and the high degree of porosity within the fibres, resulted in a high degree of elongation during the high-temperature processing. Elongations of up to 100% produced a strong preferred orientation within the fibres such that tensile moduli of up to 500 GPa were achieved. The fibres were released commercially under the trade name Thornel 25, 50 and 70, the numerical postscripts indicating the modulus in Msi.

In parallel with the work at Union Carbide, Shindo and co-workers in Japan were investigating polyacrylonitrile (PAN) as a possible precursor [12]. Despite ex-rayon fibres being the first produced in quantity [13] it was soon demonstrated that PAN was a superior starting material [14,15]. The PAN precursor increased in importance as a result of the effort of English workers sponsored by the Royal Air Force at RAE Farnborough [16,17]. Watt *et al.* showed that high-strength fibres could be obtained by oxidizing the PAN precursor while under strain without the need for a hot-stretching stage after carbonization.

All of the continuous carbon fibres produced to date have begun with organic precursors which were subsequently pyrolysed. Discontinuous carbon whiskers have been produced by vapour–liquid–solid (VLS) growth from an iron catalyst and a hydrocarbon gas [18,19]. High-modulus (HM) carbon fibres (>200 GPa) require the stiff graphene layers to be aligned approximately parallel to the fibre axis. The low shear modulus between the planes significantly decreases the fibre stiffness for any off-axis layers. Commercial carbon fibre processes develop the required orientation by plastic deformation. The preferred orientation, which can be introduced into the precursor fibres, will be retained, at least in part, upon conversion to carbon [20–22]. Axial alignment may also be developed into the fibre by high-temperature deformation [23].

The carbon fibre industry presently uses three different precursor materials: rayon and PAN fibres and pitch, either isotropic or liquid crystalline (mesophase). Rayon and isotropic pitch precursors are used to produce low modulus fibres (<60 GPa) [24–27]. Fibres from both types of precursor can be strained at high temperatures to increase their moduli substantially but the process is not operated commercially to date [23,25]. Higher-modulus fibres are made from PAN or mesophase pitch precursors. In both cases an oriented precursor fibre is spun, stabilized by slightly oxidizing to thermoset the fibres and carbonizing to temperatures in excess of 800 °C [20–22,28] to produce a carbon fibre. The fibre modulus increases with heat treatment temperature (HTT) from 1000 to 3000 °C although the relationship between modulus and HTT is non-uniform, depending very much on the precursor [6,17]. The tensile strength of all PAN-based fibres, and a number of those from mesophase pitch, more often than not maximizes at an intermediate temperature of around 1500 °C. This is due to the formation of intrafibrillar porosity at HTT > 1500 °C in PAN materials, the so-called 'Reynolds-sharp cracks' [10]. The majority of mesophase pitch precursor fibres continuously increase their strength with increasing HTT.

2.2.2 Rayon-based carbon fibres

Figure 2.2 shows the basic elements required for producing carbon filaments from rayon [10]. The first low-temperature treatment takes place typically at around 300 °C and converts the structure to a form which is stable to higher processing temperatures. The process involves polymerization and the formation of cross-links. The rayon may be subjected to a chemical treatment before the first-stage oxidation exposure. The chemical bath can be an aqueous ammonium chloride solution or a dilute solution of phosphoric acid in denatured ethanol. The chemical treatment serves to reduce the time for the low-temperature step from several hours to around 5 minutes. Of the fibre mass, 50–60% is lost to decomposition products such as H_2O, CO and CO_2 during oxidation. The carbonization step, resulting in further weight loss, is usually carried out at ≈1500 °C. The yield after carbonization is typically 20–25% of the original polymer weight. At this stage the fibres have an essentially isotropic structure. The mechanical properties of carbonized rayon are poor as a direct result of the poor alignment of the graphene layers. Stretching of the fibres during heat treatment to graphitization temperatures significantly increases both strength and modulus, but is an expensive process. The morphology of the ex-rayon carbon fibres exhibits a crenulated surface, rather like a stick of celery, which is derived from the original precursor (Fig. 2.3). The combination of poor mechanical properties, low carbon yield and expense of graphitization has meant that ex-rayon carbon fibres have generally not proved competitive in the market place, although they are used extensively in ablative technology. This is due to their poor through thickness thermal conductivity and because their composites yield high inter laminar shear strengths. Table 2.3 lists typical rayon-based carbon fibre properties. A full review of the conversion of rayon to carbon fibres is given in the book by Gill [29].

2.2.3 PAN-based carbon fibres

In the mid 1960s it was discovered that the PAN structure could be stabilized by an oxidation process which assisted controlled thermal decomposition during a second 'carbonization' stage, enabling the production of carbon fibres with superior mechanical properties to those made from rayon [30]. The late 1960s and early 1970s saw a number of companies investing in large carbon fibre production facilities and a rapid transformation from batch methods to continuous processing. Since that time many other companies have entered the field. Table 2.4 lists a number of commercial products and the precursors from which they are derived. A number of

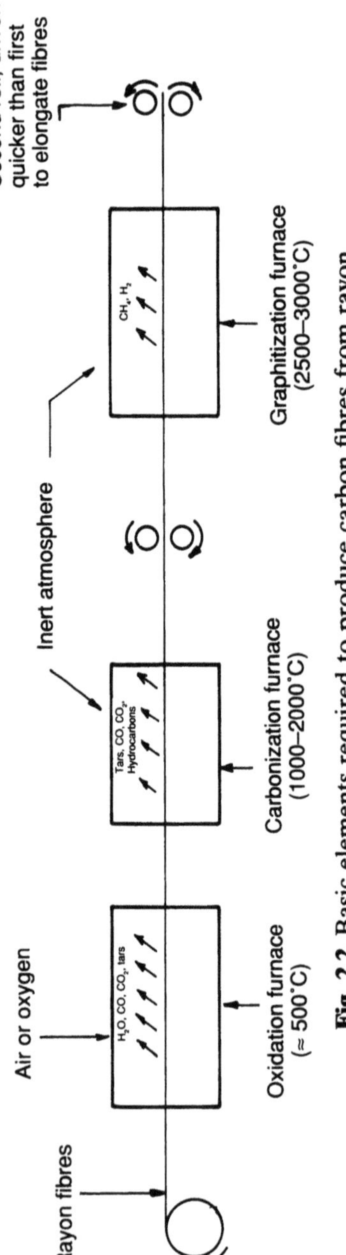

Fig. 2.2 Basic elements required to produce carbon fibres from rayon.

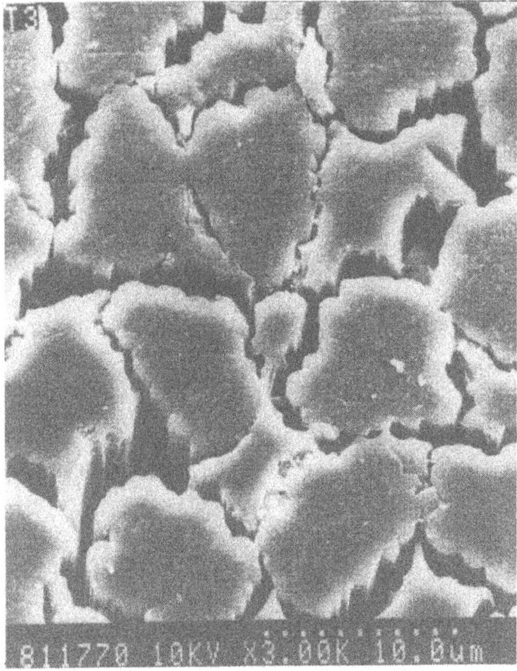

Fig. 2.3 Electron micrograph of rayon-based carbon fibres showing their crenulated surface rather like a stick of celery.

Table 2.3 Typical properties of rayon-based carbon fibres

Axial	Tensile strength	1.0 GPa
	Tensile modulus	41.0 GPa
	Elongation to break	2.5%
	Electrical resistivity	20 Ωm
Bulk	Density	1.6 g cm^{-3}
	Fibre diameter	8.5 μm
	Carbon assay	99%

manufacturers have standardized on one or two grades while the larger producers have an extensive range. The tow size (filament count) varies considerably between manufacturers, especially in the case where PAN is the chosen precursor. Tow sizes are becoming increasingly standardized at 1, 3, 6, 12 and 15k for high-tech aerospace and sports goods markets: 6 and 12k being by far the most popular. Larger tows (24k and upwards to 320k) are produced primarily for automotive applications and chopped fibre moulding compounds.

Table 2.4 Properties of a number of the major manufacturers' carbon fibre products

Manufacturer	*Product name*	*Precursor*	*Filament count*	*Density* (g cm^{-3})	*Tensile Strength* (MPa)	*Tensile modulus* (GPa)	*Strain to failure* (%)
Amoco	Thornel 75	Rayon	10K	1.9	2520	517	1.5
(USA)	T300	PAN	1,3,b,15K	1.75	3310	228	1.4
	P55	Pitch	1,2,4K	2.0	1730	379	0.5
	P75	Pitch	0.5,1,2K	2.0	2070	517	0.4
	P100	Pitch	0.5,1,2K	2.15	2240	724	0.31
Hercules	AS-4	PAN	6,12K	1.78	4000	235	1.6
(USA)	IM-6	PAN	6,12K	1.74	4880	296	1.73
	IM-7	PAN	12K	1.77	5300	276	1.81
	UHMS	PAN	3,6,12K	1.87	3447	441	0.81
Mitsubishi	K135	Pitch	2,4K	2.10	2550	540	0.5
Kasei	K139	Pitch	1,2,4K	2.12	2750	740	0.4
(Japan)							
Tonen	FT500	Pitch	2,4K	2.14	3000	490	0.61
(Japan)	FT700	Pitch	1,2K	2.17	3220	690	0.47
Toray	T300	PAN	1,3,6,12K	1.76	3530	230	1.5
(Japan)	T800H	PAN	6,12K	1.81	5490	294	1.9
	T1000G	PAN	12K	1.80	6370	294	2.1
	T1000	PAN	12K	1.82	7060	294	2.4
	M46J	PAN	6,12K	1.84	4210	436	1.0
	M40	PAN	1,3,6,12K	1.81	2740	392	0.6
	M55J	PAN	6K	1.93	3920	540	0.7
	M60J	PAN	3,6K	1.94	3920	588	0.7

The fibres derived from PAN precursors can generally be subdivided into three categories:

1. Low modulus (LM) (190–120 GPa) 'commercial' quality fibres.
2. Intermediate modulus (IM) (220–250 GPa) fibres. These fibres are of high quality, possess the highest tensile strength and strain to failure (1.2–1.4% increasing to 1.5–1.8% for specialized high strain grades) and are the favoured grade in aircraft and racing car manufacture.
3. High modulus (HM) (360–400 GPa) fibres. These fibres offer improved stiffness at the expense of strength and strain to failure (0.5–0.8%).

The prices of PAN-based carbon fibres vary considerably with grade and tow size, etc. Current (1990) prices vary, from £25 kg^{-1} for LM fibres to around £200 kg^{-1}. The output of the major manufacturers, Amoco, Hoechst, Hercules, Mitsubishi and Toray, are not disclosed but each is believed to have a capacity in excess of 100 tonnes per annum. As will be discussed later, many of the listed suppliers convert their carbon fibre product into various other forms such as woven fabrics, mats, felts and chopped fibres or even intermediate processed materials such as pultrusions, pre-moulded sheets and tubes. Additionally, a large industry exists specializing in the various secondary intermediate product forms after buying in fibres from one or more of the primary suppliers.

The reason for the variety of properties obtained from PAN-based carbon fibres is to be found in the processing parameters used in the conversion of polymer to carbon, and leads to the ability of tailoring those properties to the desired requirements. The PAN precursors to carbon fibres are selected on the basis of a fairly high degree of C–C orientation within the polymer chains. The conversion to carbon involves an initial oxidation stage, carbonization and a high-temperature 'graphitization' treatment. During each of those processes the carbon–carbon orientation is maintained along the fibre axis while competing degradative chemical reactions proceed, involving the loss of all non-carbon heteroatoms. The orientation is significantly improved during the higher temperature treatment as a result of crystallite growth and crystallite axial alignment and may be further enhanced by applying tension or stretching. The Young's modulus of the fibres is directly related to the preferred orientation of the graphene layers, hence control of that orientation will allow a tailoring of the modulus of the fibre.

Polyacrylonitrile can be spun into well-orientated polymer fibres, the nominal chemical structure of which is shown in Fig. 2.4. The chemistry of the process of conversion of PAN to carbon fibres is extremely complex, so that Fig. 2.4 is a great simplification of the actual process. Figure 2.5 shows the major steps in the continuous production process [31,32].

The fibres are first heated in an oxygen-containing atmosphere at between 200 and 300 °C in order to stabilize them for the subsequent

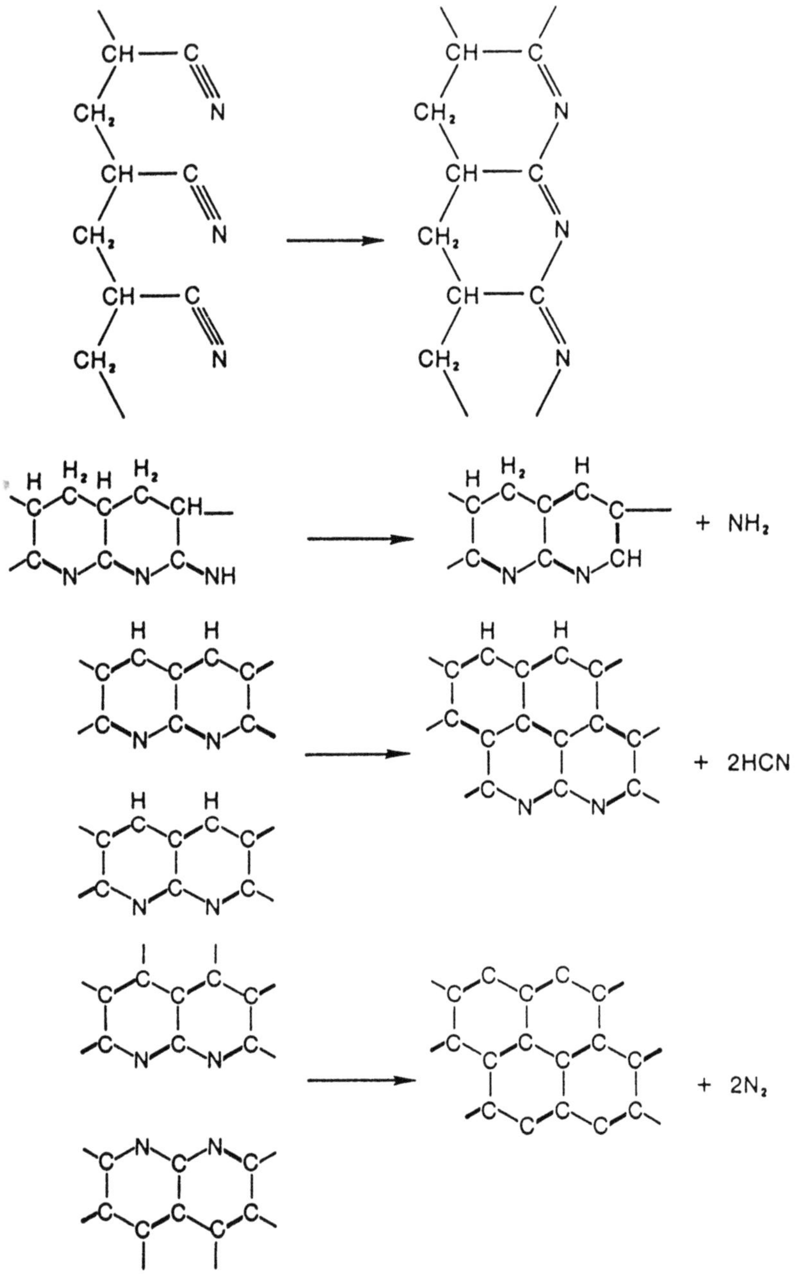

Fig. 2.4 Simplified chemical representation of the conversion of PAN to carbon fibres [32].

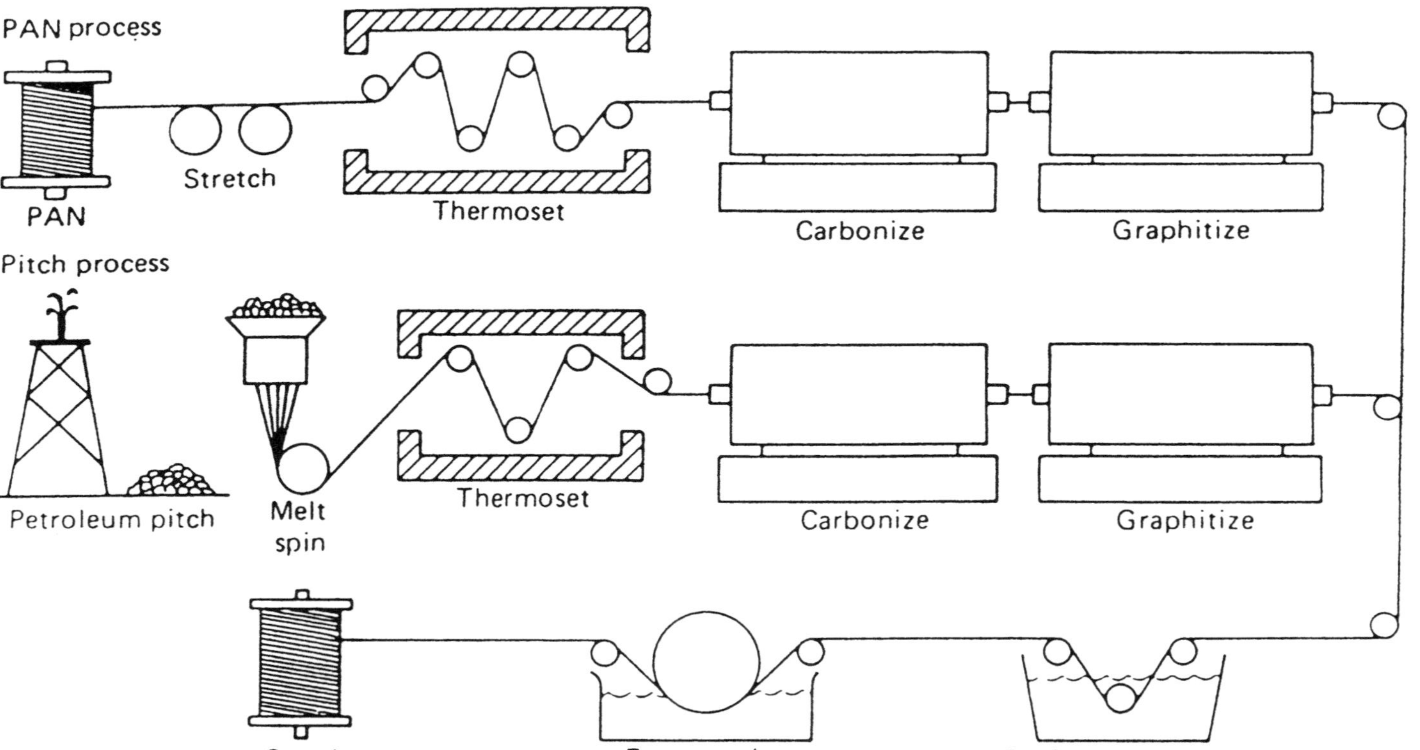

Fig. 2.5 The processing sequence for PAN and mesophase pitch-based precursor carbon fibres shows the similarity of the two processes. The PAN process obtains highly oriented carbon chains by hot stretching of the polymer chains prior to carbonization, while the high degree of orientation in pitch is a natural consequence of the mesophase (liquid crystal).

carbonization. Despite oxidation and stabilization being an extremely complex process, it is possible to identify the two most important processes which occur; (a) the nitride groups react to form a closed ring structure, and (b) oxygen aids cross-linking of the chains. The reaction of the nitride groups is extremely exothermic and must be carefully controlled. Oxidation is the essential PAN-stabilization stage since it allows the subsequent polymer degradation reactions during carbonization to proceed without collapse of the fibre or loss of orientation [33–36]. Air is mainly used as the oxidation atmosphere, although other media such as oxygen or ozone-enriched air and sulphur dioxide or nitrogen dioxide have been examined [37,38]. The orientation of the graphene layers can be maintained by 'hot stretching' the fibres, i.e. operating the process under tension.

Carbonization is carried out at temperatures of between 1000 and 1500 °C. A typical commercial process producing IM fibres would carbonize at 1000 °C followed by heat treatment to ≈1300 °C to improve the tensile strength. The carbonization stage causes fundamental changes in both chemical composition and physical properties. Around 50% by weight of the fibre is volatilized as water, ammonia, hydrogen cyanide, carbon monoxide and dioxide, nitrogen and, possibly, methane. The volume of gas evolved is around five orders of magnitude greater than that of the fibres and approximately 1000 times the volume of the carbonization equipment [30]. It is possible, therefore, to consider that the fibres are bathed in their own decomposition products during continuous processing. Carbonization is often carried out in an inert atmosphere such as 'medically pure' nitrogen or argon in order to prevent oxidation due to the ingress of air and to dilute the extremely toxic waste gases prior to extraction.

During carbonization the fibre gradually changes from a polymeric textile of low strength and modulus but high extensibility (strain to failure ($\varepsilon_f \approx 10\%$), to a brittle, high strength and modulus ceramic fibre with low extensibility ($\varepsilon_f \approx 1$–2%). The density of the fibres increases from around 1.45 to 1.7 g cm^{-3} or greater coupled with an associated reduction in fibre diameter from 10–15 to 6–9 μm. A significant longitudinal shrinkage results from the change from the more open structure of polymer chains towards the tight graphene layering of carbon atoms. The shrinkage is usually limited to less than 10% by the application of suitable tensioning. The gases evolved during pyrolysis are emitted from within the structure at various temperatures by a number of diffusion-controlled mechanisms [39]. If the alignment of C–C sequences in the polymer is to be maintained during thermal degradation it is paramount that the gaseous evolutions are strictly controlled by the manufacturing conditions. Following carbonization the fibre consists of >92% by weight of carbon. The residual nitrogen may be progressively removed between 1000 and 1500 °C.

It is clear that in proceeding from PAN precursor fibres, with a modulus of roughly 15 GPa, a complete range of modulus and associated strength

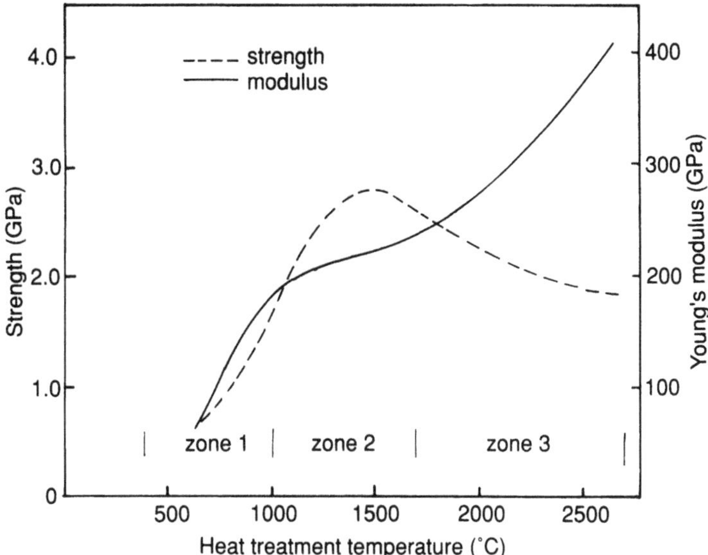

Fig. 2.6 Strength and modulus of PAN-based carbon fibres with respect to HTT.

levels may be obtained. From a commercial point of view, there is very little to be gained in competition with glass fibres ($E \approx 70$ GPa) or even polymeric fibres such as aramids (Kevlar and Twaron) with modulus values of ≈125 GPa. As a result carbon fibres settle into three groups, LM, IM and HM fibres.

Figure 2.6 illustrates typical curves for the variation of the mechanical properties of PAN-based carbon fibres with respect to HTT. The curves may be sudbivided into three zones. Zone 1 is the carbonization zone in which the pre-oxidized PAN fibres undergo the main thermo/chemical degradation losing up to 50% of their weight. Carbonization does not degrade the fibres physically and is complete at around 1000 °C. A certain amount of residual nitrogen, usually 4–8%, is present but the modulus attained (typically 200 GPa) heralds the first main level of commercial exploitation, i.e. LM carbon fibres. Zone 2 covers the region 1000–1500 °C in which very little chemical change takes place save for the expulsion of nitrogen, but where crystallite growth and further alignment begin to occur. The slope of the modulus/HTT curve is relatively shallow, such that in this region there is as much contribution from the degree of alignment in the original PAN polymer as there is from increased temperature towards the fibre modulus.

The graphene networks are aligned with respect to the fibre axis, typically making an average angle of 30%. The angle between the planes and the fibre axis is known as the **misorientation angle**, as defined in Fig. 2.7. Further improvements in the orientation, and hence modulus, may be

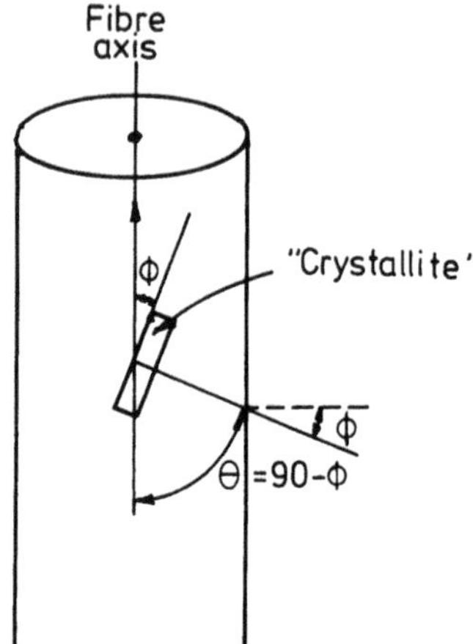

Fig. 2.7 Definition of the crystallite orientation angle θ and the corresponding 'misorientation' angle ϕ for crystallites in carbon fibres. Note: $\theta = 90 - \phi$.

achieved by the higher HTTs covered by zone 3 on Fig. 2.6 (1990–3000 °C). It is generally believed that the majority of manufacturers separate their graphitization process from the continuous oxidation and carbonization stages. They tend to avoid proceeding to this stage because of the limited markets for HM grade fibres.

'Graphitization' is again a straight-through process in an inert atmosphere under tension. There is very little evolution of gas, the major changes being in the physical structure of the fibres. Microcrystallites grow in size and the preferred orientation of the carbon basal planes is improved [39, 40]. The fibre may be considered to be transforming towards a graphitic structure. Stretching of the fibres by the application of suitable tension has been shown to aid the transformation and further improve alignment and thus modulus [41,42]. Care must be taken, however, to avoid breakage of the fibres due to overstretching.

The residence times required to complete the graphitization process are only of the order of a few minutes compared with about 1 hour for carbonization. The high-temperature technology required places a high cost premium on this grade of fibre, pushing their price above £100 kg^{-1}. Increasing the scale of manufacturing plant to produce in excess of 100 tonnes per annum would realize cost benefits, but the fundamentally low strain to failure fibres are limited to lightly loaded, stiffness-critical applications, such as satellite space frames, thus narrowing their market potential. The reasons for the low extensibility are not fully understood but are believed to be due to flaws within the fibres and the graphitic structure itself [43–45].

A preferred orientation of the graphene layers with respect to the fibre axis is essential in producing HM fibres. Tensile strength, on the other hand, is not so specifically related to orientation. In the 500–1000 °C 'zone 1' region, strength is observed to increase proportionately with modulus. In zone 2 (HTT 1000–1700 °C) strength generally exceeds 1% of the modulus value (1.2% or greater strain to failure). The tensile strength, in terms of fracture mechanics, is controlled in the main by surface and possibly internal flaw mechanisms [39,46]. In the HM zone 3, flaw mechanisms become obscure; it is thought that the graphitic structure contributes significantly to the strength [43,44,47]. A peak in fibre strength is generally observed in the 1200–1500 °C temperature range [46,48]. Although a few workers have reported a steady increase in strength with respect to HTT [30], it should be noted that even with strength increases at higher processing temperatures, the modulus increases proportionately faster, such that there is always a decrease in strain to failure in the zone 3 temperature region.

As is observed with other brittle materials, the strength of PAN-based carbon fibres is statistically complex. The wide scatter of individual filament strengths is related to the frequency and severity of contaminant flaws [46,49,50]. In general, manufacturers are able to achieve reasonable and consistent strength levels by stringent quality control. Great care is taken to ensure the spinning of clean and unflawed PAN fibres, the prevention of mechanical damage to the surface of fibres during processing, the prevention of inter-fibre sticking during oxidation and the prevention of oxidation attack during carbonization and graphitization. Improved processing has led to continual improvements in strength, reliability and strain to failure from year to year. Higher processing temperatures not only lead to higher modulus fibres but also cause densification of the fibre structure. The density of PAN-based carbon fibres varies between 1.7 and 1.9 g cm^{-3} depending on final processing temperature, with a good consistency (±0.02 g cm^{-3}) for any individual grade. Representative properties of the various grades of PAN-based fibres are given in Table 2.5.

2.2.4 Pitch-derived carbon fibres

Pitches are isotropic mixtures of polyaromatic molecules obtained as by-products of coal-tar and petroleum processing (Chapter 5). They are relatively easy to melt spin into fibres. Unfortunately, the fibres so formed are of low modulus and strength as a direct result of their isotropic structure. Their mechanical properties are not improved even when they are carbonized to high temperatures, unless there is hot stretching at very high temperatures of between 2700, and 3000 °C – a very costly and impractical process [51]. Table 2.6 lists typical properties of isotropic pitch-based carbon fibres which are generally used as relatively cheap fillers in plastics and, more recently, to improve the strength and toughness of concrete, where their resistance to environmental attack is exceptionally useful [52]. Although pitches are isotropic in character, a number of processes have

Table 2.5 Properties of carbon fibres from PAN precursors

Properties	*Low modulus (LM)*	*Intermediate modulus (IM)*	*High modulus (HM)*
Axial			
Tensile strength (GN m^{-2})	3.3	4–5	2.4
Tensile modulus (GN m^{-2})	230	270	390
Elongation to break (%)	1.4	1.7–1.9	0.6
Thermal conductivity (W $m^{-1}K^{-1}$)	8.5	—	70
Electrical resistivity (Ωm)	18	—	9.5
CTE at 21 °C ($10^{-6}K^{-1}$)	−0.7	—	−0.5
Transverse			
Tensile modulus (GN m^{-2})	40	—	21
CTE at 50 °C ($10^{-6}K^{-1}$)	10	—	7
Bulk			
Density (g cm^{-3})	1.76	1.8	1.9
Fibre diameter (μm)	7–8	6–7	4–6
Carbon assay (%)	92	96	100

Table 2.6 Properties of isotropic pitch-based carbon fibres

Property	*Value*
Axial	
Tensile strength	1.0 GN m^{-2}
Tensile modulus	41 GN m^{-2}
Elongation to break	2.5%
Electrical resistivity	20 Ωm
Bulk	
Density	1.6 g cm^{-2}
Fibre diameter	8.5 μm
Carbon assay	99%

been developed to convert the pitch into a mesophase, or liquid crystal, system (Chapter 5) [53].

Commercial pitches are complex mixtures of aromatic compounds which, when heated to temperatures of around 400 °C, undergo dehydrogenation condensation reactions to form planar aromatic molecules which aggregate into a liquid crystalline phase known as the mesophase. When the mesophase content reaches approximately 40%, a phase inversion occurs such that this highly anisotropic material becomes the continuous phase [22]. The discovery that mesophase pitch could be used as a precursor for carbon fibres led to an expectation that the process would create a low-cost,

Table 2.7 Carbon yields from the major precursors to carbon fibres [54]

Precursor	*Chemical composition*	*Carbon yield* (%)
Rayon	$C_6H_{10}O_5$	20–25
PAN	CH_2–CH–CN	45–50
Mesophase pitch	Polynuclear aromatics	75–85

high-performance fibre. It was estimated that a cost as low as £6 kg^{-1} could be achieved.

There are several reasons why the mesophase pitch process ought to yield a lower-cost high-performance fibre. Firstly, the isotropic precursor to the mesophase is extremely cheap, costing between £0.1 and 0.2 kg^{-1} compared with £0.5 kg^{-1} for acrylonitrile. Secondly, the thermosetting process does not require constant tension to be applied to the fibres because the melt spinning process inherently imparts a high degree of molecular orientation into the pitch filament as it is being spun, thus eliminating the need for tension control equipment. Thirdly, since the pitch-based carbon fibre begins with a structure closer to graphite than PAN fibre, it necessarily follows that less energy will be required to convert it to an HM axially aligned structure. One would therefore expect lower temperatures and/or shorter residence times for the carbonization of pitch-based fibres. Finally, mesophase pitch contains a far smaller percentage of heteroatoms such as nitrogen, hydrogen and other non-carbon elements than does PAN. The carbon yield of mesophase pitch is much higher than that from either rayon or PAN so the pitch process ought to be more efficient (Table 2.7) [54]. Despite the obvious cost advantages, high-performance pitch-based carbon fibres presently sell for £60 kg^{-1} and upwards. They are unable to compete with PAN-derived fibres except in the very specialized 'ultra-high modulus' ($E > 400$ GPa) market. Hughes [55] suggests that the unexpectedly high cost results from the extensive purification necessary to remove flaw-inducing particles from the precursor. The most significant cost in the production of pitch-based fibres appears to arise, however, from the difficulties incurred during processing. The problem arises because the melt spinning of mesophase pitch is far removed from 'conventional' melt spinning processes [56].

The term 'melt spinning' is, in fact, something of a misnomer due in the main to the terminology of the textile industry whence carbon fibre production has evolved. It would be far more accurate to describe the process as melt extrusion. A typical melt spinning process and the process variables which require to be controlled are shown in Fig. 2.8. The precursor is generally melted in an extruder which pumps the melt into a 'spin pack'. The spin pack contains a filter to remove solid particles from the melt. After filtration the melt leaves the bottom of the spin pack via a spinnerette.

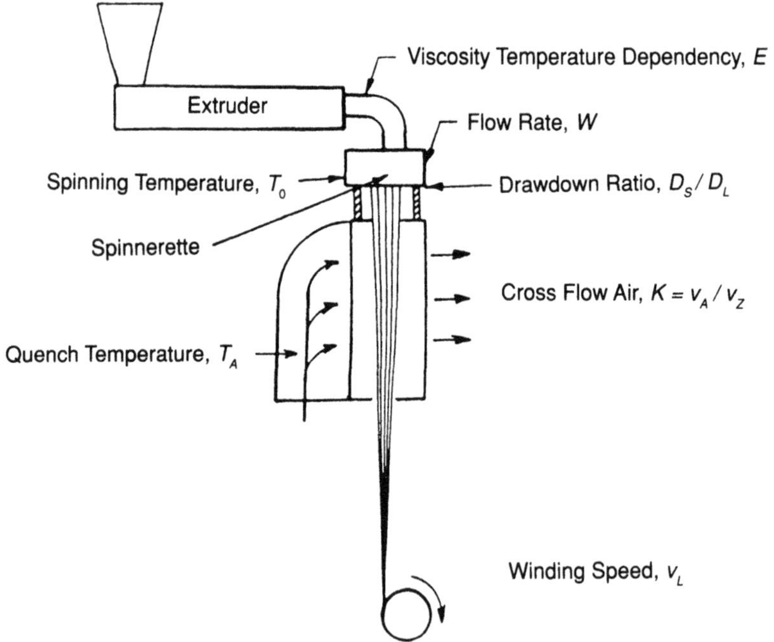

Fig. 2.8 Process variables in the melt spinning of pitch-based carbon fibres [56]. Reproduced from *Carbon* by kind permission of Pergamon Press.

The spinnerette is a plate containing a large number of individual capillaries. Melt leaving the capillaries cools and forms a filament. The solidified fibre, whose cooling is often aided by an 'air-quench' is finally wound on a spinning spool. The primary process variables are the winding speed, v_L, the precursor extrusion rate, W, the temperature at extrusion, T_0, the temperature of the air quench, T_A, the velocity of the quenching air relative to the downward velocity of the fibres, K, and the drawdown ratio D_s/D_L.

If a precursor pitch is to be melt spinnable it must be capable of forming an unbroken filament between the capillary exit of the spinnerette and the winding-up device. Should the melt temperature be too high, the extruding jet of material will disintegrate into droplets, destroying the filament. On the other hand, if the tensile stress within the filament exceeds the tensile strength of the material at any point along the threadline, the fibre will break as a result of cohesive fracture. For any given material there will be a **processing window** of spinning conditions which permit the formation of an unbroken filament. The extreme temperature dependence of the viscosity of mesophase pitch makes melt spinning a very difficult process to control. The processing window for the manufacture of pitch-based carbon fibres is extremely small [56]. For example, even at typical spinning conditions the tensile stress generated in the pitch filament is almost one-half its ultimate tensile stress (UTS), unlike polymeric fibres such as nylon which are spun under a tension of only the order of 1% of their UTS. It

Table 2.8 Properties of mesophase pitch-based carbon fibres

Property	*Low modulus*	*High modulus*	*Ultra-high modulus*
Axial			
Tensile strength (GN m^{-2})	1.4	1.7	2.2
Tensile modulus (GN m^{-2})	160	380	725
Elongation to break (%)	0.9	0.4	0.3
Thermal conductivity (W m^{-1} K^{-1})	—	100	520
Electrical resistivity (Ωm)	13	7.5	2.5
CTE at 21 °C (10^{-6} K^{-1})	—	−0.9	−1.6
Transverse			
Tensile modulus (GN m^{-2})	—	21	—
CTE at 50 °C (10^{-6} K^{-1})		7.8	
Bulk			
Density (g cm^{-3})	1.9	2.0	2.15
Fibre diameter (μm)	11	10	10
Carbon assay (%)	>97	>99	>99

has been shown that the extruding pitch filaments should be quenched slowly to improve spinnability. Also, as one might predict, larger filaments are easier to spin than the more desirable small-diameter fibres. Finally, small temperature gradients across the spinnerette will create a large variation in filament size and tensile stress due to the extreme temperature dependence of the mesophase. The melt spinning of mesophase pitch thus requires very accurate and expensive process control.

After extrusion, the fibres are thermoset by oxidation in dry air by being slowly heated to temperatures in the range 200–300 °C, followed by rapid cooling in an argon (inert) atmosphere. The fibres are subsequently carbonized and 'graphitized' to between 2500 and 2700 °C. Typical properties of mesophase pitch-based fibres are given in Table 2.8.

2.3 THE STRUCTURE OF CARBON FIBRES

The modulus of a carbon fibre is determined by the degree of preferred orientation of the graphene layers along the fibre axis. The tensile strength of the fibre is governed by both axial and radial textures along with any flaws present in the structure. Wetting of the fibres by the composite matrix and the strength of the interfacial bond are strongly influenced by the orientation of the graphene layers at the fibre surface.

2.3.1 Axial structure

A quantitative representation of the degree of orientation of graphene planes with respect to the fibre axis may be obtained by X-ray diffraction

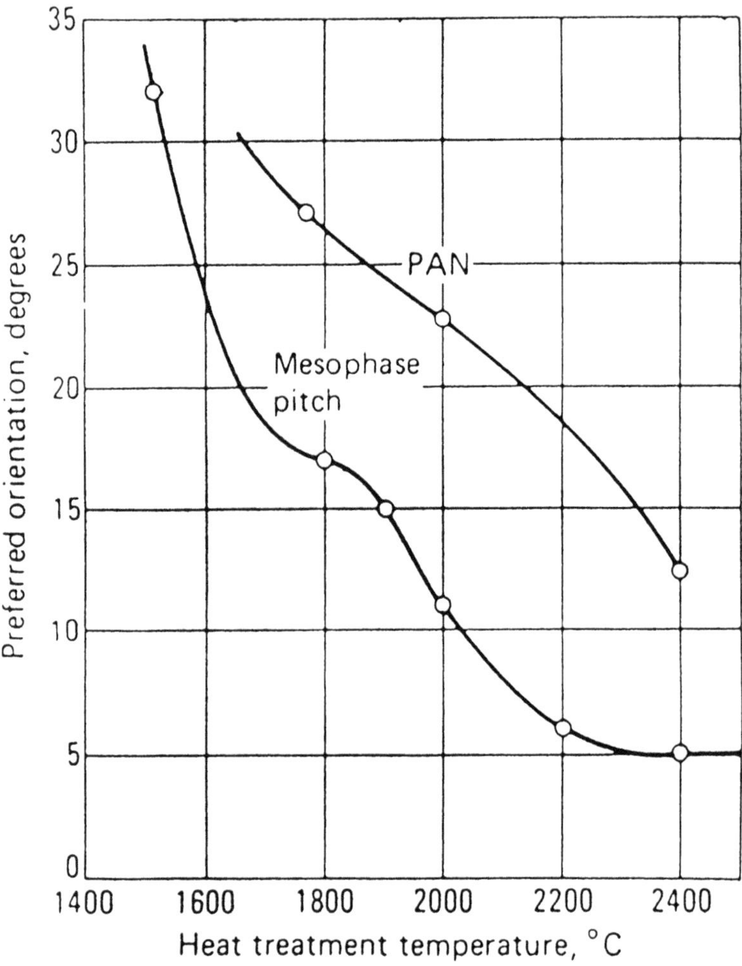

Fig. 2.9 The preferred orientation of the graphene planes is determined by HTT and the precursor type [6,58].

[57]. Figure 2.9 shows the effect of HTT on the preferred orientation of PAN and mesophase pitch-based carbon fibres [6,58]. The trends in the two curves are similar except that the PAN curve is shifted roughly 400 °C higher. The relation between preferred orientation and fibre modulus for a pitch-based fibre is shown in Fig. 2.10. Approximately 66% of the graphene layers are aligned within 15° of the fibre axis in the case of a carbon fibre with a tensile modulus of 220 GN m^{-2}. Correspondingly, for a fibre of modulus 400 GN m^{-2}, the orientation is improved such that two-thirds of the graphene layers are found to lie within 6° of the fibre axis. Similar behaviour is observed with PAN-based fibres, although it is extremely difficult to obtain preferred orientations of below 10° and thus

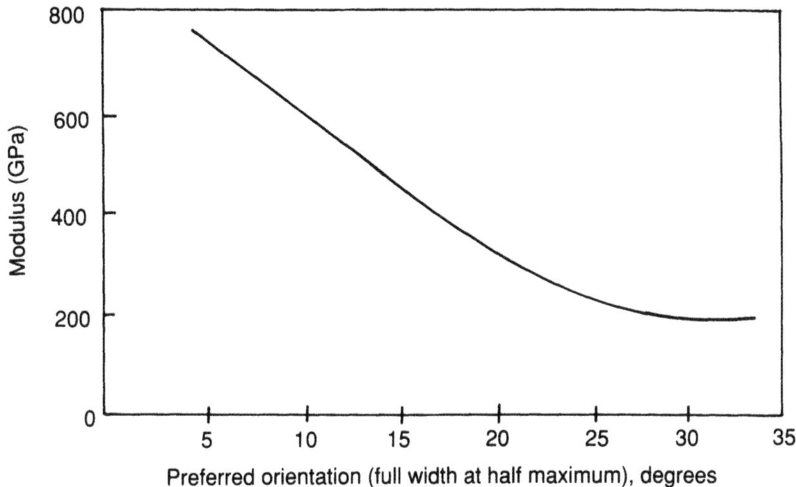

Fig. 2.10 The relationship between tensile modulus and preferred orientation for a pitch-based carbon fibre.

moduli greater than 400 GN m^{-2}. In contrast, it is significantly easier to obtain orientation down to 5° in pitch-based fibres such that fibres with moduli in excess of 800 GN m^{-2} have been produced on the laboratory scale [59]. Direct lattice imaging using high-resolution transmission electron microscopy (TEM) has shown that the graphene planes stack on top of each other rather like the pages in a book. They are ribbon-like in structure having widths of the order, typically, of a few hundred nanometres. An artist's impression of the arrangement of these planes in an ex-PAN fibre is shown in Fig. 2.11 [60]. In a longitudinal fibre cross-section, as sketched in Fig. 2.12 [61], the ribbons, which are stacks of graphene layers, would be seen to undulate and separate but the majority would be roughly aligned along the fibre axis. The presence of needle-like voids (as shown in Fig. 2.11) reduces the fibre density below the value (2.26 g cm^{-3}) one would predict for an idealized planar structure of hexagonally arrayed carbon atoms. Figure 2.12 indicates the 'crystallite' dimensions L_a and L_c, representing the extent of relatively straight portions of the lattice planes along their length, and the stacking height of the graphene planes in the ribbons respectively. Increasing the HTT allows the graphene planes to grow, with concomitant increases in L_a and L_c as the fibre tends towards a higher modulus.

2.3.2 Radial structure

A cross-section across the fibre axis, as shown in Fig. 2.10, indicates a much higher state of disorder than in a longitudinal section (Fig. 2.11).

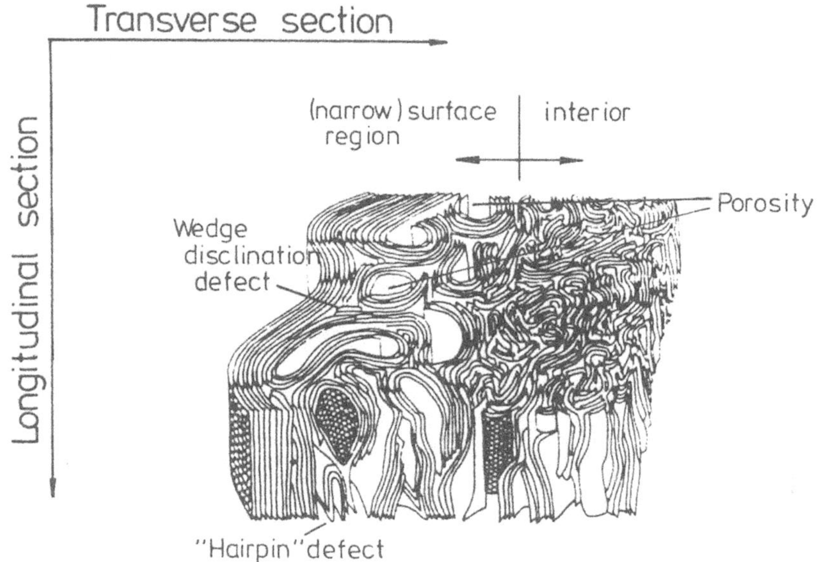

Fig. 2.11 Artist's impression of the structure of a high-strength PAN-based carbon fibre [60].

A transverse section shows the planes to be twisted over much shorter distances than the width of ribbons. Additionally, there is a random orientation of planes over the diameter of the fibre. The radial structure of the graphene planes varies with processing conditions and type of fibre. Radial texturing increases with the degree of axial preferred orientation [62,63].

Figure 2.13 shows an SEM micrograph of an ultra-high modulus mesophase pitch-based carbon fibre (Amoco P75) illustrating its radial axial texture. If the extrusion of the pitch is performed without stirring, such radial structures tend to be formed. Stirring of the pitch melt during spinning has the effect of reducing the structural ordering of the resultant fibres, as characterized by such parameters as the interlayer spacing $\tilde{C}$ and the crystallite size L_c [64].

A number of other cross-sectional structures can be achieved by stirring the pitch, such as random and onion-skin rather than radial (Fig. 2.14). The very stiff and brittle radial structured pitch-based carbon fibres such as P75 and P100 are often damaged such that a section fractures and becomes detached during handling. Fibres damaged in this way are referred to as having a 'Pac-man' section as a result of their resemblance to the computer game character (Fig. 2.14).

Polyacrylonitrile-based fibres may have contorted graphene planes and exhibit what are called 'onion-skin' structures. Low and intermediate modulus PAN fibres ($E < 350$ GN m^{-2}) have a highly contorted transverse

Fig. 2.12 Sketch of a longitudinal cross-section along the fibre axis of a PAN-based carbon fibre [61]. The in-plane and c-axis structural coherence lengths L_a and L_c are indicated.

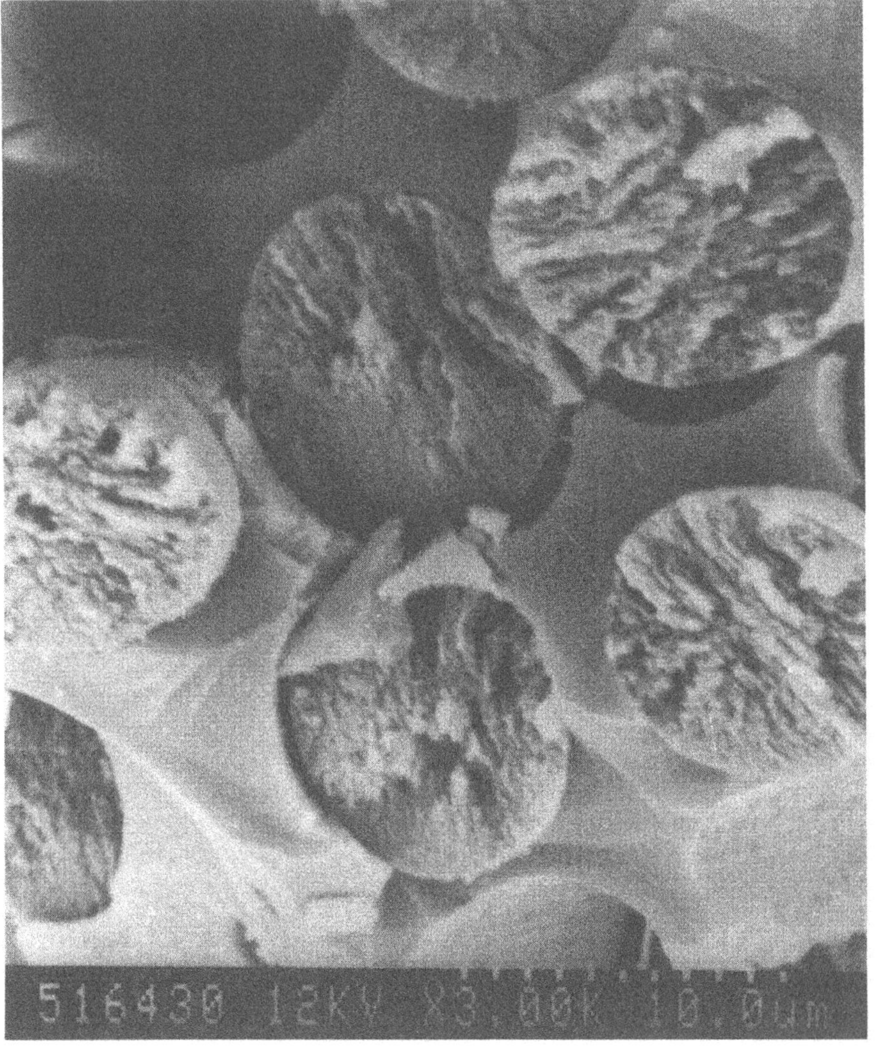

Fig. 2.13 Scanning electron micrograph of ultra-high modulus mesophase pitch-based carbon fibres showing their radial cross-section.

graphene layer structure. Higher modulus fibres have a thin, highly oriented, onion-skin surface layer about a randomly oriented, contorted layer core [65].

2.4 COMMERCIALLY AVAILABLE FIBRES

Carbon fibres are available from a variety of suppliers in a number of yarns and tows with different moduli, strengths, cross-sectional areas and

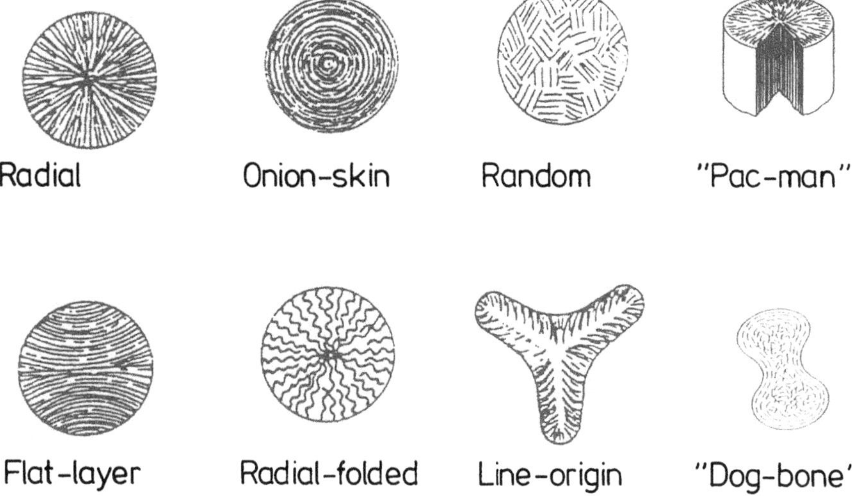

Fig. 2.14 Possible cross-sectional morphologies of high-performance carbon fibres.

shapes, twists and number of fibre ends. They may be purchased in a multitude of different forms including continuous tows, chopped and milled, woven fabrics and three-dimensional preforms (Section 2.6). The diversity and tailorability of the properties of carbon fibres are one of their advantages but, accordingly, are also a problem in that a complete evaluation is difficult and expensive to obtain. It is possible, however, to make a number of generalizations:

1. The fibres produced from PAN and mesophase pitch have significantly different combinations of properties and hence, applications. At present they cannot be considered interchangeable.
2. There are generally four classes of carbon fibre subdivided by moduli: low, intermediate, high and ultra high. Subdivision into these categories is not, unfortunately, as straightforward as, for example, metal alloys. Although the fibres made by different manufacturers may be similar they are not identical. Subtle differences in choice of precursor and processing conditions may have quite significant effects on the fibre properties and on its behaviour when combined into a composite. Constant improvements by the suppliers to their products have resulted in a degree of overlap between the categories.
3. The fibre selections that make up a particular manufacturer's grade, but have different numbers of fibres within the tows, are usually, but not always, based on the same precursor. While the fibres within each grade may be considered to possess the same properties, the fibre count or degree of twist, etc. may affect subsequent composite properties.
4. It is not possible to list all of the properties of the different suppliers'

products; Table 2.5 lists the general properties for a number of products.

5. The range of carbon fibre grades in present use is between 5 and 11 μm in diameter. The trend is towards smaller and smaller fibre diameters in order to attain improved tensile strength and processing speeds. The compressive and transverse properties of composites employing smaller diameter fibre reinforcement do not increase in proportion with tensile properties. As a result failure in compression may result at lower than predicted loads as a result of buckling.
6. Secondary load-bearing structures generally require at least 1.5% strain to failure and primary structures 2% or better. As a result, IM PAN-based grades are generally preferred for such tasks.
7. The major advantage of mesophase pitch-based fibres is that they can be made into ultra high modulus ($E > 600$ GN m^{-2}), and their use is generally limited to space structures because of their excessive cost (up to £1000 kg^{-1}).
8. Great care must be taken when selecting a reinforcing fibre for a carbon–carbon composite to avoid a situation arising in which a component is required to operate at a temperature higher than the temperature at which it was produced, when a corresponding change in properties may result. For this reason a lot of carbon–carbon materials are based on heat-treated (post-weaving) fabrics, to ensure that no shrinkage occurs throughout the process.

2.5 SURFACE TREATMENT OF CARBON FIBRES AND INTERFACIAL BONDING

Carbon fibres, especially those of HM, are extremely difficult to wet and bond with polymeric resins and virtually impossible to couple with metals [66]. In order to form a bond between fibre and matrix capable of affording an efficient stress transfer in a carbon fibre reinforced composite system, it is necessary to apply some form of surface treatment to the fibres. Such treatments serve to increase the number of active chemical groups on the fibre surface and/or roughen the fibre surface to increase the amount of mechanical 'keying' between fibre and matrix. Chemical coatings, known as 'sizes', have been extensively developed for the epoxy family of resins (the most common matrix in polymer composite systems). The size, which is applied prior to shipping, prevents fibre abrasions, improves handling and provides an epoxy-compatible interface by the inclusion of coupling agents such as titanates [67]. Other techniques include oxidation using liquid agents (nitric acid, potassium permanganate, etc.) [68], gaseous oxidation [69], gas plasma techniques [70] and a whole host of others as reviewed by Delmonte [51].

In a polymeric matrix system, it is generally accepted that a strong bond between fibre and resins is essential to achieve optimum strength and stiffness in a composite system. The degree of bonding must, however, be lessened somewhat in systems where toughness and impact resistance are of interest. Carbon–carbon composites differ significantly from their polymeric ancestors in that both matrix and fibres are brittle materials. As a result one generally aims for a weak interface in such composites so that cracks travelling through the matrix when a component is loaded may be deflected by debonding at the fibre/matrix interface. The bonding in carbon–carbon is generally considered to be of the relatively weak keying type. Recent research by the author and co-workers (Chapter 5) has shown, however, that it is possible to obtain a 'pseudo-polymeric' fibre/matrix interface under certain conditions; one of these being severe (oxidative) surface treatment of the fibres prior to impregnation.

2.6 CARBON FIBRE PRODUCT FORMS

2.6.1 Discontinuous fibres

Chopped carbon fibres, generally between 6 and 13 mm in length, are the least expensive and the most widely used as reinforcements to moulding compounds. While they may be used with all types of resin, chopped carbon fibres are mostly compounded with thermoplastic resin matrices. The fibres, which are mainly the lower cost varieties, i.e. isotropic pitch-based, and LM PAN-based, impart reduced shrinkage, greater stiffness, improved fatigue and greater resistance to electrical conductivity to the moulding compound. More finely divided milled fibres, with an average length of around 300 μm, are available, as well as longer fibres up to any specified length.

For carbon–carbon manufacture, short chopped fibres are used to manufacture fibrous webs such as felts (Fig. 2.15) and mats which are subsequently densified using CVD and/or liquid pitch impregnation/pyrolysis for the purpose of brake manufacture. Chopped fibre compression moulding compounds are also used to make brake materials but these are impregnated as continuous tows and chopped afterwards (Chapter 4). Milled carbon fibres have found no commercial exploitation in the carbon–carbon theatre to date, although it has been shown that they can be used to reduce the room temperature friability of pitch so that it may be 'prepregged' (Chapters 4 and 7) in the same way as polymer resins [71].

2.6.2 Continuous fibres

The final step in the carbon fibre production process is the **spooling** or winding of the tows on to a carrier tube (Fig. 2.16). The tows must be

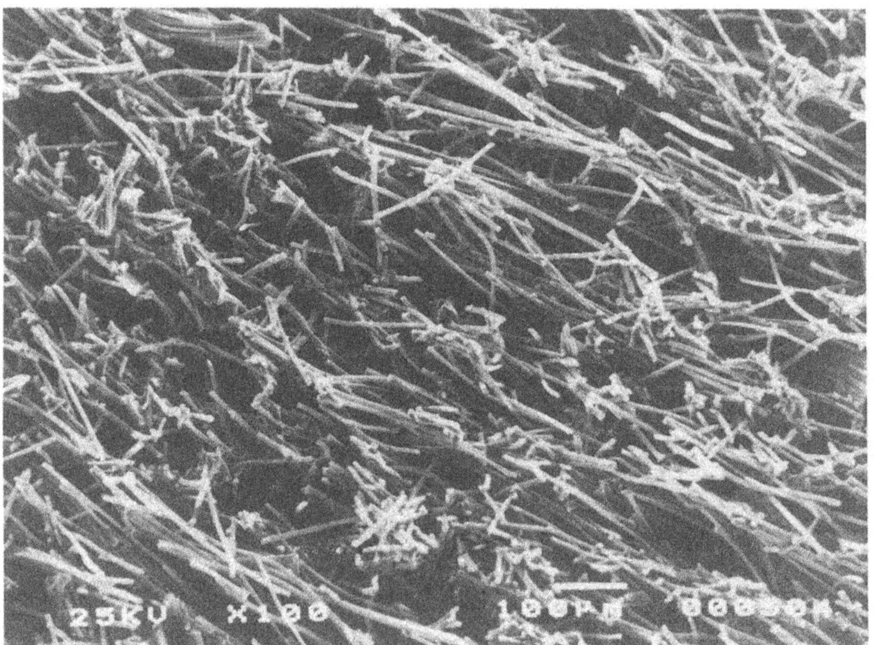

Fig. 2.15 Discontinuous, low modulus, carbon fibre felt material as used in carbon–carbon brake manufacture.

wound tight enough to form a stable package but not so tight as to impede their subsequent removal or to be damaged in the spooling process. Single tows from such spools may be used in filament winding operations (Chapter 4) or collimated to prepare unidirectional tapes, usually between 50 and 1000 mm wide, and are used to manufacture pre-impregnated sheets or prepregs (Chapter 4). Carbon fibre tows are also woven and knitted to form a number of constructions and preforms as described in the following section.

2.6.3 Woven fabrics

There are generally three reasons cited for the employment of woven products in composite structures: their ease of conformance to complex geometries, reduced manufacturing costs and improved damage resistance and tolerance. Undirectional fibre tapes have negligible strength in the direction normal to the fibres. Any attempt to stretch them in that direction to conform to the surface of double curvature tooling would therefore lead to tape splitting. The answer to that problem is to select a woven product with sufficient 'drape' to conform to the contoured surface. Fabrics, on the whole, have widths between 1 and 3 m, i.e. considerably greater

Fig. 2.16 Typical carbon fibre spools (courtesy Ciba-Geigy).

than that of tapes (50–1000 mm). It is thus possible to lay up larger areas without seams. If the fabric is close to being balanced, a single fabric ply will replace two orthogonal tape plies, thereby further reducing the amount of lay-up. This is particularly important in cases where automatic tape lay-up is not available or complexity of form precludes its use. Economic considerations alone often strongly dictate the use of fabrics especially where lamination must be done by hand. Improved laminate toughness is also an important consideration in the utilization of fabrics in brittle fibre/brittle matrix composites such as carbon–carbon.

Woven broad goods, as an intermediate product, present the fibres in a convenient format for the design engineer, resin coater (prepregger) and component fabricator to use. A great number of variations in properties are possible by combining different yarns and weaves, allowing the designer a wide range of laminate properties. The fabric pattern, or construction, is an *x*, *y* coordinate system. The *y*-axis represents the long axis of the fabric roll (typically 30–150 m) and is known as the **warp** direction. The warp yarns are known as 'ends'. The *x*-axis in the width of the roll (generally 0.9–3 m) is known as the **fill** direction and fill yarns as 'picks'. One of the defining characteristics of any fabric construction is the number of ends per unit length times the number of picks per unit length. A weave is

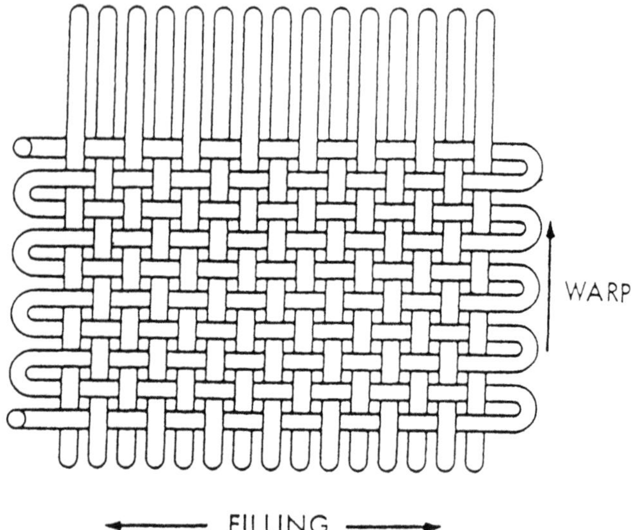

Fig. 2.17 Illustration of weaving nomenclature: warp threads are called 'ends'; filling threads are called 'picks'; fabric construction is given as ends × picks per inch (or cm); weave design (style) describes the way the fibres interlace.

described as balanced if the same yarn and weight are used in both directions. All other waves are unbalanced. The extreme cases of unbalanced weaves are known as 'unidirectional' because only a very small amount of fill yarn is used to keep what are effectively the undirectional load-carrying yarns together. The basic nomenclature of the weaving process is illustrated in Fig. 2.17 using the example of a 'plain' weave in which the yarns are interlaced at 90° to each other [72]. Fabrics in which the weave does not lie on 0/90° orientation to the length direction are known as **angle ply** fabrics.

There are basically three styles of weave used in the composites industry. They are the plain, twill and satin weaves. The plain weave, which is by far the most commonly used (Fig. 2.18), consists of the picks and ends going over and under one another, resulting in an equal amount of warp and fill yarns on either side of the fabric. The curvature, or deformation, arising as a result of the weaving is known as **crimp**. The application of a tensile load in the plane of the fabric will tend to straighten out the crimp, manifesting itself as a reduction in strength and stiffness of a fabric composite when compared to unidirectional tape of the same material. An obvious way to increase the fabric stiffness is to reduce the amount of crimp by having as much of the warp and fill fibres as straight as feasible. Twill or 'basket' weave is a variation of the plain weave in which warp and fill yarns are paired, two up and two down. Satin weaves are a family of

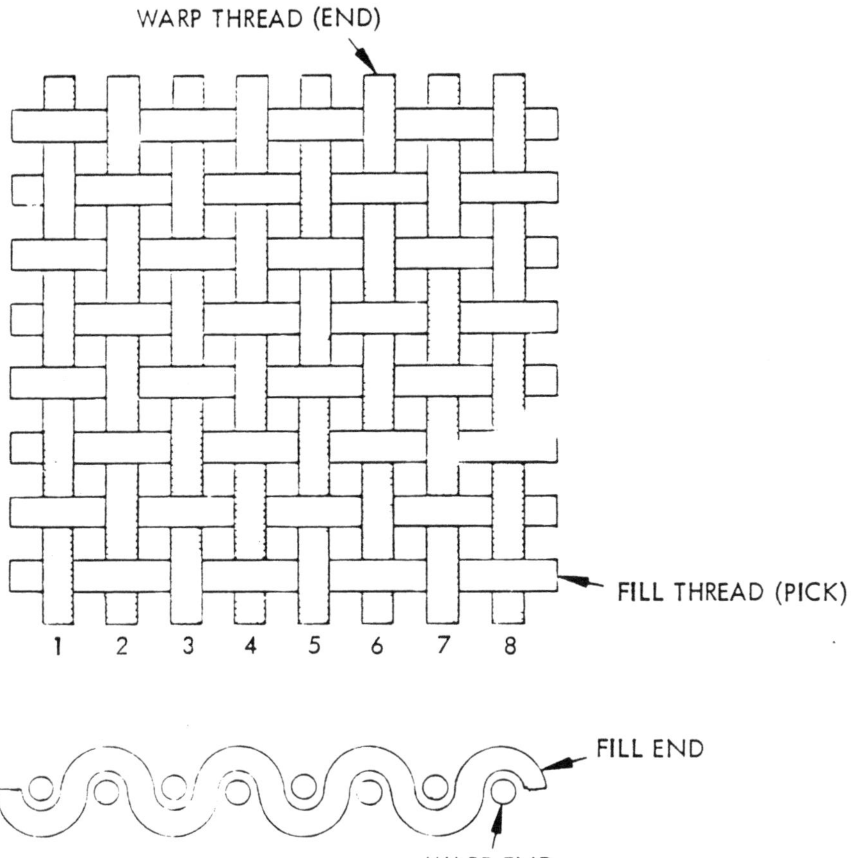

Fig. 2.18 Details of a plain weave.

constructions with a minimum of interlacing. In these weaves, the fill yarns periodically skip over several warp yarns. A **float** is the term used to describe the length of yarn between the crimped intersections. For example, consider the eight-harness satin weave illustrated in Fig. 2.19. It consists of each fill pick passing over seven warp ends and then under one end. The 'seven over/one under' pattern is repeated throughout the fabric such that no float is repeated in the same pick in one repeat of the weave. The numerical value preceding the 'HS' descriptor is always one greater than the number of warp ends over which the pick passes before crimping under a single end. The weaves obviously appear different depending on which side is viewed. When laying them up in laminates it is necessary to make a complete description to include whether each ply is laid warp or

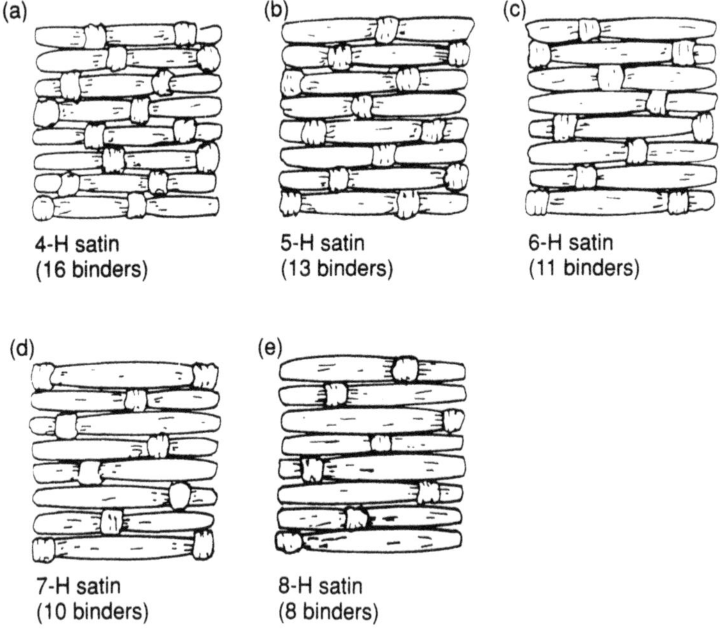

Fig. 2.20 Illustration of fabric drapability using fibre styles of satin weave [72].

fibre from the outside of the warp fibres and draw it across them, after which it is cut off. The rapier then grasps another fill yarn from a spool on the opposite side of the loom and draws it back across the warp fibres. The edges of a fabric must be reinforced to allow handling in subsequent processing. The **selvage** is the name given to the strong woven edges of a fabric (Fig. 2.21); selvages must be removed prior to lay-up.

2.6.4 Multi-directionally reinforced fabrics and preforms

The majority of applications of composite materials in aerospace and racing cars endure relatively simple mechanical loading regimes allowing the use of two-dimensionally oriented tape or fabric reinforcement. The emergence of carbon–carbon composites over the last two and a half decades has introduced a number of complicated design problems as a result of their strongly anisotropic properties. Mechanical properties are generally satisfactory in the two dimensions containing the reinforcement fabric, but are matrix dominated in the third direction, being typically an order of magnitude less than in the reinforced directions. This problem is critical in applications involving high thermomechanical loading such as re-entry components, and especially so in artefacts such a rocket motors which are required to endure an essentially isotropic stress field (Chapter 9). The result of excessive loading in the unreinforced plane is a

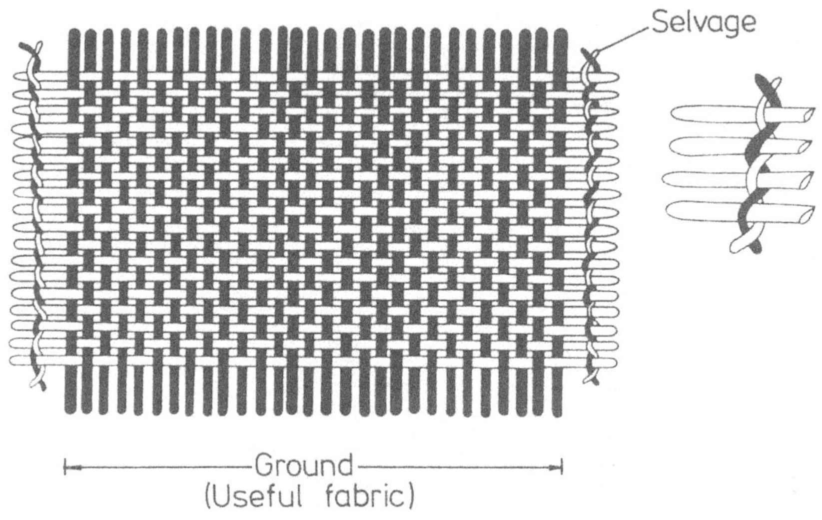

Fig. 2.21 Full-width plain weave fabric showing selvage.

delamination of the constituent plies and failure of the part under loads lower than designed for. The obvious solution to such problems, then, is to add fibre reinforcements in the third dimension, creating an (x, y, z) coordinate system of reinforcement.

The major advantage of multi-directional carbon–carbon composites is the freedom to orientate selected fibre types and amounts to accommodate the design loads of the final structural component. The disadvantages of multi-directional fabrication technology are the cost of producing the fibre preform, the size limitation on components as dictated by available equipment size and the difficulty of matrix impregnation between the three-dimensional fibre arrays [76].

The simplest type of multi-directional preform is based on a three-directional orthogonal construction and is normally used to weave rectangular, block-type preforms (Fig. 2.22). This type of preform consists of multiple yarn bundles located on Cartesian coordinates. Each of the bundles is kept straight so that the maximum structural capability of the fibre is maintained. The preforms are described by yarn type, number of yarns per site, spacing between adjacent sites, volume fraction of yarn in each direction and preform density. A number of modifications of the basic three-directional orthogonal construction are available in order to achieve more isotropic preforms. This is accomplished, as one would imagine, by introducing yarns in additional directions. Based on geometric principles, a variety of fibre orientations ranging from orthogonal, 3-directional to 11-directional reinforcements may presently be produced.

The various reinforcement patterns are described by the n-D nomenclature, where n represents the number of directions and D stands for

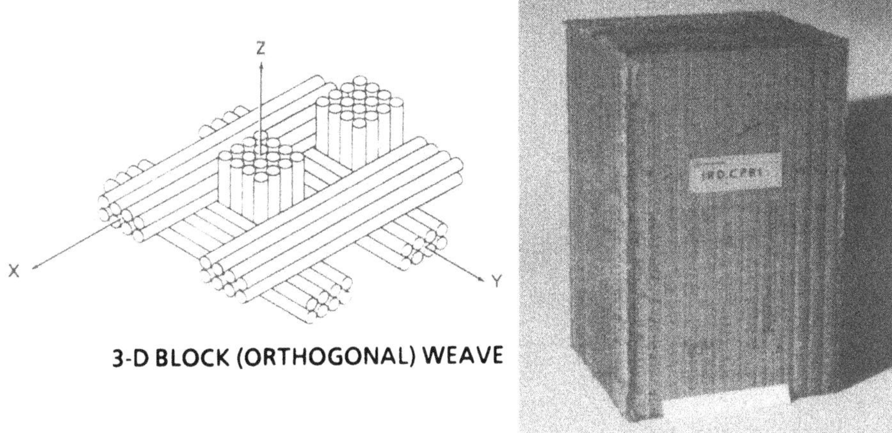

Fig. 2.22 Three-dimensional (3-D) orthogonal preform construction [77] (courtesy Hercules Inc.).

direction [77]. Thus our orthogonal reinforcement structure would be described as 3-D (not to be confused with 3-d which means 3-dimensional!). As an example, a 5-D construction can be achieved by adding two reinforcement directions that are ±45° with respect to the yarns within the *x–y* plane of the preform. Similarly, the introduction of additional diagonal yarns across the corners of the *x–z* plane would result in a 7-D reinforcement and so on. Clearly the cost of assembling preforms of this type will increase almost exponentially with complexity, such that 3-D are by far the most commonly employed.

Polar weave preforms are used to form cylinders and other shapes of revolution. They are 3-D constructions with yarns oriented on polar coordinates in the radial, axial and circumferential directions (Fig. 2.23). Preforms of this geometry normally contain 50 vol.% of fibres that can be introduced equally in the three directions. Should a specific application require unbalanced properties, a degree of variation in relative yarn distribution may be accomplished. For example, if a high hoop tensile strength is necessary, additional fibres may be added in the circumferential direction at the expense of radial and longitudinal properties.

Although originally developed as thick-walled cylinders, polar weave preforms may now be fabricated in a number of 'body of revolution' shapes such as cylinders, cylinders/cones, and convergent/divergent sections. A two-step process may be used to form non-axisymmetric shapes such as leading edges (Fig. 2.24) and conical/rectangular transitions. The first stage involves the weaving of a preform of an appropriate geometry. The preform is then placed in a metal die, deformed into the required shape and impregnated with a suitable resin to ensure geometric stability during the

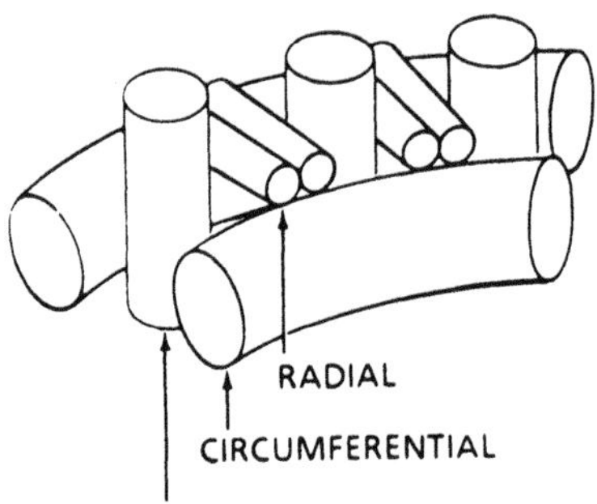

Fig. 2.23 Three-dimensional polar weave preform geometry [77] (courtesy Hercules Inc.).

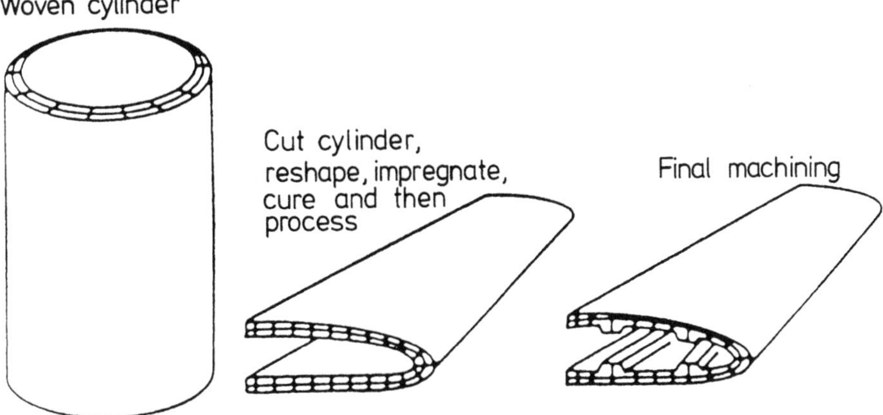

Fig. 2.24 Deformation of a 3-D cylindrical preform to produce a leading edge (courtesy Hercules Inc.).

remainder of the densification process. Depending on the shape being formed, it may or may not be necessary to slit the preform. Formation of a leading edge as shown in Fig. 2.24. would require slitting, whereas a simple deformation from conical to conic/rectangular would not. At present, polar weave preforms may be produced up to 2.1 m in diameter and 1.3 m in length. Wall thicknesses vary from 6.4 to 200 mm and there is a minimum inside diameter constraint of ≈75 mm as a result of the space requirements of the weaving mechanism.

Angle (or warp) **interlocks** are multi-layered fabrics in which the warp yarns travel from one surface of the fabric to the other. Up to eight layers of fabric may be held together, creating a thick 2-D fabric as shown in Fig. 2.25(b). Should higher in-plane strength be required, additional 'stuffer' yarns may be added to create a quasi-3-D fabric (Fig. 2.25(a)). Despite angle interlock being economical to produce on commercially available weaving equipment, its use is limited because it is unavailable in closed geometries. The production of cones and cylinders using such fabrics would be inherently weak because of the requirement of joints and their attendant plane of weakness.

Both **stitched fabric** and **needled felt** may be considered 3-D preforms although, despite having reinforcement in all three dimensions, the amount of fibre in the inter-ply direction is more often than not negligible. The inter-ply properties of both these fabrics are, thus, seldom significantly better than the matrix-dominated properties of 2-D composites.

The majority of multi-directional preforms used in carbon–carbon manufacture are represented by the orthogonal or polar constructions or by some modification of these constructions. The techniques used to manufacture the preforms include weaving dry yarns [78], piercing fabrics

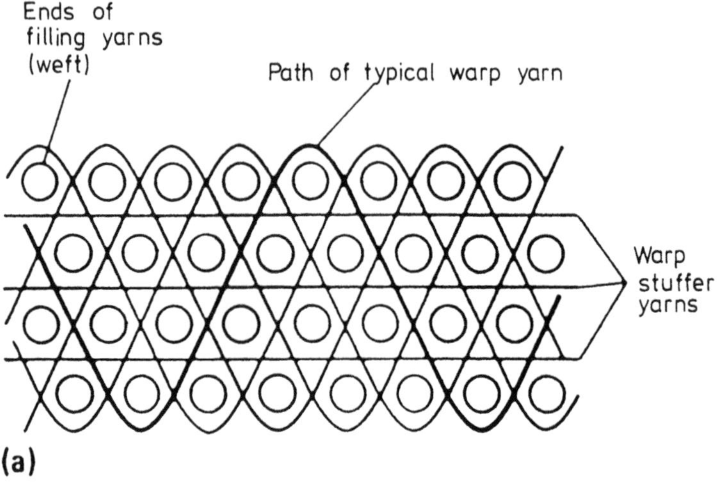

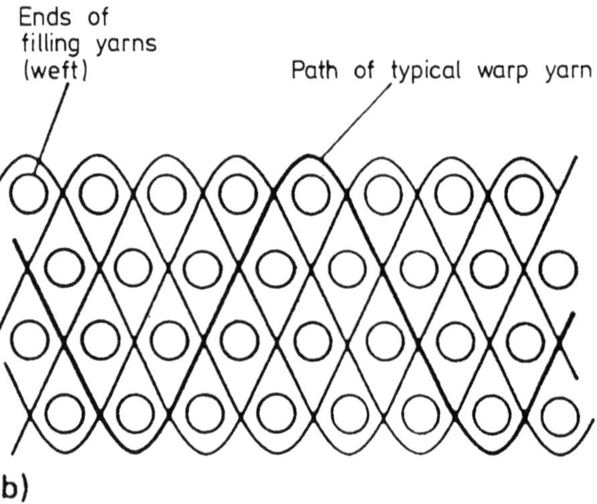

Fig. 2.25 Geometry of angle-interlock fabric (a) with and (b) without added stuffer yarns.

[79], assembling resin-rigidized yarns [80] and modified filament winding technology [81]. Infiltration or impregnation of the carbon matrix becomes increasingly difficult with increasing thickness of preform and complexity of reinforcement. Chemical vapour deposition (Chapter 3) can be used only on relatively thin sections (up to a few centimetres). Vacuum impregnation with polymeric resins followed by pyrolysis (Chapter 4) is a possible method, but by far the most efficient is the hot isostatic pressure impregnation carbonization (HIPIC) process using petroleum or coal-tar pitch as a matrix precursor (Chapter 5).

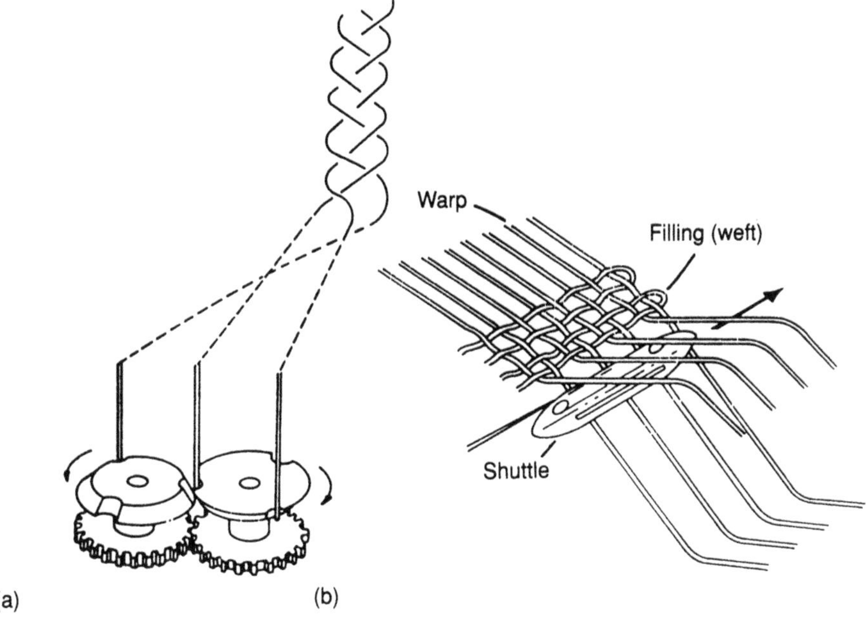

Fig. 2.26 Fabric techniques: (a) braided; (b) woven.

2.6.5 Braiding

Braiding is a textile process known for its simplicity, in which two or more systems of yarns are intertwined in the bias direction to form an integrated structure. Braided structures differ from woven fabrics in the method of yarn introduction into the fabric and in the manner by which the yarns are interlaced (Fig. 2.26). Braiding is similar in many ways to filament winding (Chapter 4). Dry fibre tows or prepreg tapes may be braided over a rotating and removable mandrel in a controlled manner to form a variety of shapes, fibre orientations and fibre volume fractions. Braiding cannot achieve as high a fibre volume fraction as filament winding, but braids can assume more complex shapes. The interlaced nature of braids provides a higher level of structural integrity, essential for ease of handling, joining and damage resistance. The low interlaminar properties of composites can be lessened by use of a three-dimensional braiding process [82]. The application of braided materials in composite engineering materials was initiated in the late 1970s [83]. The current trend in braiding technology is to expand to large-diameter braiding, develop more sophisticated techniques for braiding over complex-shaped mandrels, multi-directional braiding, production of near-net-shape preforms (Fig. 2.27) and the extensive use of computer-aided design and manufacturing (CAD/CAM).

A typical braiding machine (Fig. 2.28) consists of a track plate, spool carrier, former and a take-up device. In some cases a reversing ring is used

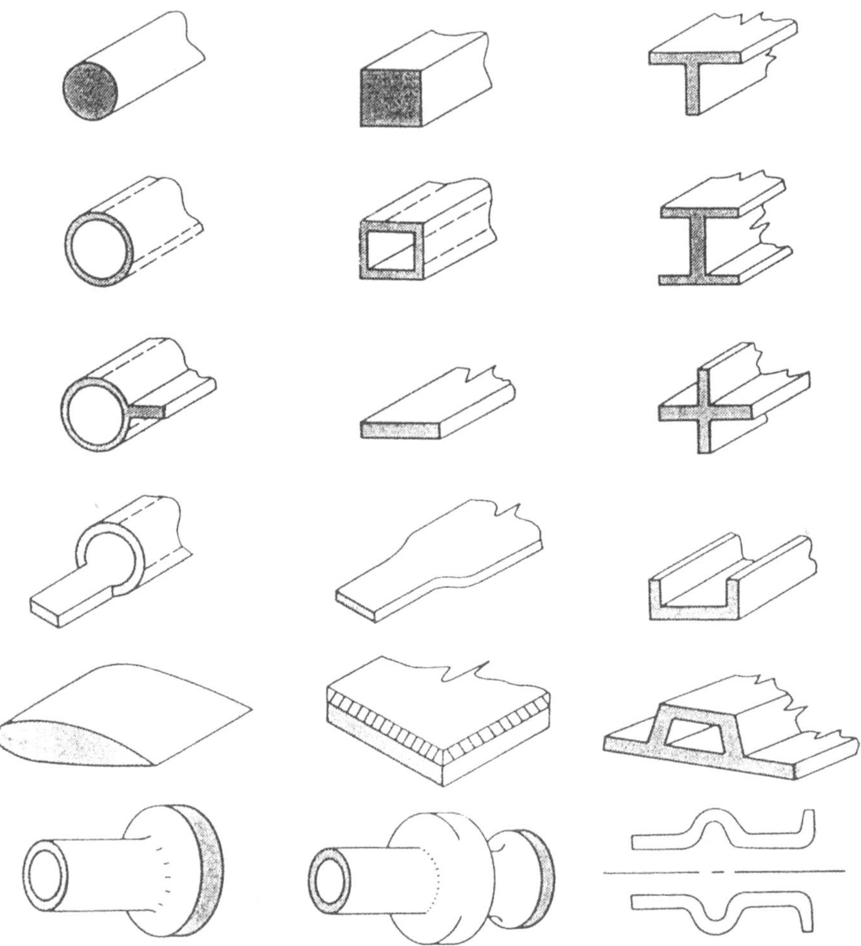

Fig. 2.27 Net shape structures produced by three-dimensional braiding.

to ensure uniform tension on the braiding yarns. The resulting braid geometry is defined by the braiding angle, θ, which is half the angle of the interlacing between yarn systems, with respect to the braiding (or machine) direction. The tightness of the braided structure is reflected in the frequency of interlacings. The distance between interlacing points is known as the pick spacing. The width or diameter of the braid (flat or tubular) is represented as *d*. Should longitudinal reinforcement be required, a third system of yarns may be inserted between the braiding yarns to produce a triaxial braid with $0° \pm \theta$ fibre orientation. If there is a need for structures having more than three yarn thicknesses, several layers (plies) of fabric can be braided over each other to produce the required thickness. For a higher level of through-thickness reinforcement, multiple track braiding

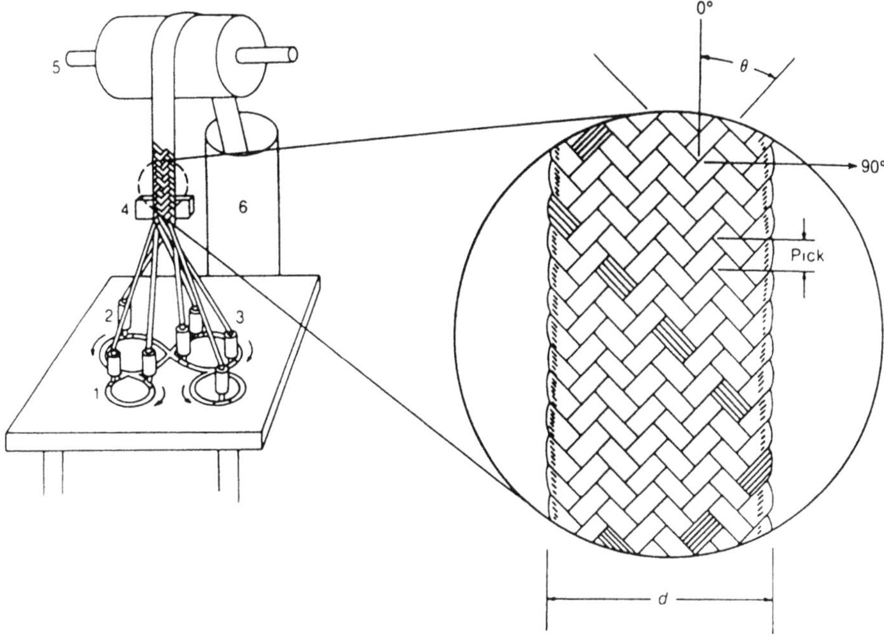

Fig. 2.28 Flat braider and braid. 1, track plate; 2, spool coiner; 3, braiding yarn; 4, braiding point and former; 5, take-off roll with change gears; 6, delivery can.

Table 2.9 Braiding classifications

Parameter	*Levels*		
Yarn axes	Biaxial	Triaxial	Multiaxial
Dimension of braid	Two-dimensional	Three-dimensional	Three-dimensional
Shaping	Formed shape	—	Net shape
Direction of braiding	Horizontal	Vertical	Inverted vertical
Construction of braid	1/1	2/2	3/3

pin braiding, or three dimensional braiding can be used to fabricate structures in an integrated manner. The various criteria and braiding classifications are shown in Table 2.9. A braided structure having two braiding yarn systems with or without a third laid-in yarn is defined as two-dimensional braiding. When three or more systems of braiding yarns are involved in forming an integrally braided structure it is known as three-dimensional braiding. The three-dimensional braiding system can produce structures in a wide variety of complex shapes as shown in Fig. 2.27. The structures may be produced as thick as desired by proper selection of the sizes of the yarn bundles. Fibre orientation can be chosen and 0° longitudinal reinforcements can be added as desired.

2.7 FOOTNOTE

Carbon fibres may be made in a variety of grades and forms each with their own characteristics. The ability to vary the properties considerably by varying processing conditions and choice of precursor allows considerable 'tailoring' of the reinforcement phase within a carbon–carbon composite. While this tailorability of reinforcement properties is clearly of advantage in forming a composite, it also creates a severe problem when trying to categorize, evaluate and standardize properties. The following three chapters will show how similar effects are available when introducing the matrix phase. It is thus possible to see how carbon–carbon is not a single material but rather a family or class of materials whose properties can be varied depending on how and from what they are made.

Polyacrylonitrile and pitch represent the major categories of carbon fibres. Both have a unique set of advantages and disadvantages, but PAN-based fibres are the dominant fibres today and are expected to remain so for the foreseeable future. They posses an excellent balance of mechanical properties, good handleability and an extensive data base. New products are continually being introduced with the emphasis on higher strength and higher modulus. As the world-wide demand for carbon fibre grows there is an increasing emphasis on improved manufacturing technology in order to meet competitive carbon-fibre pricing and lower the capital requirements for new capacity.

Pitch-based carbon fibres develop a high modulus far more easily than those derived from PAN. They are an excellent reinforcing fibre if stiffness, coefficient of thermal expansion or conductivity are important. The key process development area for both economics and properties is improved mesophase technology in spinning and the handling of exceedingly brittle as-spun fibres. Continued success in pitch technology (quality, price, performance, etc.) will lead to increased consumption into applications with low (industrial) and ultra-high (space) modulus requirements. At present there is only one commercial supplier in the West (Amoco in the USA) and a number in Japan. One would therefore expect to see a number of others over the next 3–5 years, perhaps in Europe.

REFERENCES

1. Griffith, A. A. (1920) *Phil. Trans. Roy. Soc. London*, 221A.
2. Savage, G. M. (1989) *Metals and Materials*, **5**, 285.
3. Savage, G. M. (1988) *Metals and Materials*, **4**, 544.
4. Duthie, A. C. (1987) *Proc. PRI. Conf. Polym. in Def.*, Bristol, 18–20 March.
5. Stein, B. A. and Johnson, N. J. (1985) *Proc. ASTM Symp. on Toughness of Composites*, Houston, Texas, March 1985.

6. D'Abate, G. D. and Diefendorf, R. J. (1985) *Proc. 17th Biennal Conf. on Carbon*, Am. Carbon Soc., p. 390.
7. Berman, R. (1979) in *The Properties of Diamond* (ed. J. E. Field), Academic Press, New York.
8. *Carbon*, **20** (5), 445 (1982).
9. Edison, T. (1880) US Patent 223,898.
10. Milanski, J. V. and Katz, H. (eds) (1987) *Handbook of Reinforcements for Plastics*, Van Nostrand Reinhold Amsterdam.
11. Fitzer, E. (1989) *Carbon*, **27** (5), 621.
12. Shindo, A. (1961) *Osaka Kogyo Gijitsu Shikuko*, **12**, 110,119.
13. Bacon, R., Pallozza, A. A. and Slosarik, S. E. (1966) *Proc. Soc. Plastics Ind. 21st Tech. and Mang. Conf., Sect 8E.*
14. Shindo, A. (1961) *Rept* Covt. Ind. Res. Inst., Osaka, 317.
15. Shindo, A. (1964) *Carbon*, **1**, 391.
16. Watt, W., Philips, L. N. and Johnson, W. (1966) *The Engineer*, **221**, 815.
17. Standage, A. E. and Prescott, R. (1966) *Nature* (London), **211**, 169.
18. Hughes, T. V. and Chambers, C. R. (1889) UK Patent 405,480.
19. Oberlin, A., Endo, M. and Koyama, T. (1976) *J. Cryst. Growth*, **32**, 335.
20. Johnston, W., Phillips, L. N. and Watt, W. (1964) UK Patent 1,110,791.
21. Johnston, W., Phillips, L. N. and Watt, W. (1968) US Patent 3,412,062.
22. Singer, L. S. (1977) US Patent 4,005,183.
23. Bacon, R. and Schalamon, W. A. (1967) *Proc. 8th Biennial Conf. on Carbon*, Am. Carbon Cttee.
24. Hawthorne, H. (1971) *Proc. 1st Int. Conf. on Carbon Fibres*, Plastics Ind., p. 81.
25. Soltes, W. (1961) US Patent 3,001,981.
26. Abbott, W. (1962) US Patent 3,053,775.
27. Otani, S. (1965) *Carbon*, **3**, 213.
28. Singer, L. S. (1972) Netherlands Patent 239490.
29. Gill, R. M. (1972) *Carbon Fibres in Composite Materials*, published for the Plastics Institute, London, by ILIFFE Books, London.
30. Watt, W. and Perov, B. V. (eds) (1985) *Strong Fibres*, Amsterdam.
31. Dresselhaus, M. S., Dresselhaus, G., Sugihara, K., Spain, I. L. and Goldberg, H. A. (1988) *Graphite Fibres and Filaments*, Springer-Verlag, Heidelberg.
32. Henrici-Olivé, G. and Olivé, S. (1983) in *Industrial Developments* (*Advances in Polymer Science*, **51**), Springer-Verlag, Heidelberg, p. 1.
33. Hay, J. N. (1968) *J. Polym. Sci.* **A1** (6), 2127.
34. Watt, W. and Green, J. (1971) *Proc. Int. Carbon Fibre Conf.*, Plastics Inst., London, Paper 4.
35. Clarks, A. J. and Bailey, J. F. (1973) *Nature*, **243**, 146.
36. Fitzer, E. and Muller, D. J. (1975) *Carbon*, **13**, 163.
37. Toray Inc. (1970) German Patent 1,958,361.
38. Raskovic, V. and Marinkovic, S. (1978) *Carbon*, **16**, 351.
39. Watt, W., Johnson, D. J. and Parker, E. (1974) *Proc. 2nd Int. Carbon Fibre Conf.*, Plastics Inst., London, Paper 1.
40. Johnson, D. J. (1971) *Inst. Carbon Fibre Conf.*, Plastics Inst., London, Paper 8.

41. Rose, P. G. (1977) *Kohlenstoff und Aramidfaser Verstärkte Kunststoffe*, VDI-Verlag GmbH, p. 9.
42. Johnson, W. (1971) *Proc. Third Conf. Ind. Carbons and Graphites*, Soc. Chem. Ind., London, p. 447.
43. Reynolds, W. N. (1971) *Proc. Third Conf. Ind. Carbons and Graphites*, Soc. Chem. Ind., London, p. 427.
44. Reynolds, W. N. and Sharp, J. V. (1974) *Carbon*, **12**, 103.
45. Moreton, R. and Watt, W. (1974) *Nature*, **247**, 360.
46. Johnson, J. W. (1969) *J. App. Polym. Symp.*, **9**, 229.
47. Stewart, M. and Feughelman, M. (1973) *J. Mat. Sci.*, **8**, 1119.
48. Watt, W. and Johnson, W. (1969) *CASI Trans.*, **2**, 81.
49. Johnson, J. W. and Thorne, D. J. (1969) *Carbon*, **7**, 659.
50. Moreton, R. (1969) *Fibre Sci. Tech.*, **1**, 273.
51. Delmonte, J. (1981) *Technology of Carbon and Graphite Fibre Composites*, Van Nostrand Reinhold, New York.
52. Ohama, Y. (1989) *Carbon*, **27** (5), 729.
53. Volk, H. F. (1977) *Proc. Symp. on Carbon Fibre Reinforced Plastics*, Bamberg, FRG, 11 May 1977.
54. Riggs, J. P. (1985) in *Encyclopedia of Polymer Science and Engineering*, **2**, Wiley, New York, p. 640.
55. Hughes, J. D. H. (1987) *J. Phys. D: Appl. Phys.*, **20**, 276.
56. Edie, D. D. and Dunham, M. G. (1989) *Carbon*, **27** (5), 647.
57. Ruland, W. (1968) *Polym. Preprints, Polymer Chemistry Div. Am. Chem. Soc.*, **9** (2), 1368.
58. Le Maistre, C. W. and Diefendorf, R. J. (1973) *SAMPE Quart.*, **4**, 1.
59. Diefendorf, R. J. (1988) in *Engineered Materials Handbook*, **1**, *Composites*, ASM Int. p. 49, Amsterdam.
60. Guigon, M., Overlin, A. and Desarmot, G. (1984) *Fibre Sci. Tech.*, **20**, 177.
61. Fordeaux, A., Perret, R. and Ruland, W. (1971) *Proc. Int. Conf. on Carbon Fibres, their Composites and Applications*, London, Plastics and Polymer Conference Supplement no. 5, Plastics inst., London, p. 57.
62. Tokarsky, E. W. and Diefendorf, R. J. (1975) *Polym. Eng. Sci.*, **15** (3), 150.
63. Diefendorf, R. J. and Tokarsky, E. W. (1975) *Report AFMI-TR-72–133*, parts I–IV, US Air Force Mats. Lab.
64. Bright, A. A. and Singer, L. S. (1979) *Carbon*, **17**, 59.
65. Butler, B. L. and Diefendorf, R. J. (1970) *Proc. 10th Carbon Composite Tech, Symp. Am. Soc. Mech. Eng.*, p. 109.
66. Donnet, J. B. and Ehrberger, P. (1977) *Carbon*, **15**, 143.
67. Monte, S. J., Sugarman, G. and Seeman, D. J. (1977) *Proc. 32nd Ann. Tech. Conf.*, SPI, RP/C Inst. Washington, DC, Feb. 1977.
68. Barr, J. (1974) US Patent 3,791,840.
69. Nyo, H., Heckler, D. L. and Hoernshcemeyer, P. L. (1979) *SAMPE, 24th Nat. Symp.*, **14**, 51, May 1979.
70. De Lolhis, N. J. (1979) *SAMPE J.*, **15** (3), 10.
71. Savage, G. M. unpublished results.
72. Bailie, J. A. (1989) Woven fabric aerospace structures, in *Structure and Design* (eds C. T. Herakovich and Y. M. Tarnopoloski), Metals Park, Ohio, North-Holland, pp. 353–91.

73. Ishkiawa, T. and Chou, T. W. (1982) *J. Comp. Mat.*, **6**, 2.
74. Ishikawa, T. and Chou, T. W. (1983) *AIAA J*, **21** (12), 1714.
75. Levin, J. (1975) *SAMPE Quart.*, 20.
76. McAllister, L. E. and Taverna, A. R. (1971) *Proc. 73rd Ann. Mtg Am. Ceram, Soc.*, Chicago.
77. *Fabrication of Composites* (*Handbook of Composites*, **4**) (eds A. Kelly and S. T. Mileiko (1986), Amsterdam.
78. Barton, R. S. (1968) *SPE J.*, **4** (May), 31.
79. McAllister, L. E. and Taverna, A. R. (1972) *Proc. 17th Nat. SAMPE Symp.*, Paper IIIA-3, AZ, USA.
80. Lamicq, P. (1977) *Proc. AIAA/SAE 13th Prop. Conf.*, Paper 77–882, Orlando.
81. Mullen, C. K. and Roy, P. J. (1972) *Proc. 17th Nat. SAMPE Symp.*, Paper IIIA-2, AZ, USA.
82. Sanders, L. R. (1977) *SAMPE Quart.*, **38**.
83. Post, R. J. (1977) *Proc. 22nd Nat. SAMPE Symp.*, p. 486.

Gas Phase Impregnation/Densification of Carbon–carbon and other High-temperature Composite Materials

3

3.1 THE CVD PROCESS

Chemical vapour deposition (CVD) is a process in which a solid product nucleates and grows on a substrate, by decomposition or reaction of gaseous species, and involves the heating of a fibre preform in a gaseous environment so that the matrix is deposited from the gas phase. The technology developed to date allows fine control over the composition and morphology of the solid deposit. Various processes have been used for the production of thin film semiconductor devices for the communications industry and the production of hard, abrasion-resistant coatings on cutting tools and coatings on radioactive pellets. The CVD techniques have also been widely used for the preparation of oxidation- and wear-resistant coatings to carbon–carbon composites. Well-processed CVD-derived composites generally possess excellent mechanical properties as a consequence of the slow, steady build-up of matrix material around the fibre network. The CVD method has proven especially useful for the production of ceramic matrix composites, where melt-processing techniques are inapplicable and conventional powder-processing methods – such as those used for the production of monolithic ceramics – result in serious fibre degradation. The major drawback of CVD is the very slow rate of deposition, leading to large material/energy inputs and a high final cost.

The patent literature on inorganic composite materials (including carbon and ceramic matricies) is the subject of a recent monograph [1]. Several workers active in the field refer to CVD densification of composites as chemical vapour infiltration (CVI). The purpose of this nomenclature is to distinguish the deposition of material in fibre preform from the simple layer deposition techniques used in the semiconductor and coatings industries. It will become apparent throughout the text that the deposition of material within the pore system of a fibre preform imposes severe kinetic limitations on industrially operated CVD processes.

3.2 PHYSICO–CHEMICAL PRINCIPLES OF THE CVD PROCESS

The CVD process is, in principle, a very simple one. A gas (or mixture of gases) comes into contact with a hot surface, and reacts to deposit the thermodynamically stable phase. In practice, however, the mechanistic and kinetic features of CVD may be extremely complex.

3.2.1 Thermodynamics

Any simple hydrocarbon gas will become thermodynamically unstable with respect to carbon at elevated temperatures. Carbon deposition usualy results in a volume increase for the system such that the $T\Delta S$ entropy term in the free energy equation overcomes any endothermic enthalpy term. Simple free energy calculations can be performed to predict the onset of carbon depositions for the various reactions, for example:

$$CH_{4(g)} \rightarrow C_{(s)} + 2H_{2(g)}. \tag{3.1}$$

For simple C_2 species, the order of stability is $C_2H_6 > C_2H_4 > C_2H_2$. Methane is one of the most stable hydrocarbons. Temperatures in excess of 550 °C are required before carbon deposition is thermodynamically favourable. Even at this temperature, the rate is very slow, necessitating higher temperatures for measurable deposition rates. The widespread availability of methane and its excellent diffusion properties have, nevertheless, made it the most common raw material for carbon matrices in CVD-fabricated composites. The processes are generally operated at low pressures, or by adding inert diluent gases such as H_2, He, N_2 or Ar to the gas stream in order to improve diffusion conditions by increasing the mean free path of the gas molecules.

The CVD of ceramic matrices is less straightforward than carbon. Oxide ceramics can be deposited from a single alkoxide precursor (such as SiO_2 from $Si(OEt)_4$, or TiO_2 from $Ti(O^iPr)_4$ for example), but generally two reacting precursor materials are needed. Some representative reaction mixtures are shown:

$$SiCl_4/NH_3/H_2 \rightarrow Si_3N_4 + HCl \quad (3.2)$$

$$CH_3SiCl_3/H_2 \rightarrow SiC + HCl \quad (3.3)$$

$$BCl_3/CCl_4/H_2 \rightarrow B_4C + HCl \quad (3.4)$$

$$TiCl_4/BCl_3/H_2 \rightarrow TiB_2 + HCl \quad (3.5)$$

$$AlCl_3/CO_2/H_2 \rightarrow Al_2O_3 + HCl. \quad (3.6)$$

The reactions are by no means stoichiometric. The gas mixtures can generate other solid phases depending on the conditions employed in the experiment (temperature, pressure and mole fractions). If one assumes that the CVD process operates at equilibrium, it is possible to perform thermodynamic calculations to predict the equilibrium solid phase for a given set of experimental conditions. A number of computer programs have been written to perform such tasks [2]. Figures 3.1 and 3.2 show examples of these calculations for both SiC and TiB_2 deposition respectively. Figure 3.1 illustrates the relative amounts of Si, C and SiC produced as a function of the inlet H_2/CH_3SiCl_3 concentrations for deposition at 927 and 327 °C. Although CH_3SiCl_3 would appear to produce SiC in a simple $CH_3SiCl_3 \rightarrow SiC + 3HCl$ reaction, thermodynamic analysis suggests that considerable carbon deposition also occurs unless hydrogen is added [3] as is clearly visible in Fig. 3.1. Figure 3.2 shows the complex array of phases which it is possible to deposit from a $TiCl_4/BCl_3/H_2$ system at 654 °C and 0.84 bar [4]. To allow a simple depiction, one corner of the phase diagram is represented by the (50% H + 50% Cl) variable.

A pure TiB_2 deposit will only be obtained by careful control of inlet reactant mole fractions. Theoretical calculations of this nature are generally reliable but kinetic constraints often result in the formation of different phases. Workers at the Oak Ridge National Laboratories, for example, performed a number of computer optimization calculations on the conditions required to generate a TiC/SiC composite from $TiCl_4/CH_3SiCl_3/H_2$, but found that in practice they generated a $TiSi_2$/SiC material [5].

3.2.2 Mechanism and kinetics

A model for the different processes that occur during CVD has been developed by Spear [6]:

1. forced flow of reactant gases into reaction vessel;
2. diffusion of reactants through laminar flow boundary layers around substrate;
3. adsorption of reactants on surface of substrate;
4. reaction of adsorbed reactants to give solid products and adsorbed gaseous products;

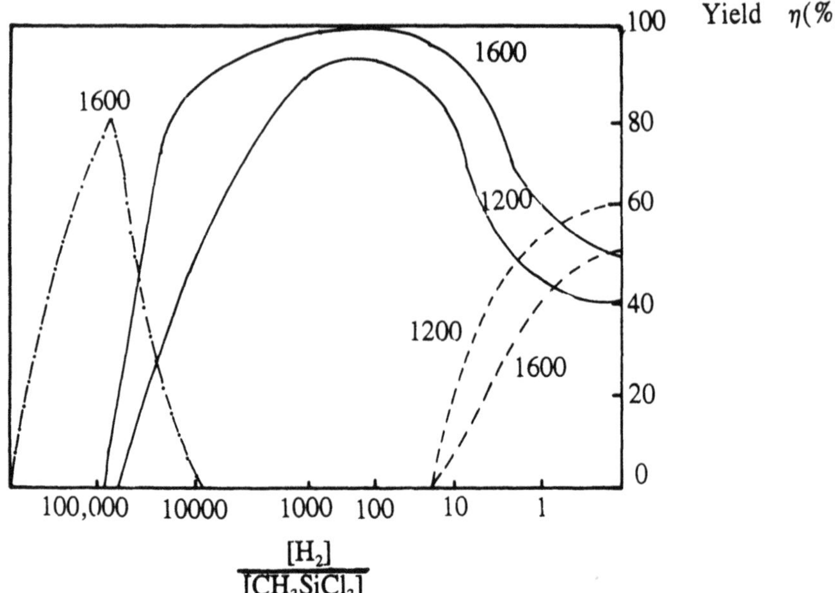

Fig. 3.1 Silicon carbide deposition from $CH_3\ SiCl_3/H_2$. (—) ηSiC; (- · -) ηSi; (- - -) ηC.

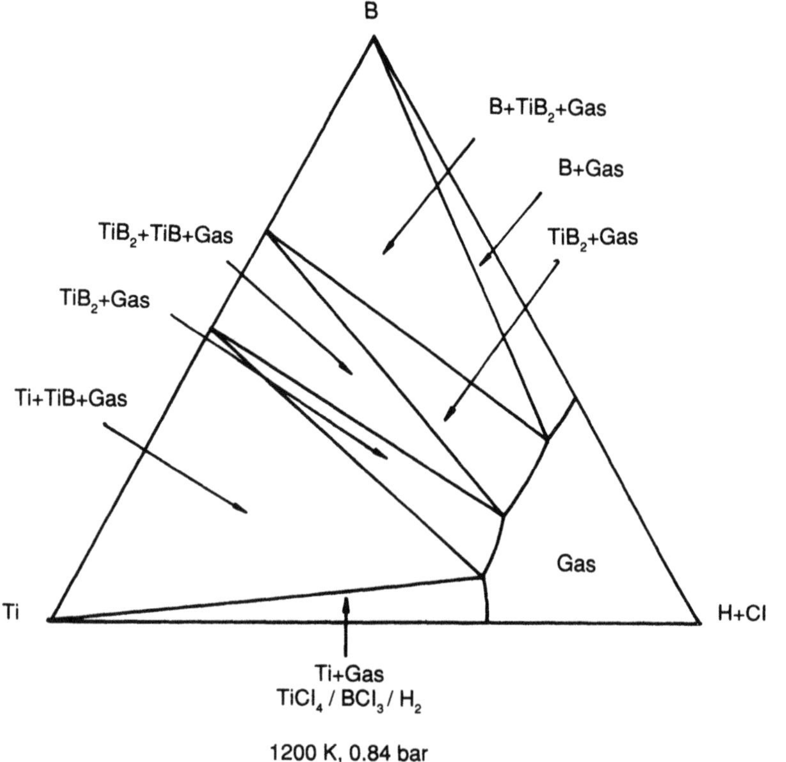

Fig. 3.2 Titanium diboride deposition from Ti/Cl/B/H mixtures.

5. desorption of adsorbed gaseous products;
6. diffusion of gaseous products through a boundary layer region;
7. forced flow of gaseous products through reaction vessel exit.

Any attempt to uncouple fully the overall kinetics of the CVD process in terms of these seven stages would obviously be a Herculean task. In practice the CVD process tends to be controlled by the surface reaction kinetics at low temperatures and pressures, whereas at high temperatures and pressures diffusion control dominates.

The deposition of carbon is particularly complex. A range of chemical processes can occur, dependent on the experimental conditions. The CVD process described above is essentially heterogeneous – the entire reaction occurs at the surface. In the case of carbon deposition from methane, this would correspond to a series of dehydrogenation/polymerization reactions taking place on the surface to produce graphitic materials. At high temperatures and pressures, however, pre-reaction occurs in the gas phase, generating fine, globular soot particles which agglomerate on the surface to form a low-density, mechanically weak carbon. A thorough discussion of the mechanism of carbon deposition and the morphologies of the deposits has been written by Bokros [7]. Analysis of the gaseous products from methane pyrolysis has shown that a range of aromatics are formed, indicating the complex dehydrogenation and polymerization reactions that occur [8].

Workers at the Sandia National Laboratories have made a considerable attempt to understand the chemical processes involved in the CVD of a carbon matrix around a carbon preform [9]. They suggest that methane decomposes via acetylenic or aromatic (C_6) species. The relative importance of these two routes depends on the experimental conditions. Below 1250 °C, benzene was found to be the predominant species, decomposing to give a smooth laminar carbon, characterized by a high density. Above 1250 °C, acetylene appears to predominate, with a tendency to produce an isotropic carbon. Around 1250 °C, where a mixture of acetylenic and aromatic species occurs, an intermediate rough laminar carbon results. The relation between CVD conditions and the carbon microstructure and the resulting physical properties will be explored at a later stage (Section 3.4.2).

To date, no analysis has been made of the chemical processes involved in the CVD of ceramic materials. One important point is that sometimes low-temperature pre-reaction occurs in the gas phase, giving a deposit with poor mechanical properties. This is particularly important in the CVD fabrication of Si_3N_4 using $SiCl_4/NH_3$. Reactions occur forming a range of smokes/'soots' unless the two reactants are kept in discrete inlet systems to ensure that a reaction may only occur on the substrate. No detailed process conditions/deposit microstructure correlations have been performed for CVD-fabricated composite matrices, although Nuhara and co-workers have performed careful studies on grain size/morphology of solid plates

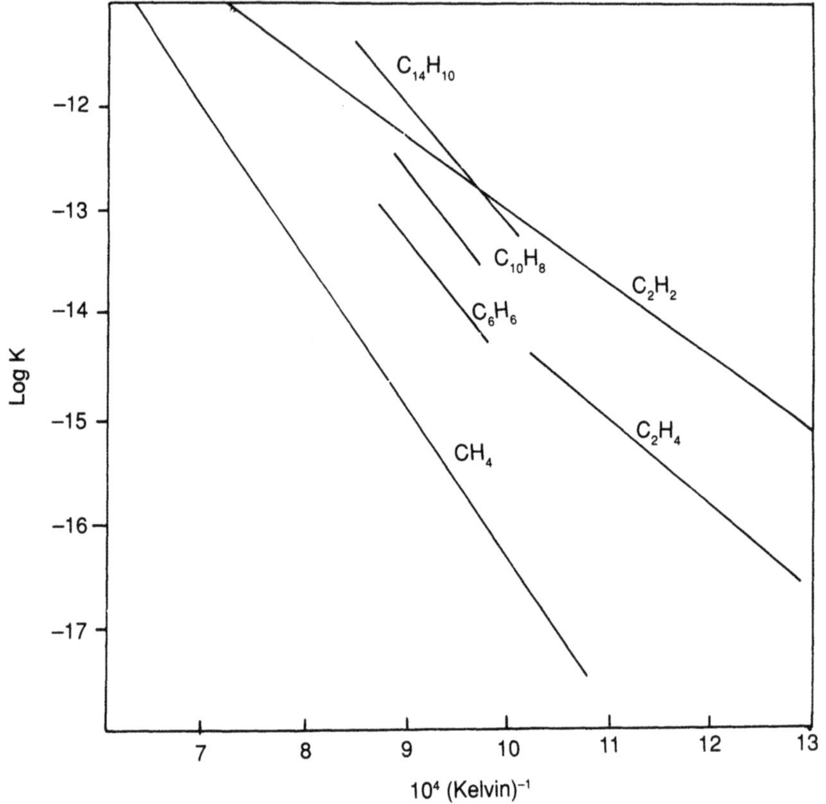

K=Rate constant of CVD carbon laydown g cm^{-2} s^{-1} Pa^{-1}

Fig. 3.3 Dependence of carbon deposition rates on temperature/type of hydrocarbon precursor during the CVD process.

(up to 5 mm thick) of CVD-fabricated SiC, Si_3N_4 and B_4C, as a function of experimental conditions [10].

The kinetics of carbon deposition have been studied on a wide range of substrates by Tesner and co-workers in the former Soviet Union [11]. They concluded that rates of deposition were a complex function of reactor temperature, pressure, gas phase composition/flow rate and substrate geometry and type. Of particular interest is the dependence of deposition rate on the type of hydrocarbon precursor. Figure 3.3 shows the rate constant/temperature graph for a range of hydrocarbons displaying Arrhenius-type kinetics over the temperature regions studied.

While empirical studies on the influence of experimental conditions on CVD kinetics are very valuable, it is important to realize how complex the behaviour can be. Van der Brekel and Lersmacher at the Philips Laboratories in Eindhoven have demonstrated the difference in behaviour of 'hot wall' and 'cold wall' reactor systems for carbon deposition at a given temperature [12]. 'Hot wall' systems rely on radiative heating from an external furnace, such that the reactor wall and reactant gases are heated up as well as the substrate. 'Cold wall' systems, on the other hand, rely on internal heating of the substrate, generally via inductive or resistive heating.

Observations on carbon deposits formed in both reactors at the same set of conditions showed that, at certain temperatures, sooty isotropic deposits are formed in the 'hot wall' system as a result of homogeneous gas phase reaction. In contrast, laminar deposits were found in the 'cold wall' reactor since the 'cold wall' configuration avoids excessive gas pre-heating.

3.2.3 CVD in a pore system

The preceding discussion of the CVD process was confined to deposition on a flat surface. Fabrication of composite materials requires deposition of the desired matrix inside and around a fibre preform – whether it be a woven continuous fibre structure, or a mat of short fibres or whiskers. The fibre preform will have a well-defined initial pore structure, dependent on fibre form, content and arrangement. It is essential that matrix material be deposited throughout the pore structure, if a strong dense composite is to result. This applies severe constraints to the CVD process. The reactants must diffuse through the boundary layer of laminar flow around the preform, diffuse into the pores and then adsorb and react. The products must be desorbed, and diffuse back out along the same route (pore and boundary layer). If the surface chemical reaction needed to produce a solid deposit occurs rapidly with respect to the diffusion processes, deposition will occur near the mouth of the pore rather than along it, rapidly sealing off the pores. Closed porosity is created which serves to concentrate mechanical stress and is thus detrimental to the mechanical performance of the composite (Fig. 3.4(a)). If, on the other hand, conditions are chosen such that the surface reaction rate is a good deal slower than the diffusion rate, deposition can occur evenly along the length of the pore, to give a well-densified material as shown in Fig. 3.4(b).

Several factors affect the rates of the surface chemical reactions and the diffusion processes involved. The kinetics of deposition, previously discussed, are well documented for carbon. Diffusion in the pores can be of two types: bulk diffusion, where pore size is irrelevant, and Knudsen diffusion, where the pore size is important due to the effect of molecule–wall collisions. With the relatively coarse pores present in fibre preforms, it is likely that bulk diffusion predominates. In bulk diffusion regimes the diffusion rate is influenced by the temperature, pressure, molecular weights and collision cross-sections of the molecular components involved. Combination of diffusion kinetics and surface reaction kinetics is well known in the catalysis engineering literature. An expression can be derived for

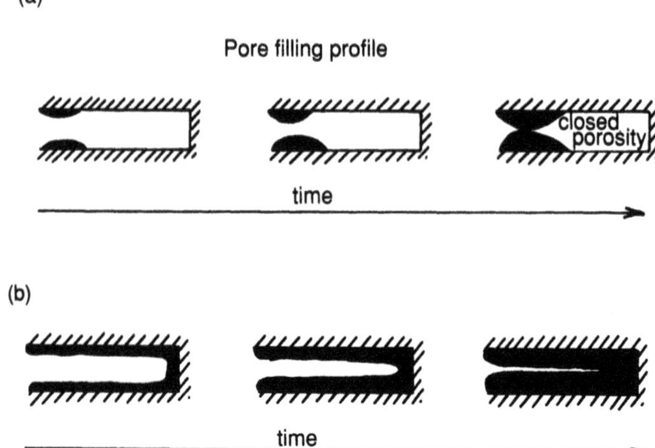

Fig. 3.4 The balance of diffusion and surface reaction kinetics; idealized depictions: (a) surface reaction rate >> diffusion rate; (b) diffusion rate >> surface reaction rate.

the ratio of the reaction rate at a point along the pore to the rate at the mouth of the pore, in terms of a dimensionless number called the Thiele modulus (ϕ) which is defined as

$$\phi = L\ 1/2(4k/dD_e), \tag{3.7}$$

where L is the length of the pore, d the diameter of the pore, k the rate constant for surface reaction, D_e the diffusion constant for reacting gases.

Figure 3.5 shows the ratio of the deposition rate at a given distance into a pore to the rate at the pore mouth, as a function of that distance. Curves corresponding to different experimental conditions (different values of ϕ) are shown. At high values of ϕ, corresponding to rapid surface kinetics in comparison to diffusion, the reaction rate is much faster at the pore mouth (the ratio of reaction rates drops sharply as the distance from the pore mouth increases). Under these conditions, pore 'plugging' occurs. At low rates of ϕ, corresponding to slow surface kinetics, the reaction rate is almost uniform throughout the pore, leading to successful, in-depth pore filling. Unfortunately, this experimental constraint, vital to successful CVD densification of a composite, means that processing times can be very long indeed, as will be discussed more fully in the section on isothermal methods of CVD processing (3.3.2).

The problem of uneven deposition along a capillary pore has been experimentally verified by Diefendorf and Sohda [13]. Capillaries of 400–1000 μm diameter and 38 mm length were bored into a graphite block. Carbon CVD experiments were subsequently performed using methane as a source gas at different temperatures, pressures and flow rates. At the end of each deposition run the thickness of deposited carbon was

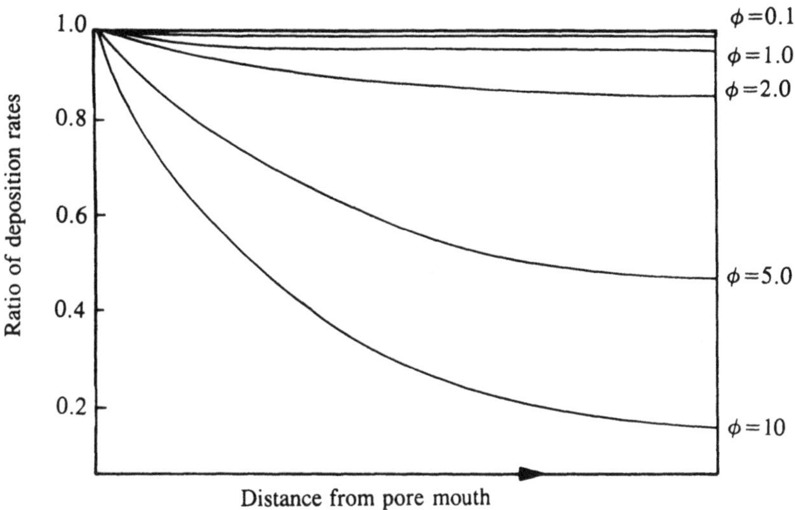

Fig. 3.5 Theoretical deposition/pore depth profiles; the role of the Thiele modulus.

measured as a function of the distance from the pore mouth. Figures 3.6(a)–(c) show different deposition rate/length profiles for independent variation of temperature, pressure and flow rate. At high temperatures, pressure and flow rates, the deposition rates were faster near the mouth of the pore, as had been theoretically predicted by the high value of the Thiele modulus (ϕ). The experiments further emphasized the need for careful selection of process parameters when performing a CVD densification of a composite. Naslain *et al.* [3] performed similar experiments on the silicon carbide infiltration of a porous carbon–carbon skeleton. Their results, discussed in Section 3.5.2, provide yet more evidence of this delicate balance. A comparison of theoretical predictions and experimental study of SiC/Si_3N_4 impregnation profiles has been carried out by Fitzer and Hegen in Karlsruhe [14].

3.3 EXPERIMENTAL CVD TECHNIQUES

3.3.1 Introduction

The pore structure of a fibre preform clearly places considerable constraints upon the process conditions that can be employed in a conventional (isothermal) CVD process. In order to remove the diffusion constraints, a number of novel experimental approaches have been developed. A brief review of each technique will now be given, together with some comment on the relative merits and disadvantages present in that technique.

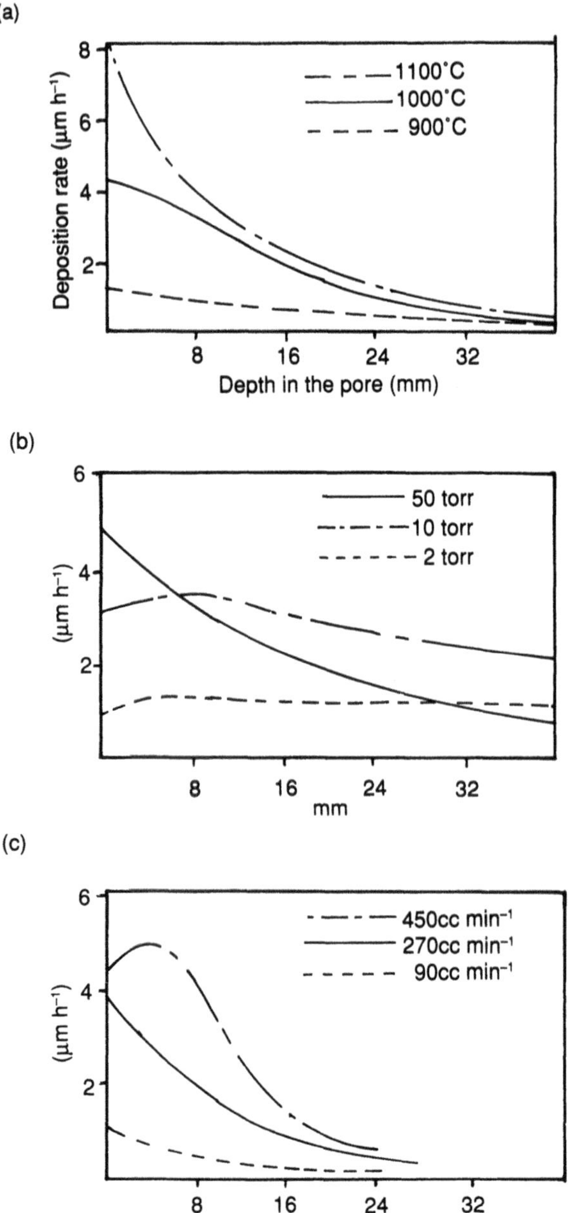

Fig. 3.6 Experimental deposition/pore depth profiles: (a) The influence of temperature; (b) The influence of pressure; (c) The influence of flow rate.

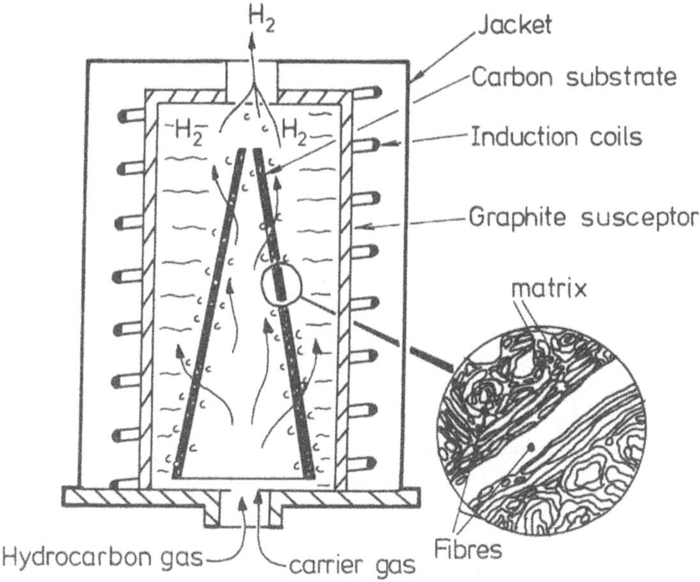

Fig. 3.7 The isothermal method of CVD processing.

3.3.2 The isothermal method

The isothermal method is a simple technique whereby a substrate is placed into an even temperature furnace and the reactant gases passed over it (Fig. 3.7). One thus relies on diffusion in and out of the pores. To avoid sealing the pores off at the mouth, the surface reaction rate must be kept lower than the diffusion rate. Unfortunately this means that the rate of weight gain is very slow, and process times become very long. An additional problem is that the rate of densification slows down as the amount of porosity decreases, making full densification virtually impossible. Figure 3.8 shows the weight/time graph for the densification of a porous carbon–carbon billet with boron nitride (from BF_3/NH_3) as measured by Naslain and co-workers at Bordeaux/Société Européenne de Propulsion (SEP) [15]. The graph clearly shows the rate of weight gain to drop as the composite approached full densification. Inevitably, selection of isothermal CVD experimental parameters reflects a compromise between obtaining well-densified materials and working at an economically viable deposition rate.

The isothermal process, despite its fundamental limitations, remains in widespread use for composite production. The process is readily amenable to scale-up, with large furnaces capable of processing several objects simultaneously. In practice, carbon–carbon billets are processed for up to a week, removed from the furnace, the surface machined to remove the surface blockage of the pores, and then redensified in the CVD furnace.

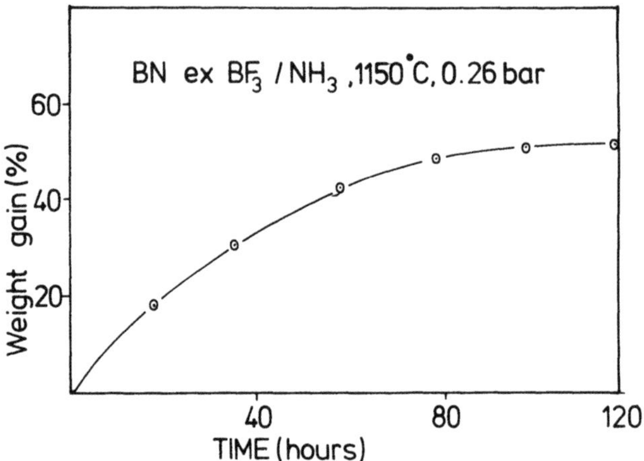

Fig. 3.8 Weight gain/time graph for the densification of a carbon skeleton with boron nitride.

This cycle sometimes has to be repeated three or four times. As a result, processing can often take up to a month or more, despite the use of low deposition pressures and diluent gases to minimize pore blockage by aiding diffusion.

3.3.3 The thermal gradient method

The thermal gradient method is another diffusion-controlled process. A temperature gradient is deliberately set up across the fibre preform, so that surface deposition of the material, and hence overcrusting, is avoided. In practice, a fibre preform is applied to a mandrel (often by winding). The mandrel (generaly solid graphite) acts as a susceptor material for inductive heating. A typical experimental set-up is shown in Fig. 3.9. By choosing the appropriate temperature, thermal conductivity of the preform and precursor gas flow rate, one can ensure that deposition occurs only on the hot mandrel/fibre interface region. In this way, the deposited matrix moves progressively out towards the outside surface of fibre preform.

There are two significant experimental points in discussing thermal gradient methods. Firstly, the fibre preform must possess a low thermal conductivity to allow the thermal gradient to be established; secondly, high gas flow rates must be used (generally a mixture of methane and nitrogen is used). These two factors ensure that a large thermal gradient occurs across the fibre preform, with a gas flow sufficient to cool the exterior surface of the preform. Temperature differences of up to 500 °C across a 1 cm thick preform have been observed [16]. Carbon felts are commonly used as substrates as they have considerably lower thermal conductivities than woven fabric structures.

The presence of a thermal gradient allows considerably higher deposition

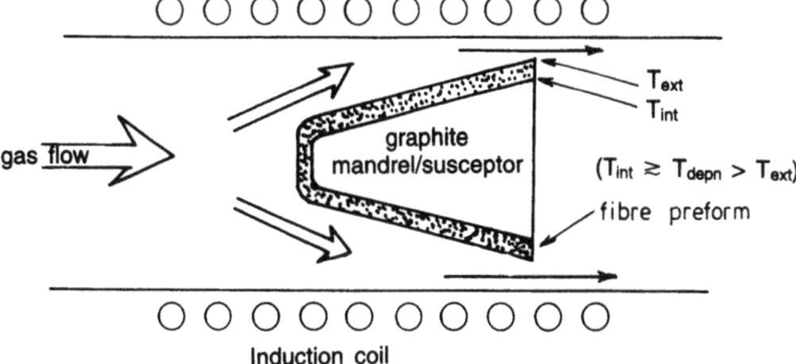

Fig. 3.9 The thermal gradient method.

rates to be attained than with 'conventional' isothermal techniques. It has been suggested that an increase in rate of around an order of magnitude can be obtained [17]. Other advantages include the lower incidence of overcrusting and the ability of the process to run at atmospheric pressure. The machining of samples and the need for low-pressure vessels and pumping apparatus commonly associated with the isothermal method can thus be eliminated. A great deal of work has been carried out at the Sandia National Laboratories on thermal gradient CVD processing of filament-wound carbon–carbon rocket nose cones, with a conical piece of graphite being used as the former and the susceptor mandrel. The disadvantages of the thermal gradient method are the limitation to single-item processing and difficulties in scaling up this process.

3.3.4 The pressure gradient method

The pressure gradient method relies on forced flow of the precursor gas mixture through the pore system of a fibre preform, removing the combined diffusion/surface reaction limitation of the isothermal CVD method. A fibre preform is sealed into a gas-tight unit, and placed into a heated region. The resistance to gas flow of the fibre preform causes a pressure gradient to be set up. Unlike both the isothermal and thermal gradient methods, where the deposition rate slows down as the process progresses, the deposition rate increases during the pressure gradient process because the pressure gradient increases as the pores are filled up. The deposition rate is found to be proportional to the pressure drop across the preform. A typical experimental set-up is shown in Fig. 3.10.

Although high deposition rates can be attained, the pressure gradient method suffers from severe drawbacks. The method is confined to single-item processing and is dependent on robust high-temperature/pressure seals. In addition, overcrusting and pore blockage may occur. Removal of the component and machining are necessary to ensure in-depth infilling. The pressure gradient method is not in widespread commercial use.

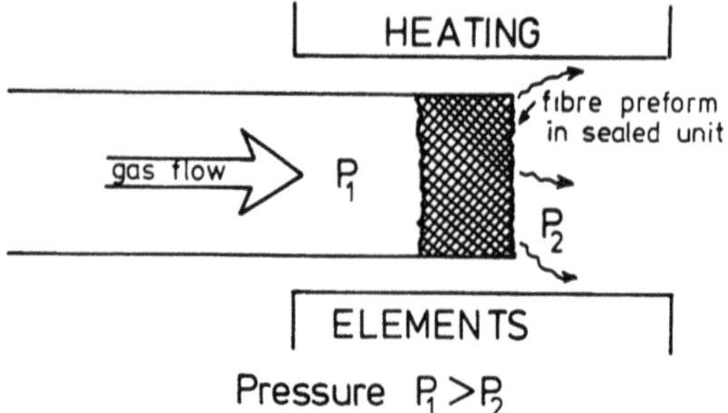

Fig. 3.10 The pressure gradient method.

3.3.5 Pulse CVD methods

An alternative method for removing diffusion limitations is the pressure/vacuum pulse infiltration technique. Here the hot reaction vessel is cycled between atmospheric pressure and a rough vacuum (generally a few torr) so that reactant gases are forced deep into the pore structure of the preform. The precursors are allowed to react and the gaseous products pumped out, ready for infiltration with fresh reactant in the next cycle. This 'forced diffusion' technique allows rapid, in-depth pore filling.

Japanese workers have recently published a study of pulse CVD infilling of a porous carbon skeleton with titanium nitride, derived from a $TiCl_4/N_2/H_2$ mixture [18]. They found that careful adjustment of the vacuum/gas cycle times and process temperatures was necessary to ensure an even impregnation profile.

While the pulse CVD process offers a direct way of overcoming diffusion limitations, the practical difficulties in setting up a rapid pressure cycled reactor have meant that the method is not used commercially in carbon–carbon production. It is possible, however, that the increasing capabilities in computer process control will make pulse methods more attractive.

3.3.6 Miscellaneous methods

Trinquecoste and co-workers have investigated DC plasma-enhanced CVD processes for carbon–carbon production [19]. They found that application of a plasma increased the efficiency of the process. The temperature could be reduced from 1050 °C in a normal process to 850 °C in the plasma-enhanced process, with similar quality materials being produced. It was also found that lower methane flow rates could be used with the plasma process.

Workers at Le Carbone-Lorraine have described a process for speeding up CVD process times in carbon–carbon fabrication using 'aerogels'. A carbon fibre preform was impregnated with silica using sol-gel processing.

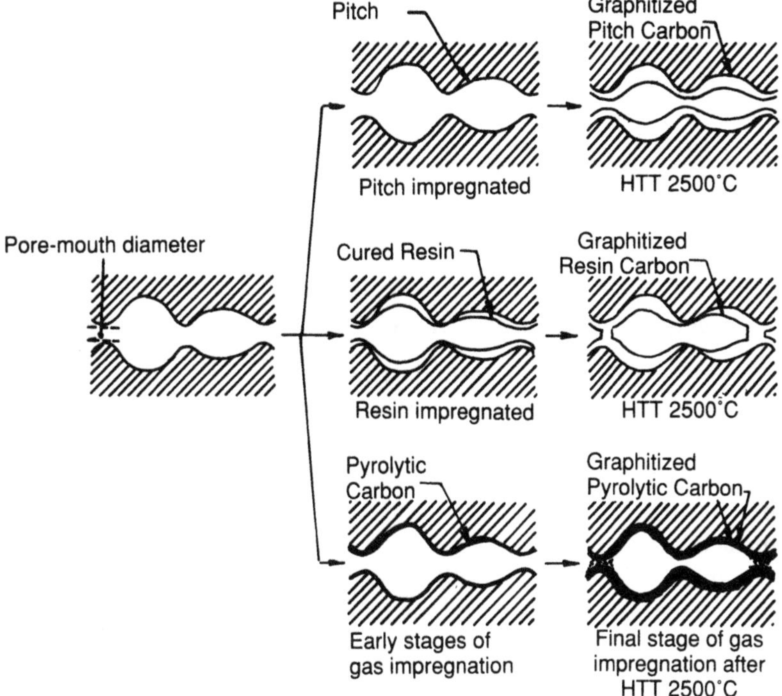

Fig. 3.11 Carbon pore-filling mechanisms.

The gel was dried under critical pressure–temperature conditions to provide a highly microporous silica aerogel around the carbon fibre preform [20]. Deposition of carbon (from methane at 1000 °C, 0.01 atm) led to the densification of the composite in a much reduced processing time. The silica aerogel was removed by sublimation of the silica at 2500 °C.

3.4 CVD PROCESSING OF CARBON–CARBON COMPOSITES

Carbon–carbon composites are generally processed in one of three ways (or a combination):

1. CVD processing;
2. multiple impregnation–pyrolysis using thermosets (e.g. phenolics) (Chapter 4);
3. multiple impregnation–pyrolysis using thermoplastics (e.g. pitch) (Chapter 5).

Figure 3.11 depicts the pore-filling mechanisms for the three methods [21].

Table 3.1 Comparison of matrix formation techniques

Material	*Density* (g cm^{-3})	*Tensile strength* (MPa)	*Flexural strength* (MPa)	*Shear strength* (MPa)
Resin impregnated with pyrolysis	1.65	82.7	68.9	27.6
CVD	1.50	120.6	142.6	51.7

3.4.1 Comparison of carbon–carbon matrix formation techniques

In theory, CVD-derived carbon is likely to possess greater mechanical strength than either the needle-like cokes derived from mesophase pitch pyrolysis or the amorphous glassy carbons derived from thermosetting resin pyrolysis [17]. The steady build-up of the matrix carbon around the fibres during the CVD process does not generate the severe fibre–matrix interfacial stresses seen with the phenolic resin impregnation–pyrolysis method. In fairness, it should be pointed out that although CVD-processed composites have excellent properties, the majority of composites are produced at least partially via impregnation–pyrolysis routes for the sake of process economics.

A number of studies have directly evaluated the relative merits of different carbon matrix formation techniques for the same fibre preform. McAllister and Taverna compared resin impregnation–pyrolysis cycles to CVD as a processing technique for the same 3-D preform. The mechanical properties of the composites are summarized in Table 3.1 [22].

Although the properties of the CVD-processed composite are superior to the resin-based composite, the authors do comment that some of the disparity can be explained by the absence of a graphitization stage in the CVD fabrication. Graphitization generally increases the oxidation resistance and density of a composite, but reduces its strength.

In another more thorough study, Mullen and Roy fabricated 3-D carbon–carbon rings by a range of different processes [23]. The variation of mechanical properties with process technique is shown in Table 3.2. Several points of interest arise from these data. The two methods of CVD used, isothermal and pressure gradient, resulted in very different products being produced. The pressure gradient densified material had a relatively low density (1.28 versus 1.59 g cm^{-3} for the isothermal method product), but possessed a higher breaking stress. It is also significant to note that the supplementary use of a phenolic impregnation–pyrolysis cycle to densify the CVD-produced composites resulted in considerable increase in density (1.59 $\rightarrow$ 1.73 g cm^{-3} for the isothermal sample, 1.28 $\rightarrow$ 1.58 cm^{-3} for the pressure gradient material), but did not appear to be of benefit mechanically. This is somewhat at odds with later findings. In general, the authors

Table 3.2 Mechanical property process relationships for 3-D carbon–carbon hoops [23]

Impregnation technique	*Density* ($g\ cm^{-3}$)	*Tensile strength* (MPa)	*Tensile modulus* (GPa)	*Strain to failure* (%)	*Compressive strength* (MPa)	*Compressive modulus* (GPa)	*Strain to failure* (%)
Phenolic	1.62	118.5	70.3	0.18	52.9	20	0.59
High-melt pitch	1.64	94.4	106.1	0.08	—	—	—
Low-melt pitch	1.65	128.2	64.1	0.05	—	—	—
Isothermal CVD	1.59	113.7	77.2	0.15	103.1	33.8	0.48
Isothermal CVD/phenolic	1.73	106.8	77.9	0.13	73.9	27.6	0.40
Differential pressure CVD	1.35	136.4	68.2	0.20	115.1	23.4	0.72
Differential pressure CVD–graphitized	1.28	130.2	61.3	0.20	—	—	—
Differential pressure CVD–phenolic	1.58	128.2	64.1	0.20	82.7	18.6	0.65

Table 3.3 Mechanical property data for Le Carbone-Lorraine CVD processed carbon–carbon [24]

		Aerolor 32	*Aerolor 33*
Density (g cm^{-3})		1.6–1.7	1.8–1.9
Flexural strength:	*XY* (MPa)	80	40
	Z (MPa)	80	60
Tensile strength:	*XY* (MPa)	70	70
	Z (MPa)	70	110
Tensile modulus:	*XY* (GPa)	50–70	70
	Z (GPa)	50–70	120
Compressive strength:	*XY* (MPa)	50–80	100
	Z (MPa)	50–80	100
CTE	(°C^{-1})	3	4
Thermal conductivity	(W m^{-1} c^{-1})	10	120

suggested that the better mechanical properties of the CVD-processed composites were a result of the superior fibre–matrix interactions.

A more recent study by Girard and Slonia at Le Carbone-Lorraine provides further information on the use of an impregnation/pyrolysis step after CVD processing [24]. The mechanical property data of two products, Aerolor 32, produced by CVD processing of a 3-D fibre preform, and Aerolor 33, produced by a combined CVD-impregnation/pyrolysis method on a similar preform, are shown in Table 3.3.

In contrast to the results of Mullen and Roy, the data in Table 3.3 convincingly show the densification of a CVD-processed composite by impregnation–pyrolysis to improve the mechanical properties. The tensile and compressive properties benefit particularly, although the flexural properties are diminished. Of particular interest is the significant increase in thermal conductivity on impregnation–pyrolysis of the CVD-fabricated composite. The thermal conductivity of the composite is clearly influenced by the residual porosity.

Two studies have considered the effect of a CVD processing step after impregnation–pyrolysis processing of a carbon fibre preform. McAllister and Lachmann compared the properties of phenolic-based 3-D carbon–carbon materials, with and without additional CVD processing, with the properties of pitch-derived composites [25]. Table 3.4 shows that the CVD process markedly improves the strength of the phenolic-derived material, presumably by removing the residual porosity that acts to increase stress concentration in the matrix material (and hence reduce breaking stresses). It is noticeable, however, that the mechanical properties of the two pitch-derived composites (one from a cinnamalydehyde–indene synthetic

Table 3.4 Comparison of matrix formation techniques for 3-D carbon–carbon composites [25]

Material	*Flexure strength* (MPa)	*Flexure modulus* (GPa)	*Compressive strength* (MPa)	*Compressive modulus* (GPa)
Phenolic	89.0	27.6	56.5	7.6
Phenolic and CVD	108.9	24.1	73.0	6.9
CAI	102.7	32.4	50.3	6.9
LTV pitch	153.6	32.4	71.7	10.3

pitch (CAI), the other from an LTV proprietary pitch, see Chapter 10) are better than the phenolic derived composites (especially the LTV synthetic pitch). The LTV pitch was specifically designed to give good fibre/matrix interfacial properties, and the success of the design is evident from the mechanical property data.

Hill and co-workers at AWRE, Aldermaston studied thermal gradient CVD as a method of densifying and strengthening unidirectional (ID) phenolic-derived composites [26]. When untreated carbon fibres were used in the phenolic prepreg, very little fibre/matrix adhesion was observed. As a result, the material had, on pyrolysis, little adhesion between the carbon fibres and glassy carbon matrix. The CVD densification of this 'first pass' material resulted in considerable pore infilling, especially at the fibre/matrix interface, and gave a considerable improvement in longitudinal flexural properties. The authors comment that using CVD-derived pyrolytic carbon to fill the spaces around the fibre results in composites capable of dissipation of fracture energy by fibre pull-out. Pyrolytic carbon possesses the optimum shear properties, and adhesion to both the phenolic-derived glassy carbon and the untreated carbon fibre surface, to ensure good fracture toughness.

Finally, Seibold at McDonnell Douglas has evaluated CVD (CH_4 at 1100 °C) as an intermediate process cycle in the multiple impregnation–pyrolysis processing of high-density 3-D carbon–carbon materials [27] (up to 15 cycles used). He found that application of initial CVD process steps limited the final density attainable after multiple cycles to 1.83 g cm^{-3}. The density limitation was caused by the creation of closed porosity during the initial CVD step. Intermediate CVD processing was found to be no more effective than conventional resin impregnation–pyrolysis for further densification. Seibold concluded that CVD processing is not worth while.

A proviso should be added, however, that by operating the CVD methane cracking process at 1100 °C, one might well be handicapping the final result by sealing in closed pores. The use of more 'gentle' experimental conditions might well result in better composite properties.

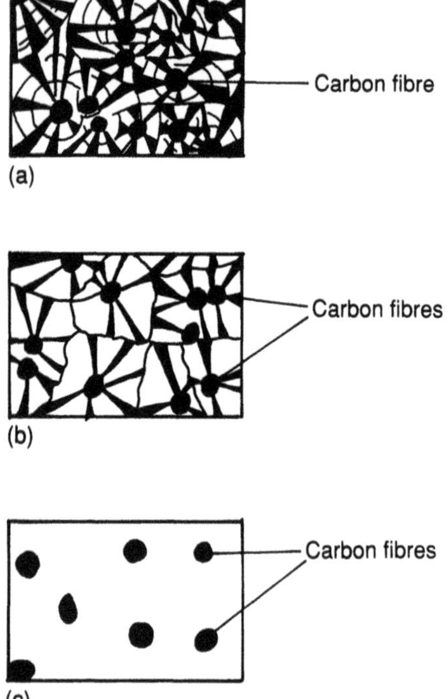

Fig. 3.12 CVD-derived carbon microstructures: (a) rough laminar; (b) smooth laminar; (c) isotropic.

3.4.2 Carbon matrix microstructure: the influence of process conditions

Early work on pyrolytic carbon deposition in fluidized beds established that several different morphologies of CVD-deposited carbon can be produced. Three types of carbon microstructure are commonly seen in CVD composite fabrication experiments:

1. smooth laminar
2. rough laminar
3. isotropic

These microstructures are characterized by their optical activity under polarized illumination; schematic depictions of their optical activities are shown in Fig. 3.12.

Several research groups have performed experimental studies to elucidate the experimental conditions which control the microstructure of the deposited carbon by careful variation of pressure, temperature, gas flow rates and precursor nature, etc. Trinquecoste and his team studied the microstructures of carbons derived from methane over a range of conditions [28], while Kimura *et al.* studied the microstructures of propane-derived carbons [29]. Both groups observed transitions from one microstructure to another as conditions changed.

Liebermann and Pierson at the Sandia National Laboratories have developed a chemical model for the CVD of carbon which rationalizes the observed microstructure/experimental correlations. They postulated that methane decomposes to give either ethyne (acetylene) (C_2H_2) or benzene (C_6H_6). These two species act as precursors to the elemental carbon formed in the deposit. As previously stated, acetylenic precursors are thought to be responsible for isotropic carbon, whereas aromatic precursors, such as benzene, are thought to be responsible for smooth laminar carbon [9]. Thermodynamic calculations based on experimental pressures, temperatures and gas flow/composition were performed and the types of carbon deposit successfully correlated with the equilibrium predicted ratios of acetylene/benzene concentrations. The results and conclusions are summarized as follows [30]:

1. smooth laminar – low deposition temperature, high partial pressure CH_4, no added H_2, $[C_2H_2] : [C_6H_6] < 5$;
2. rough laminar – intermediate deposition temperature, intermediate partial pressure, some added H_2, $[C_2H_2] : [C_6H_6] > 5, < 20$;
3. isotropic – high deposition temperature, low partial pressure CH_4, large amount of added H_2, $[C_2H_2] : [C_6H_6] > 20$.

Liebermann and Pierson carefully studied the microstructures of the CVD matrices produced over a wide range of conditions using the thermal gradient CVD technique (the thermal gradient technique allows considerable variation of experimental parameters without causing the sealing of pores). They noticed that the matrix microstructure generally changes across a sample. The changes were always smooth laminar → rough laminar, or rough laminar → isotropic, consistent with the model developed. In fact, it was found to be very difficult to deposit carbon in one particular microstructure throughout the preform. Deviations from their theoretical model occurred whenever the conditions were a long way from equilibrium, such as at high flow rate.

Given that the CVD process offers some degree of manipulation of the matrix microstructure, what is the optimum type of microstructure for a carbon–carbon composite? Granoff and co-workers concluded that a rough laminar was the preferred microstructure. Isotropic carbon was found to be relatively low in density and the smooth laminar carbon was shown to be prone to thermal stress microcracking [31]. Liebermann *et al.* were subsequently able to fabricate a PAN-derived carbon felt/CVD carbon composite with a matrix consisting of 90% rough laminar carbon. The material was found to have excellent thermal and mechanical shock resistance [32]. Yasuda and co-workers have fabricated carbon–carbon materials using the pressure gradient method. Under a variety of different experimental conditions, they produced two types of composite matrix: a combination smooth/rough laminar matrix, and a combination rough laminar/isotropic one. The rough laminar/isotropic matrix was found to possess superior strength and stiffness. The smooth/rough laminar matrix composite, on the other hand, possessed excellent toughness which was attributed to the energy-absorbing properties of the network of thermal stress cracks in the smooth laminar matrix, leading to a less brittle failure mode.

It is worth noting that the temperature and pressure gradient methods allow considerable freedom in the selection of experimental conditions for CVD carbon deposition. A choice of microstructure can therefore be made. By contrast, the isothermal process requires operation under quite a narrow range of conditions in order to satisfy the diffusion rate/surface reaction rate criteria previously discussed. As a result, the choice of microstructure is very much narrower. Under typical industrial operating conditions (20 torr CH_4 at 1050 °C) the matrix possesses a predominantly smooth laminar microstructure. Raising the temperature to promote the rough laminar structure would result in the formation of undesired closed porosity.

3.4.3 Fibre preform geometry/CVD process relationships

The nature of the fibre substrate, be it a woven long-fibre multidimensional structure or a compacted carbon felt substrate, will greatly influence the final mechanical properties of the composite. In addition, each fibre substrate will have a defined pore size distribution, which will affect the process conditions required for successful densification.

Conventional composite theory leads one to expect the mechanical properties of a composite to increase with increasing volume fraction of reinforcing fibre. This is not necessarily the case with CVD-processed carbon–carbon. One might predict that low fibre content materials can be processed to higher densities with the high density of pyrolytic carbon relative to carbon fibre. A carbon felt preform of initial density of 0.1 g cm^{-3} for example, can be processed to a density of 2.06 g cm^{-3}, whereas a long-fibre 2-D preform of density 1 g cm^{-3} (58% v/v) can be processed to a density of only 1.6 g cm^3. A high initial fibre content means that the initial pores are a good deal smaller and therefore more prone to sealing, thus causing poor mechanical performance. Kotlensky and Bauer found that the mechanical properties of a series of CVD-processed eight-harness satin weave preforms did not increase as the fibre volume increased from 34 to 70%. High fibre content materials generally exhibited lower densities and lower flexural strengths than the low fibre content preform-derived composites [33].

Collaborative work between teams at the Sandia and Oak Ridge National Laboratories has further illustrated the effect that initial porosity can have on the properties of CVD-processed carbon–carbon materials [34]. They prepared two types of filament-wound preform – one with wound yarn (each yarn consisting of up to 2500 carbon fibres) and the other with the same wound yarn arrangement with additional short chopped fibres sprayed on to the preform during the filament winding stage. These chopped fibres served to reduce considerably the macropore size between the wound yarns. The two types of preform were densified by thermal gradient techniques and subjected to mechanical and pore size testing. The addition of chopped fibres was found to reduce considerably the number of large pores and, accordingly, boost the mechanical performance of the composite.

Kotlensky identified different mechanical property characteristics of CVD-processed carbon–carbons produced by different CVD processes [17]. Comparison of the isothermal and temperature gradient CVD methods for the formation of a matrix around filament-wound yarn preforms showed the isothermal method to be superior at filling the small pores in between the individual fibres; the thermal gradient method preferentially filled the large pores. One study has evaluated the effect of the CVD process type used on the mechanical properties of the resulting composite. Using a carbon felt substrate, the isothermal method was found to produce the highest flexural strength materials [16] even though thermal gradient processed materials possessed equivalent densities. It appears that the residual porosity left in the thermal gradient processed material is more detrimental than that remaining in the isothermally processed composite.

Finally, it is instructive to consider how scale-up can cause the need for a change of process conditions. Workers at the General Electric Company noted that it was necessary to change the CVD conditions (to a longer deposition time and lower operating pressure) in order to achieve successful densification of a 20 × 20 × 30 cm fibre preform in comparison with a 10 × 10 × 20 cm preform. It is noteworthy that CVD processing is only really cost-effective for relatively thin preforms – a thick preform places too many demands on the diffusion-controlled stage.

3.5 CVD PROCESSING OF CERAMIC MATRIX COMPOSITES

3.5.1 Introduction

The excellent high-temperature strength, oxidation resistance and toughness of well-processed ceramic matrix composites make them extremely attractive materials for high-temperature environments. A specific example of an area where ceramic matrix composites might be deployed is in the fabrication of jet turbine blades. The insidious increase in temperature of the nickel alloys in current use, place restrictions on the operating temperature of the turbine. Processing by CVD represents a viable method for matrix formation – the relatively low temperatures involved do not cause fibre degradation, processing to high densities can be achieved and the process yields very pure materials (if the thermodynamic constraints discussed in Section 3.2.1 are considered). For the high-temperature refractory materials, such as SiC and TiC, other composite fabrication techniques based on hot pressing/sintering or melt infiltration are not applicable, and CVD is the preferred fabrication technique. One of the great advantages of CVD fabrication for carbon–carbon over rival processing routes is its versatility. The same equipment can be used to deposit carbon, ceramic or hybrid matrices and to produce oxidation-protective coatings.

3.5.2 Hybrid carbon fibre/ceramic matrix compounds produced by CVD

A major contribution to this field has been made by the group led by Naslain at CNRS (Bordeaux)/SEP, which have made great strides in understanding the fabrication processes and mechanical behaviour of ceramic matrix composites [35]. A great deal of this work has been concerned with upgrading the mechanical and oxidation resistance properties of CVD-produced carbon–carbon. Partially densified carbon fibre–CVD carbon materials were densified with a number of CVD-derived ceramics to different levels of final residual porosity. Systematic studies of the mechanical, chemical and thermal properties as a function of the ceramic content were carried out.

The first published study concerned the formation and properties of C–C/SiC and C–C/TiC composites [3]. Thermodynamic calculations were used to set the inlet flow conditions to produce pure SiC (from CH_3SiCl_3/H_2) and TiC (from $TiCl_4/CH_4/H_2$). A low-temperature, low-pressure experimental set-up was utilized to ensure in-depth deposition throughout the pore system. The experimental conditions (temperature, pressure, flow rate) were optimized by the use of Si microprobe analysis on cross-sections of materials produced under different conditions. Silicon microprobe analysis showed that if the deposition rate was too high (due to temperature, pressure or flow rate being too high), overcrusting occurred rapidly, with very little SiC being deposited within the inner part of a cylindrical 2-D carbon–carbon preform. The rate of SiC deposition was found to be slightly greater than that of the carbon deposition; the rate of TiC deposition was, however, much slower.

All of the studies were based on a 2-D PAN-derived carbon fibre preform, densified with CVD carbon to a residual porosity of around 35%. These preforms were then densified with varying amounts of SiC or TiC up to full densification, and the mechanical and chemical properties studied as a function of the ceramic content. Linear increases in compressive strength (σ_{11}^{c}) and modulus (E_{11}^{c}) were noted with increasing SiC or TiC contents. Of particular interest are the mechanical properties of the fully densified C–C/SiC and C–C/TiC composites in comparison with a fully

Table 3.5 Mechanical properties of carbon/ceramic hybrid matrix composites

Ratio of carbon fibre to ceramic matrix		E_{11} (GPa)	σ_{11} (MPa)
C–C	—	10	100
C–C/SiC	5 : 7	51	310
C–C/TiC	5 : 7	48	330

densified carbon–carbon composite (since all of the materials are based on similar PAN–carbon fibre preforms, with 40% of the total composite volume being fibre). The mechanical property data are shown in Table 3.5 [3].

The mechanical properties are clearly improved by partial substitution of a carbon matrix with either of the carbides. The composite materials fail in different modes depending on the volume of carbide present. At low carbide content, failure occurs by delamination, whereas at higher carbide content failure occurs by combined fibre/matrix fracture.

A more complete mechanical analysis of the C–C/TiC system is given in a follow-up paper by the same authors [36]. The oxidation resistance of these materials was studied and correlated with oxidation conditions (temperature and partial pressure of oxygen) and degree of carbide infiltration. The C–C/TiC composites initially showed a weight gain on oxidation, due to the formation of TiO_2, followed by a loss in weight as oxidation of the carbon occurred. The transition from weight gain to weight loss regimes occurred more rapidly as the temperature increased. As expected, increasing the level of TiC infiltration increased the oxidation resistance. The C–C/SiC composites exhibited the best oxidation resistance of all. The fully densified C–C/SiC materials showed reasonable oxidation resistance up to 1600 °C, at which temperature the SiC began to oxidize and vaporize.

A similar study has been performed on a series of C–C/B_4C materials [37]. A C–C/B_4C composite, with a matrix carbon : matrix carbide ratio of 5 : 7, was found to possess a compressive strength of 310 MPa, i.e. similar to the C–C/SiC and C–C/TiC materials. The degree of oxidation resistance was found to be on a par with the C–C/TiC material, but inferior to the C–C/SiC material.

To complete the series of ceramic substitution materials, Naslain and co-workers prepared and studied a series of C–C/BN hybrid matrix composites using similar techniques [15]. The boron nitride was deposited using low-pressure CVD from a BF_3/NH_3 mixture. The mechanical properties of C–C/BN composites were found to be inferior to those of the related C–C materials, a consequence of the low modulus of BN. The oxidation resistance of the C–C/BN materials was, however, found to be superior to the C–C analogues. Above 900 °C, weight increases occurred

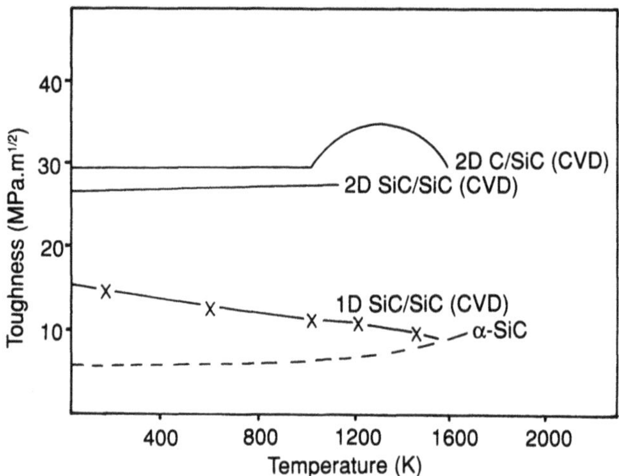

Fig. 3.13 The temperature dependence of fracture toughness of silicon carbide composite materials.

due to the formation of B_2O_3. At higher temperatures, the oxidation resistance of the C–C/BN material diminished, due to the volatilization of B_2O_3 which becomes appreciable at temperatures above 1100 °C.

3.5.3 CVD Ceramic Matrix Composites

Despite a well-established technology for producing CVD ceramic coatings, relatively little work has been done on the fabrication of ceramic matrix composites. The best studied matrix system is silicon carbide from which excellent results have been obtained. A 2-D SiC–SiC composite has entered commercial production [38]. The material is unusual in that the matrix (CVD-SiC) is stiffer than the fibres (Nicalon SiC fibres). The SiC fibres generally used are produced by the pyrolysis of a spun carbosilane fibre and contain substantial inclusions of SiO_2 and C. The toughness/temperature curves for the SEP range of SiC matrix composites are shown in Fig. 3.13. Of note is the high toughness of the SiC/SiC and C/SiC materials relative to sintered, monolithic SiC (carborundum). The SiC–SiC composite is reputed to possess the highest toughness yet observed for any ceramic material [35].

Observation of the fracture pattern of the SiC–SiC composites explains their high toughness. Stress relief occurs via matrix microcracking, allowing for the tensile strain to failure to be increased from around 0.08% in monolithic SiC to 0.6% in the composite. The process conditions used by SEP in the fabrication of their composites have been disclosed in a patent [39]. The furnace operating temperature is 900 °C, and the partial pressures of H_2 and $MeSiCl_3$ and 15 are 5 torr respectively.

Fitzer and Gadow have performed several studies on SiC matrix composites, comparing CVD as a matrix fabrication technique with reaction bonding [40]. Reaction bonding is a fast and simple technique. A porous carbon body is dipped into molten silicon and heat treated to form a silicon carbide matrix. Fitzer concluded that CVD-fabricated composites possessed far superior mechanical properties (up to 50–80% better) than the reaction-bonded materials. He also showed that extemely high flexural strengths could be achieved using CVD-derived SiC fibres, rather than the more widespread organometallic polymer-derived 'Nicalon' SiC fibres. At high temperatures, however, the 'Nicalon' SiC fibres degraded slightly to weaken the interfacial fibre–matrix bond, so that the 'Nicalon' SiC-derived composites exhibited a less brittle, tougher fracture mode. Fitzer has also performed several studies on the use of CVD for strengthening SiC and reaction-bonded Si_3N_4 monoliths [14]. While in-depth infiltration of the SiC body with SiC was quite simple, the rapid formation of Si_3N_4 (from $SiCl_4/NH_3$), even at low temperatures, meant that the porous Si_3N_4 body could not be effectively densified.

Workers at the Oak Ridge National Laboratories in the USA have developed a novel pressure/temperature gradient process for CVD fabrication of ceramic matrix composites [41]. In this method, a fibre preform is placed in a gas-tight die, so that a pressure gradient can be applied across it. The unit is then placed into one end of a hot zone, to provide a combined pressure/temperature gradient. The experimental set-up is shown in Fig. 3.14. Considerable success was achieved in densifying SiC fibre (or whisker) preforms with SiC (from CH_3SiCl_3/H_2) and Si_3N_4 (from $SiCl_4/NH_3$); a SiC/SiC composite was processed in 12 h, a considerable improvement on conventional isothermal method process times (i.e. 150 h for Fitzer/Gadow) [40]. 1-D and 2-D SiC–SiC composites have been made using this SiC–SiC method. The 1-D composites were found to possess a flexural strain to failure of 1%. Examination of the fracture faces showed that extensive fibre pull-out had occurred, indicative of a well-processed material with the optimum level of fibre–matrix adhesion. It is interesting to note that the Oak Ridge workers have recently used the temperature/pressure gradient method to make lightly densified, porous ceramic whisker/ceramic matrix composites which they have successfully tested as high-temperature filter materials [42].

An interesting extension of the Oak Ridge CVD programme resulted from the accidental production of $TiSi_2$/SiC dispersed phase composite. The aim was to produce a TiC/SiC material from a feed of $TiCl_4/CH_3SiCl_3/H_2$ but kinetic factors precluded this aim. The fracture toughness of the $TiSi_2$/SiC material was found to exceed greatly that of pure SiC. The authors suggest that the $TiSi_2$/SiC material might be very useful as an oxidation-resistant coating for carbon–carbon composite materials.

Finally, Naslain and co-workers have recently reported the CVD

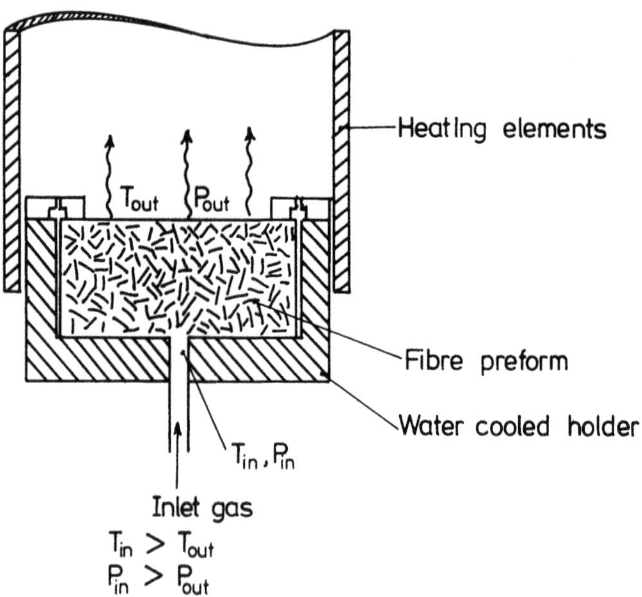

Fig. 3.14 The Oak Ridge National Laboratory combined pressure/temperature gradient CVD technique [41].

fabrication of an Al_2O_3–Al_2O_3 composite [43]. The alumina matrix was formed from an $AlCl_3/CO_2/H_2$ feed stream at working conditions of 975 °C and 2.5 kPa. The composite, densified to a 10% residual porosity, showed a very brittle mode of fracture, consistent with the strong fibre–matrix adhesions present.

3.6 A BRIEF SURVEY OF COMMERCIAL CVD COMPOSITE FABRICATION PROCESSES

Fabrication of carbon–carbon materials by CVD has existed for nearly 20 years, in competition with the multiple resin impregnation–pyrolysis method. By contrast, CVD fabrication of ceramic composites is less widespread. The choice of CVD or impregnation–pyrolysis as a matrix fabrication method is influenced by many factors. The generally advantageous mechanical properties of CVD-derived carbon–carbon in comparison with thermoset-derived impregnation–pyrolysis materials can be offset by the long process times required. The CVD method tends to work best for relatively thin preforms, where the diffusion of gases to and from the middle of the preform is not too difficult.

In the UK, the Dunlop Aviation Division make carbon–carbon via CVD, the product being aimed almost entirely at the aircraft brake market (brake materials make up 63% of the total end market for carbon–carbon at present). The carbon–carbon is made via a low-pressure isothermal technique. After a slow build-up of matrix (up to a month), the material is graphitized at 2500 °C and then machined for use. The key to commercially viable production obviously lies in a well-integrated production system and the use of large CVD furnaces to facilitate economies of scale. Dunlop have achieved the former by buying in PAN fibres for carbonization, lay-up and processing on-site, and the latter by a large-scale furnace investment programme. Arguably the world's leading nation in CVD composite fabrication technology is France, notably through SEP and Le Carbone-Lorraine. The former produce three types of composite material: SEPCARB, a carbon–carbon product aimed mainly at the brake materials market, SEPCARBINOX, a carbon–silicon carbide material designed for use in propulsion systems where the oxidation is too severe for conventional carbon–carbon, and CERASEP, a silicon carbide–silicon carbide material whose outstanding oxidation resistance and toughness have led to applications in engines and aerospace. Le Carbone-Lorraine has pioneered the use of CVD technology for the production of 3-D carbon–carbon with very high densities, used mainly for its excellent ablation properties. The two main German carbon companies (Schunk and Sigri) appear to make their carbon–carbon materials by resin impregnation–pyrolysis.

In the USA, a large amount of work on CVD fabrication of carbon–carbon has been carried out at the Sandia National Laboratory, mainly aimed at fabricating filament-wound missile nose cones. A large number of companies have used CVD for carbon–carbon fabrication – the Super-Temp division of BF Goodrich, LTV, Hitco, Bendix, ABS, GE and Avco [44]. It is hard to generalize on what the accepted route for C–C manufacture is in the US C–C industry; CVD is used either (a) when the product is manufactured on multiple-scale economics or (b) the outstanding properties of CVD carbon–carbon are necessary. It is significant to note that Du Pont have recently chosen to enter the high-temperature composites business by signing a technology licence agreement with SEP [45]. They aim to begin production by the early 1990s, with a view to supplying various US government aerospace and defence programmes. In view of the wide-ranging CVD fabrication technology for carbon–carbon that must be available in the USA, it appears likely that Du Pont are particularly interested in the CVD technology for silicon carbide materials, rather than carbon–carbon, alone.

3.7 SUMMARY

The fabrication by CVD of carbon–carbon composite materials offers considerable control over the matrix composition. The microstructure of the CVD-derived material can be controlled by the experimental parameters – temperature, pressure, gas phase composition and flow rate, and even the fibre preform geometry. The CVD method has considerable advantages over the other fabrication techniques in that the same equipment may be used to deposit non-carbon oxidation inhibitors and coatings and to densify ceramic matrix composites as well as carbon and graphite. The successful densification of a composite material using CVD necessitates great care in the selection of experimental conditions which will allow the deposition of matrix material throughout the pore system of the fibre preform. It is a slow process that requires considerable skill. The conventional isothermal process, in which the fibre preform is placed in a uniform temperature/pressure furnace, must be operated under conditions where the surface nucleation and growth rates are slower than the gaseous diffusion rates, in order to avoid 'plugging' of the surface pores in the fibre preform. In practice, therefore, the equipment must be operated at low temperatures and pressures, thus depositing material at a low rate.

Experimental techniques exist to overcome the diffusion/surface kinetics limitations of the isothermal CVD process by the application of thermal gradients and/or forced gas flows to the fibre substrate. The majority of such methods are, unfortunately, not amenable to production scale-up. The isothermal method remains the most economical method for large-scale CVD production of composites despite being beset with the problem of long processing times.

The CVD method produces good quality carbon–carbon materials. The technique is especially suited to the production of relatively thin artefacts, such as braking materials, where the diffusion problems are less severe. Many of the problems associated with the resin impregnation–pyrolysis route (Chapter 4), such as the severe thermal stresses generated in the composite during pyrolysis, are avoided with the CVD technique. The widespread use of CVD-fabricated carbon–carbon materials has to date been severely hindered by the cost when compared to material produced by other routes.

In many respects, CVD is the best developed of any of the production methods for high-temperature composites, especially carbon–carbon. It has proved to be the best method for the fabrication of long-fibre reinforced silicon carbide materials. The commercial SiC–SiC product, CERASEP, exhibits outstanding toughness for a ceramic material and seems to be on the threshold of widespread use as a material capable of operating in oxidizing environments up to 1350 °C.

REFERENCES

1. Bracke, P., Schurmans, H. and Verhoest, J. (1984) *Inorganic Fibres and Composite Materials*, (*EPO App. Tech. Series*, **3**), Pergamon Press, Oxford.

2. Besmann, T. M. (1977) *Oak Ridge Nat. Lab. Report*, no. TM 5775 (April 1977).
3. Naslain, R., Rossignol, J. Y., Hagenmuller, P., Christin, F., Heraud, L. and Choury, J. J. (1981) *Rev. Chim. Miner.*, **18**, 544.
4. Randich, E. and Gerlach, T. M. (1983) *Chem. Tech.*, **102**, 2.
5. Stinton, D. P., Lackey, W. J., Lauf, R. J. and Besmann, T. M. (1984) *Ceram. Eng. Sci. Prod.*, **5**, 668.
6. Spear, K. E. (1982) *Pure Appl. Chem.*, **54**, 1297.
7. Bokros, J. C. (1969) *Chem. Phys. Carbon*, **5**, 1.
8. Noles, G. T. and Liebermann, M. L. (1975) *J. Chromatog.*, **114**, 211.
9. Liebermann, M. L. (1972) *Proc. 3rd Int. Conf. on CVD*, **3**, 95.
10. Nuhara, K. (1984) *Am. Ceram. Soc. Bull.*, **63**, 1160.
11. Tesner, P. A., (1984) *Chem. Phys. Carbon*, **19**, 65.
12. Van der Brekel, C. H. J. and Lersmacher, B. (1983) *Proc. Eur. Conf. CVD*, **4**, 321.
13. Diefendorf, R. J. and Sohda, Y. (1983) *Ext. Abstr. Prog., 17th Biennal Conf. on Carbon*, **17**, 31.
14. Fitzer, E. and Hegen, D. (1979) *Angew. Chem. Int. Ed.*, **18**, 295.
15. Hannache, H., Quenisset, J. M., Naslain, R. and Heraud, L., J. (1984) *Mat. Sci.*, **19**, 202.
16. Kotlensky, W. V. (1971) *Materials 71, SAMPE 16th Nat. Symp. Exhib.*, **16**, 257.
17. Kotlensky, W. V. (1973) *Chem. Phys. Carbon*, **9**, 173.
18. Sugiyama, K. and Nakamura, T. (1987) *J. Mat. Sci. Lett.*, **6**, 331.
19. Lachter, A., Trinquecoste, M. and Delhaes, P. (1985) *Carbon*, **23**, 111.
20. Pajonk, G. A. and Teichner, S. J. (1986) *Aerogels* (*Springer Proc. Phys.*, Vol. **6**), Springer-Verlag, Heidelberg, p. 193.
21. Filzer, E. (1987) *Carbon*, **25**(2), 163.
22. McAllister, L. E. and Taverna, A. R. (1971) *10th Biennal Conf. on Carbon*, Bethlehem, Pa, Paper no. FV-40.
23. Mullen, C. K. and Roy, P. J. (1972) *Proc. 17th Nat. SAMPE Symp.*, III-A-2.
24. Girard, H. and Slonia, J. P. (1978) *Proc. 5th Int. Conf. on Carbon and Graphite*, **1**, 483.
25. McAllister, L. E. and Lachmann, W. L. (1983) in *Fabrication of Composites* (*Handbook of Composites*, Vol. **4**) (eds A. Kelly and S. T. Mileiko, Amsterdam, ch. 3, p. 109.
26. Boyne, L., Hill, J. and Turner, K. (1975) *Proc. 5th Int. Conf. CVD*, **5**, 577.
27. Seibold, R. W. (1977) *Ext. Abstr. 13th Biennal Conf. on Carbon*, **13**, 404.
28. Delhaes, P., Trinquecoste, M., Pacault, A., Goma, J., Oberlin, A. and Thebault, J. (1984) *Chimie Physique*, **81**, 809.
29. Kimura, S., Yasuda, E., Takase, N. and Kasuga, S. (1981) *High Temp, High Pressure*, **13**, 193.
30. Pierson, H. O. and Liebermann, M. L. (1975) *Carbon*, **13**, 159.
31. Granoff, B. Pierson, H. O. and Schuster, D. M. (1973) *Carbon*, **11**, 177.
32. Liebermann, M. L., Curlee, R. M., Braaten, F. H. and Noles, G. T. (1975) *J. Comp. Mat.*, **9**, 337.
33. Bauer, D. W. and Kotlensky, W. V. (1973) *SAMPE Quart.*, **4**, 10.
34. Brassell, G. W., Horak, J. A. and Butler, B. L. (1975) *J. Comp. Mat.*, **9**, 228.
35. Naslain, R. (1986) *J. Phys. Colloq.*, Cl, **47**, 703.
36. Rossignol, J. Y., Quenisset, J. M. and Naslain, R. (1987) *Composites*, **18**, 135.

37. Naslain, R. and Langlais, F. (1985) *Proc. 21st Univ. Conf. Sci. Ceramics*, Penn. St. Univ.
38. SEP (1987) promotional literature.
39. SEP (1988) Fr. Demande, 2, 401, 888.
40. Fitzer, E. and Gadow, R. (1986) *Am. Ceram. Soc. Bull.*, **65**, 326.
41. Caputo, A. J. and Lackey, W. J. (1984) *Ceram. Eng. Sci. Proc.*, **5**, 654.
42. Sheppard, L. M. (1987) *Adv. Mat. Processes*, **73**, 7.
43. Colmet, R., Lhermitt-Sebire, I. and Naslain, R. (1986) *Adv. Ceram. Mat.*, **1**, 185.
44. Royse, S. (1987) *The Chemical Engineer*, September, p. 38.
45. *Advanced Materials Newsletter*, 22 June 1987, p. 9.

Thermosetting Resin Matrix Precursors

4

4.1 GENERAL CONSIDERATIONS

Thermosetting resins are used as matrix precursors in carbon–carbon composites because they are relatively easy to use to impregnate fibres, and a large technology base exists from their use in 'conventional' composites processing. The resin impregnation/carbonization route is extremely flexible. Large structures, often with complex geometries, can be manufactured using all of the methods proven in the composites field, e.g. filament winding, prepreg, hand lay-up or pultrusion. In general, thermosetting resins polymerize at low temperatures (<250 °C) to form a highly three-dimensionally cross-linked non-softening amorphous solid. When pyrolysed, the resins form a glassy, isotropic carbon [1] which does not graphitize at temperatures up to 3000 °C.

Carbon yields from resins that cyclize, condense and are readily converted to carbon usually range from 50 to 60% by weight [2]. The low density of the carbon formed (≈1.5–1.6 g cm^{-3}) can limit the density of the final carbon composites but there are many applications where a non-graphitic matrix is desirable. There is one exception to the normal behaviour in carbon–carbon densification. Shrinkage stresses in the vicinity of the fibres can cause glassy carbon materials to graphitize [3] at high temperatures (>2500 °C).

The following is a summary of some of the features which need to be considered when selecting a thermosetting resin for densification processing of carbon–carbon:

1. Yields are in the range 50–70% by weight. Experimental data indicate that carbon yields are not increased by the application of pressure during carbonization.
2. Carbon matrix structures are glassy and do not graphitize at temperatures up to 3000 °C.

3. Stresses applied (or induced) during heat treatment can lead to a graphitic microstructure.
4. In order to attain usable densities and hence properties, components formed by this route must be reimpregnated/recarbonized to minimize the porosity produced during pyrolysis.

4.2 ISOTROPIC CARBON

When thermosetting polymers are heated in an inert atmosphere reactions take place such that one or two possibilities generally occur; the polymer chains may degrade into small molecules with the products being evolved as gases leaving little or no carbon behind, or the carbon–carbon chains remain intact, while heteroatoms are driven off and the carbon chains coalesce with neighbours. The material does not pass through a plastic or liquid state. The final carbon is isotropic if the precursor material is isotropic and can be anisotropic if the precursor is anisotropic. In either case the carbon is non-graphitizable, even when heated to 3000 °C. The primary aim of the thermoset resin precursor route to carbon–carbon is to convert the thermoset polymer matrix of a laminated carbon fibre component to carbon by means of a pyrolysis process. Thus it is resins exhibiting the latter of the two possible responses to heat treatment which are of concern. The carbonization of such non-decomposing, cross-linked thermoset polymers under carefully controlled conditions yields a hard non-graphitizing carbon material, historically known as amorphous carbon because of its non-crystalline, isotropic structure. The material is also known as 'vitreous' carbon as a result of its high lustre and similarity to glass in both appearance and fracture [4].

Vitreous, amorphous, glassy or more properly, isotropic carbon is the carbonaceous residue, glass-like in appearance, from the pyrolysis of a polymeric resin precursor [5]. The material is typified by a low density of around 1.5 g cm^{-3} [6]. A variety of defects identified in the material have been attributed to residual impurities that are presumably present in the parent polymer [7]. The presence of these defects is thought to be responsible for the highly variable strengths observed in isotropic carbon.

The fracture strength is found to vary depending on the quality and heat treatment temperature (HTT) of the carbon. Fractographs of this type of carbon which have failed in flexure (generally accepted as the most useful method of measuring the mechanical properties of the material) exhibit a smooth, featureless, conchoidal fracture surface similar to that of glass [8] (Fig. 4.1).

The mechanical properties of isotropic carbons are directly related to their structure which consists of a three-dimensional random network of stacks of graphene layers appearing rather like tangled ribbons, with three

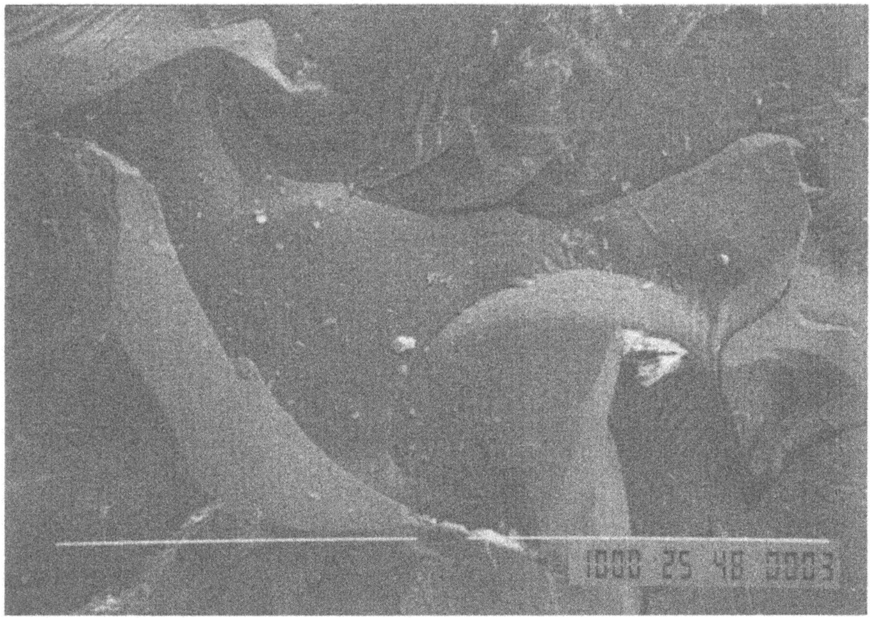

Fig. 4.1 Electron micrograph showing the conchoidal, featureless, brittle fracture face of an isotropic carbon formed from the pyrolysis of a thermosetting resin.

types of bonding therein. Between closely spaced parallel ribbons only weak van der Waals' type forces exist, which can be neglected. Within the ribbons the C–C bond is the stiffest component of the system. This is only noticed, however, with a high degree of preferred orientation as is the case in carbon fibres. In a random configuration of ribbons, the response to an applied stress is related primarily to the ease with which segments can rotate with respect to their environment, which is resisted by carbon bonding between the ribbons. Considering its non-crystalline structure, the Young's modulus and hardness of isotropic carbon are very high. This is due in the main to the presence of cross-links between the ribbon-like bundles of graphene layers which prevent the layers from shearing past one another [9]. Isotropic carbon, because of its high hardness, relatively high strength and chemical inertness to body fluids [10], has been employed as a structural material in biomedical implants such as artificial heart valves [11]. The material is somewhat limited when load bearing is required, because of its extreme brittleness [12]. Mechanical failure is almost completely controlled by brittle fracture. The failure occurs with negligible plastic deformation, without prior warning (due to the rapid rates of crack propagation) and involves low-energy absorption [13]. Isotropic carbon is much stronger than most other carbon or graphites. It has a hardness value of between 6 and 7 on Mohs' scale. When using the Vickers Diamond

Indentation equipment, the indentation disappears almost as soon as the load is removed, suggesting that the material retains some of its elastic properties after heat treatment [14].

The hardness and high strength of the material suggest the structure includes a degree of tetrahedral (sp^3) cross-links between the hexagonal layers. This is considered likely in view of the presence of such cross-links in the starting material and the isotropy of properties irrespective of the method of fabrication of the polymer precursor. The inter-ribbon cross-links are strained and so are less thermally stable than the trigonal (sp^2) bonds within the graphene layers. They are thus progressively ruptured at high temperatures and thought to be responsible for viscoelastic behaviour observed at around 2500 °C [8].

4.3 CARBON YIELD FROM POLYMERS

The number of thermoset polymer impregnants that could be used to process carbon–carbon composites is almost limitless. When all the process and property requirements are considered, the selection is reduced to relatively few materials. In the selection of the matrix precursor the following characteristics must be considered:

1. viscosity;
2. cure conditions;
3. carbon yield;
4. shrinkage during carbonization;
5. matrix microstructure.

It is necessary to achieve good impregnation of the preform structure, controlled cure of the resin and bonding of the fibre/matrix interface.

One of the most important prerequisites for an efficient carbon–carbon process is a matrix precursor with a high carbon yield. The carbon yield of a polymer can be found by heating a known weight of the polymer in an inert atmosphere at a given temperature, and weighing the carbon residue. The carbon yield is given simply by the ratio of the weight of the carbon residue to the initial weight and is expressed as a percentage figure. The conversion efficiency of the resin is defined as the ratio of the weight of carbon residue to the amount of carbon in the original resin.

Many long-chain polymers break down completely into gaseous products. Madorsky [15] showed that polyethylene at one extreme and PTFE at the other leave no carbon residue whatsoever. Polyvinyl fluoride $(CH_2CHF)_n$ and polytrifluoroethylene $(CHFCF_2)_n$ also leave very little residue. By contrast $(CF_2CH_2)_n$ has a carbon yield of 30% which accounts for all of the carbon originally present in the polymer, i.e. a conversion efficiency of 100%. Mackay [16] determined the carbon yield for a whole range of

Table 4.1 Carbon yields of various thermoset precursors

Precursor	*Carbon yield* (%)
Phenolic resins	50–55
Furan resins	50–60
Oxidized polystyrene	55
Polyvinyl alcohol	50
Polyacrylonitrile	44
Polyvinylidene chloride	25
Cellulose	20
Epoxy resins	5
Polystyrene	5
Polymethylmethacrylate	5

polymers (Table 4.1) and came to the conclusion that the carbon yield of a polymer cannot be predicted. Linear polymers such as polystyrene or poly-*p*-xylene, despite a high degree of aromacity, produce low carbon yields. Linear polymers of furan resins and polyphenylene oxide, in which ring structures are interspersed with ether linkages, provide high carbon yields. The conversion efficiency for furan resins, for example, is 91%.

Carbon yield does not seem to be influenced by whether the polymer is thermoplastic or thermosetting, linear or cross-linked, but rather by whether it is capable of cyclization, ring fusion or chain coalescence at the onset of carbonization. In general, the resins should have a high degree of aromacity, or similar ring structures, and high molecular weight. There should be no more than one carbon atom between aromatic rings otherwise chain scissions will take place leading to volatilization of fragmental parts, as witnessed in the low yield from poly-*p*-xylene. Nitrogen, if present, should be in the ring structure and not in the chain. For example, polybenzimidazole is more stable and provides higher carbon yields than amine-cured epoxy resins. Other elements such as sulphur do not affect the stability of the polymers but do result in lower carbon yields and low conversion efficiency. All phenolic resins, except those in which the *para*-position is blocked by either phenyl or methyl groups, are good carbon-forming materials.

Very few resins have carbon yields of over 40% on pyrolysis, thus the choice of matrix precursor is severely limited. Many types of thermosetting resin have been studied for potential use as matrix precursors for carbon–carbon composites. Carbon yield and X-ray diffraction data on several resins that exhibit high carbon yields are given in Table 4.2 along with some of their chemical formulae/structures in Fig. 4.2. Even though the chemical structures of these resins are quite different, they all produce an isotropic, 'glassy' carbon residue as shown by X-ray diffraction data. Data on graphite are included for comparison.

Table 4.2 Characteristics of carbonized high-yield thermoset resins

	Carbon yield at 800 °C	*X-ray diffraction data after HTT of 2700 °C*	
Resin precursor	(%)	L_c (Å)	L_a (Å)
Furan	50–60	75	3.41
Phenolics	50–60	132	3.40
Polyimide	60	75	3.44
Polybenzimidazole	73	40	3.45
Polyphenylene	85	54	3.44
Biphenol resins	65	83	3.43
*Natural graphite	100	—	3.36

Polyvinylchloride

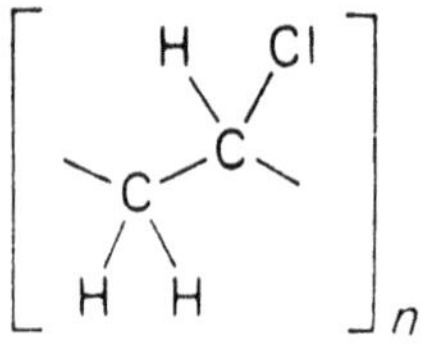

Polyvinylidenechloride

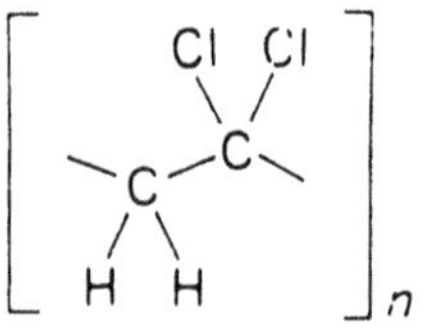

Cellulose

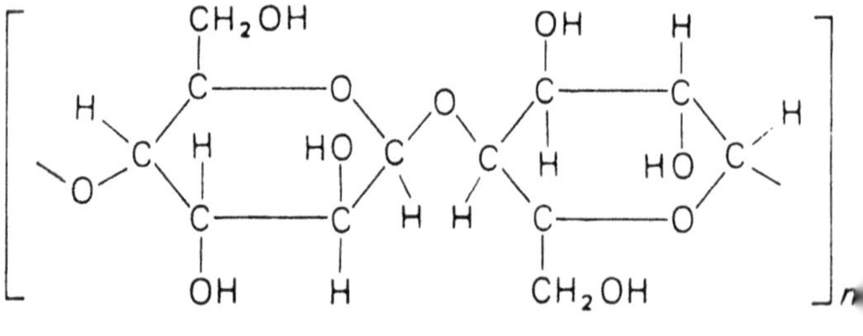

Phenolic resin

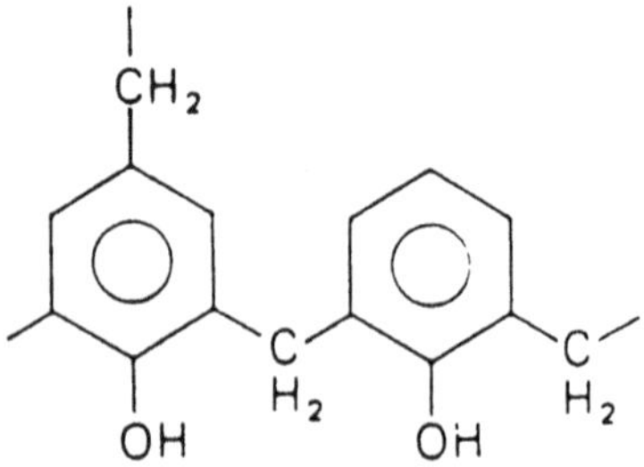

Furfuryl alcohol

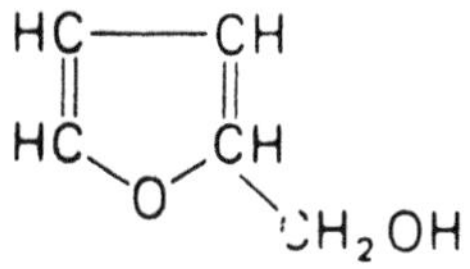

Polyacrylonitrile

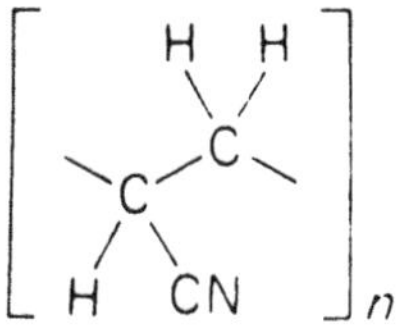

Kapton (Polyimide)

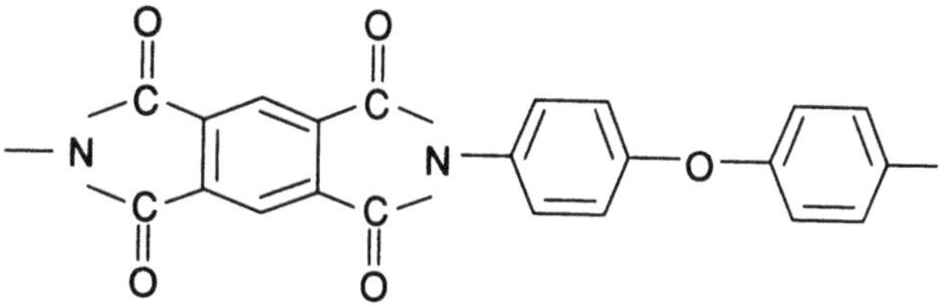

Polyphenylene

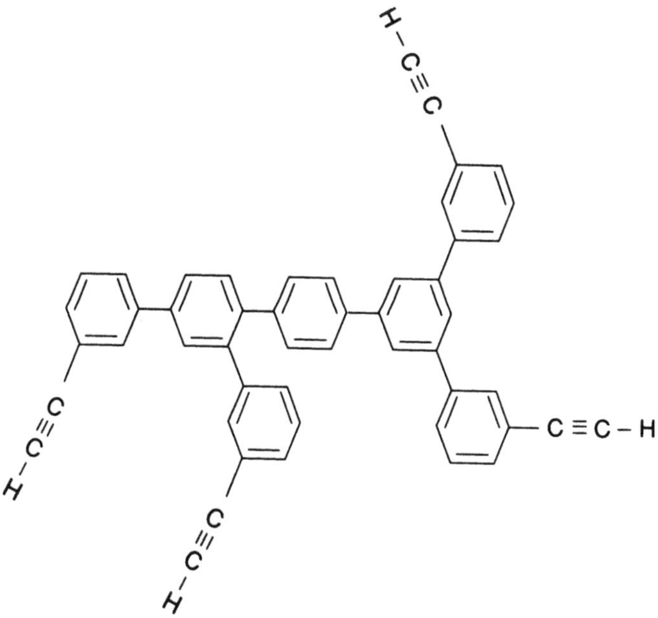

Fig. 4.2 Chemical structure of a number of thermoset resin carbon matrix precursors.

4.4 CARBONIZATION OF POLYMERS

Along with a high carbon yield and ease of impregnation of the fibres, there are three further requirements for a suitable matrix precursor. Firstly, the carbonization shrinkage of the matrix should not damage the carbon fibre skeleton. Secondly, the porosity formed during pyrolysis of the resin must be open and accessible to further impregnation. Only under such preconditions can further improvement of density and mechanical properties be achieved by subsequent processing cycles. Finally, the thermoset precursor will not have a T_g much below the onset of decomposition/carbonization, otherwise the material will carbonize in a rubbery state and the pyrolysis gases will explode the porosity. Phenolic resins, for example, have a T_g greater than the onset of carbonization temperature (due to the high primary covalent and secondary N bond cross-link density) and are thus ideal.

In principle, the resin impregnation process consists of forming a near net shape component of carbon fibre reinforced, high carbon yield, resin by standard composite technologies. The component is cured and post-cured according to the instructions of the resin supplier, followed by carbonization in an inert atmosphere to around 1000 °C. Optimization of the curing and post-curing stages is of great significance since they have a profound effect upon the carbon yield of the polymer and the mechanical properties of the carbon produced [13]. Efficient curing allows the polymerization of the resin to tend towards completion, thus reducing the number of low molecular weight compounds that would be liable to evaporate during pyrolysis and lower the carbon yield. Post-curing induces three-dimensional cross-linking in the resin which is carried over into the structure of the isotropic carbon. Cross-linking further reduces carbon loss during pyrolysis and, by restricting the movement of graphene fibrils under loading, improves the mechanical properties. Furthermore, post-curing causes the water by a condensation mechanism produced during polymerization which (but not fully desorbed) to come out of the cured structure. If the water is left in, there will be very high pore pressures in the laminate as it is carbonized. When the temperature exceeds 100 °C the vapour pressure of the water in the laminate will exceed the ambient pressure used in carbonization and will eventually exceed the interlaminar strength of the component and cause processing delaminations. There is, however, an optimum degree of cross-linking for each resin system. If the amount of cross-linking becomes too great, large thermal stresses can build up during carbonization which lead to premature failure. The shrinkage of the resin during curing and carbonization produces slit-shaped pores which require to be refilled by liquid matrix precursor for subsequent recarbonization if the density and mechanical properties of the porous carbon–carbon skeleton are to be improved.

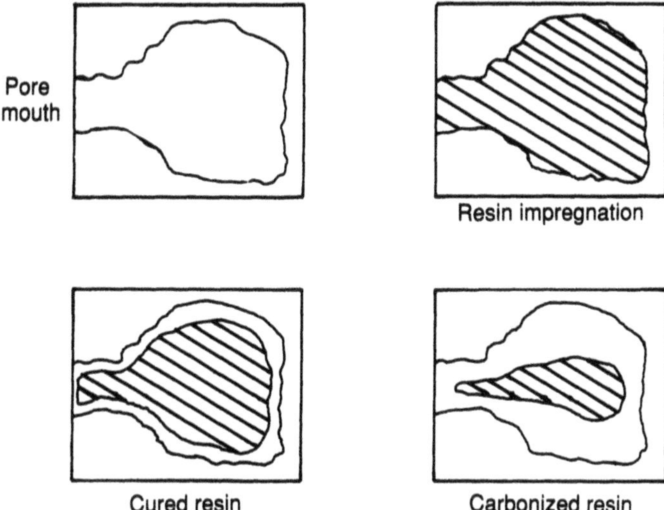

Fig. 4.3 Schematic mechanisms of pore filing and blockage during the resin densification of carbon–carbon composites [17].

Figure 4.3 shows a schematic diagram of the mechanisms of pore filling and pore blockage by liquid re-impregnation. The chemical vapour deposition (CVD) technique may also be employed [17]. In the liquid densification method, the component is generally vacuum impregnated with a resin or pitch and then recarbonized. The resin may be the same as that used in the initial lamination, a different resin, a combination of resins or a combination of resin and pitch. To aid penetration into the bulk of the sample, the resins are often diluted with a solvent to lower the viscosity; the solvent must be evaporated prior to curing and carbonization. Figure 4.4 shows how the density of the composite increases rapidly with the first two or three reimpregnation cycles but then, reaches a plateau such that there is little to be gained from carrying out more than five cycles. The redensification process approximately follows a '1/2 power law'. The reason for this is that some of the surface pores become blocked after the first few cycles, thus reducing the efficiency of subsequent treatements. Despite the fact that liquid reimpregnants are often considerably 'thinned down' using solvents, it is considerably difficult to densify areas in the centre of thick components as a result of the blocking effect of the densified outer plies of the laminate. This problem is analogous in many ways to that of surface crusting hindering the densification of thick artefacts by CVD (Chapter 3). The inability of the multiple impregnation cycles to fill/heal the porosity and shrinkage cracks formed in a composite during carbonization results in a relatively large number of cracks and voids present in the microstructure of 'fully dense' composites (Fig. 4.5).

The suitability of carbon matrices for further impregnation has been

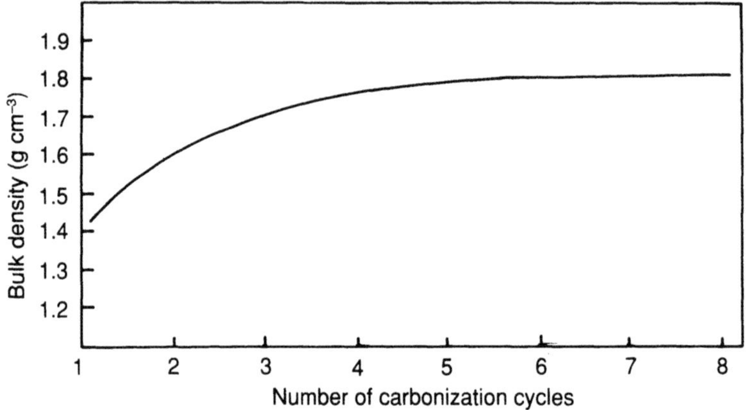

Fig. 4.4 Effect of reimpregnation/recarbonization cycles on the density of carbon–carbon composites fabricated by the thermoset resin route.

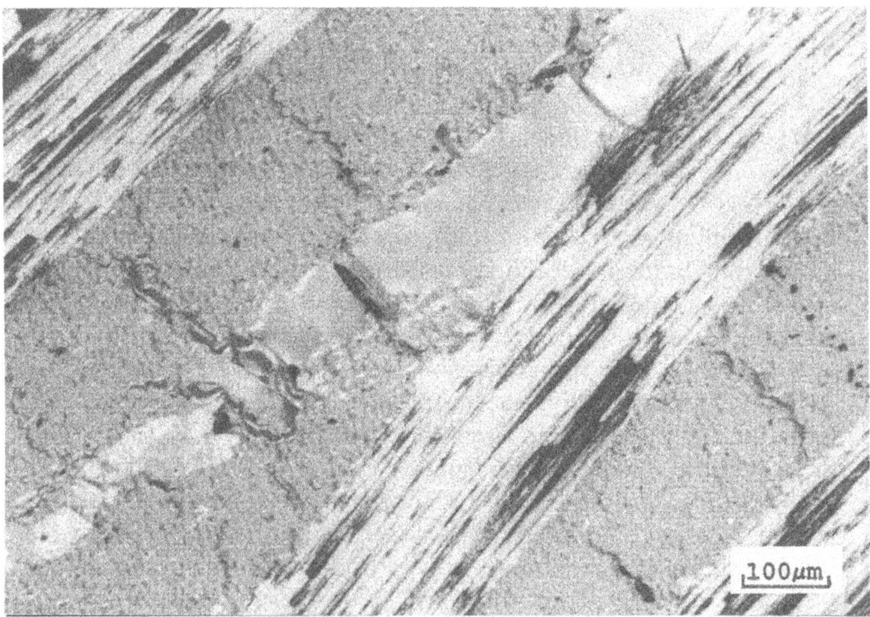

Fig. 4.5 Polarized optical micrograph of carbon–carbon composite made from carbon/phenolic precursor, reimpregnated five times with a pitch/furan blend. Note the lined but open shrinkage cracks and pores, illustrating the inefficiency of the reimpregnation technique in the interior of thick samples.

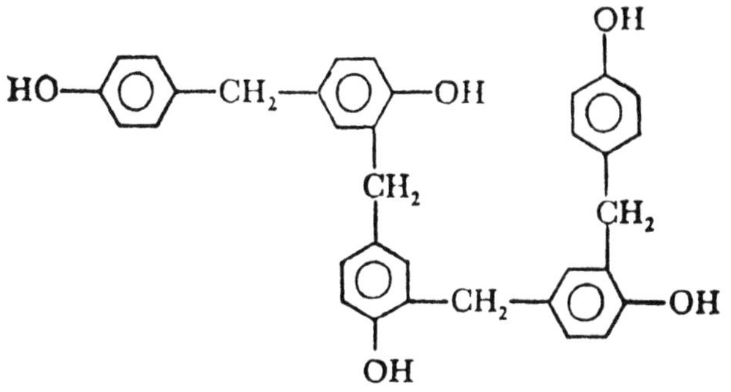

Fig. 4.6 Phenolic 'Novolac' chemical structure.

studied in detail by Fitzer and Gkogkdis [18]. The resins fall into two distinct groups with quite different pyrolysis behaviour. Resins such as polyimides and polyphenylene achieve high carbon yields in excess of 75% and produce high-strength composites after only one cycle of impregnation and carbonization. This results from the high yield and low isotropic shrinkage without damage to the fibre. Unfortunately, the mechanical properties achieved after the first carbonization cannot be enhanced by further densification cycles because the carbon formed consists mainly of closed pore systems [19,20]. The second group of resin precursors, represented by phenolics and furans, form porous carbon matrices. The composites possess poor mechanical properties after the first carbonization as a result of the inefficient stress transfer from matrix to fibres. Large improvements in mechanical properties can be achieved by repeated impregnation and carbonization cycles. For example, the flexure strength, measured in three-point bending, can be more than doubled after five cycles.

The problem with the majority of thermosetting resins is that they are not available in a form suitable for effective impregnation of porous preforms. Furthermore, many of the high carbon yield resins form an isotropic carbon with closed rather than open porosity, so that further densification and consequent improvement of composite mechanical properties are not possible. Use of high-yield resins will reduce the number of densification cycles required to achieve desired density. A high resin precursor raw-material cost may be borne since reducing the number of recarbonization cycles significantly lowers the processing costs. Unfortunately, many of the high-yield systems investigated to date have involved polymers whose prices approach those of precious metals and their use cannot therefore be justified! A brief review of high carbon yield resin systems is given in Section 4.8. The problems of processability, formation of closed porosity and cost have meant that only two resin systems, phenolics and furans, have been commercially exploited to any significant extent in carbon–carbon processing. In general, phenolic resins are used for the initial impregnation /carbonization cycle, while furan resins and resin/pitch mixtures are used for further densification.

4.4.1 Phenolic resins

Phenolic resins are a group of polymers in which the chain consists of a phenolic group interspersed with, for example, a methylene bridge. They are formed in a condensation polymerization between, for example, phenol and formaldehyde. The presence of excess formaldehyde produces methylene cross-links between the chains, again by a condensation mechanism. Phenolic resins may be produced by two methods, one of which is a single-stage process, the other involving two stages.

In the single-stage process phenol is reacted with an excess of formaldehyde so that the phenol : formaldehyde (P : F) ratio is less than 1. The mixture is heated with an alkaline catalyst, such as sodium hydroxide or ammonia. The polymerization is halted at an early stage so that the resin is converted to the A- or B-stage. The A-stage resin, known as a Resol, is a short chain, low molecular weight, linear polymer which is completely soluble in the alkaline solution. The B-stage resin, or Resitol, is a relatively long linear polymer with a slight degree of cross-linking between the chains. The compound is insoluble in alkaline solution, but readily soluble in organic solvents. When cooled, the B-stage resin becomes hard and brittle but on heating resoftens. A- and B-stage resins are used as adhesives, castings, plastics and as matrices in composites. Prolonged heating to higher temperatures results in extensive cross-linking and the formation of the C-stage resin, a hard and rigid infusible and insoluble product [21].

The two-stage process begins with a mixture in which the P : F ratio is greater than 1. The material is heated in the presence of an acid catalyst resulting in a linear chain condensation product (referred to as Novolac [22] which contains methylene linkages almost quantitively [23,24]. In the structure, *ortho*- and *para*-links occur at random as shown in Fig. 4.6. The Novolac, like Resitol, is fusible and soluble in organic solvents. It is generally ground to a powder and mixed with hexamine ($N_4(CH_2)_6$) which acts as a cross-linking agent via methylene linkages thus converting the Novolac to a hard infusible solid [25].

The response of a phenolic resin to heat treatment depends on the degree of cross-linking. When a weakly cross-linked material is heated slowly, low molecular weight compounds, such as unreacted phenol, short-chain polymers and water are evolved between 100 and 350 °C. Above 500 °C small quantities of carbon monoxide and methane are evolved. If the material is highly cross-linked water is not evolved until above 400 °C and the amount of low molecular weight polymer volatilized is considerably reduced. It is difficult to follow the detailed rearrangements of the chemical

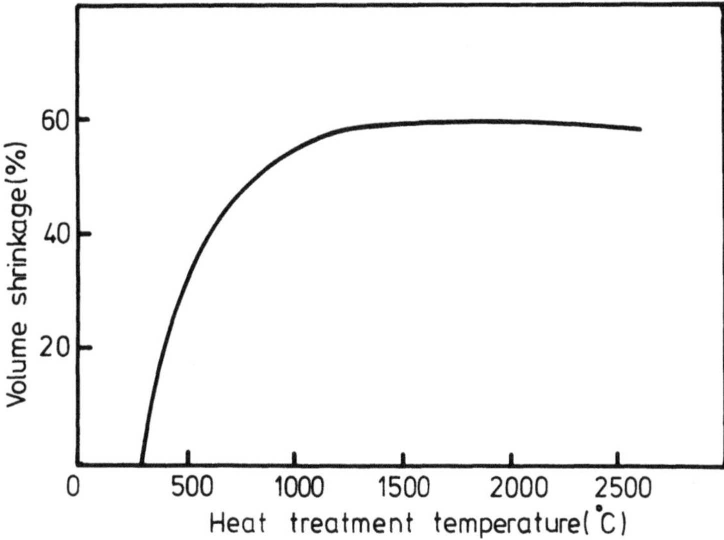

Fig. 4.7 Volume contraction of phenolic resins during pyrolysis.

structure of the resin during carbonization. A number of hypothetical interpretations of the complex reactions which take place, often simultaneously, are available [26–28] however. Fitzer *et al.* [29] suggest that between 100 and 300 °C polymerization proceeds, forming long-chain, cross-linked polymers. Above 300 °C oxygen is liberated in the form of water and carbon monoxide or carbon dioxide. At around 500 °C the structure has rearranged into a highly unsaturated aromatic residue capable of forming a cross-linked aromatic network. The carbonization of pheonolic resin may be followed qualitatively by observing contraction with respect to temperature. A typical curve (Fig. 4.7) of volume contraction versus time exhibits three features: around 40% contraction below 500 °C due to the removal of water, abrupt contraction of approximately 10% between 600 and 700 °C and a steady contraction at higher temperatures as a result of dehydrogenation.

4.4.2 Furan resins

Furan resins are made from furan,

```
  H           H
   \         /
    C ——— C
    ||      ||
H — C       C — C — H
     \     /     ||
        O        O
```

which is derived from waste vegetable matter [30]. The compound is produced on a large scale from oathulls, which give a 10% yield when obtained as a by-product in the manufacture of Quaker Oats [31].

CNO $\xrightarrow{33\ \%\ NaOH}$ COOH + CH_2OH

Furan resins [32] are produced by the resinification of furfuryl alcohol in the presence of an acid catalyst, usually with heat, until the desired degree of polymerization is achieved. The polymerization, which is exothermic, produces a thermosetting resin. Heat is required before the process will commence and the reaction can be stopped at any point by cooling the resin and catalyst mixture. The chemical reaction involves the alcohol group from furfuryl alcohol and an active hydrogen from the ring of an adjacent molecule to give a difurfuryl alcohol. Intermolecular dehydration (stage 1) is followed by formation of a linear polymer (stage 2) and the final reaction stage, again in the presence of an acid catalyst, produces a highly cross-linked, infusible resin [33].

H_2O

CH_2 OH H + $\xrightarrow{1}$ CH_2

2

CH_2 $\{ CH_2 \}_n$ CH_2OH

Commercial furan resins are low-viscosity liquids which have been produced by the action of heat and an acidic catalyst on furfuryl alcohol. The acid catalyst is neutralized before a viscous stage is reached, which could prevent effective neutralization in order to obtain a storage-stable product. As a result, the commercial product still contains large amounts of unreacted monomer.

The thermosetting of the liquid intermediate resin is accomplished by the addition of an acidic catalyst, usually followed by the external application of heat. In the thermosetting process, any monomer present will largely be polymerized (condensed), with the splitting out of water which will vaporize under the prevailing conditions. A small amount of the monomer will undoubtedly vaporize and be lost before it can be chemically condensed. Volatilization of both the water and the low molecular weight compounds will tend to create porosity in the thermoset product.

Furan resins have been termed 'reactive' compounds [34]. The patent literature describes a great number of suitable catalysts. Mineral acids, strong acids, Lewis acids and acyl halides are very active. Industrially, control of the exothermic reaction is achieved by regulation of catalyst concentration and temperature. The mechanism of resinification has been inferred, in part at least, by identification of the intermediate products. The initial and predominant reaction is intermolecular dehydration leading to higher molecular weight condensation products in which furan rings are linked together by methylene bridges in a linear chain. To a lesser extent, furfuryl ether is also formed. Formaldehyde, resulting from thermal decomposition, is always found in the reaction mixture. Some hydrolytic fission can also be inferred based on infra-red (IR) evidence.

Experimental evidence [35] indicates that formaldehyde formed during resinification may, under appropriate conditions, recondense with the intermediate products, and this is one mechanism by which the chains can become chemically cross-linked. Finally, cross-linking may occur by a mechanism involving the nuclear double bonds during the later curing stages. This explains the high degree of chemical inertness of the thermally set resin. Figure 4.8 illustrates the linkages most probably present in cured furan polymers; it is not intended to represent the unit polymer itself.

Data for the determination of the pyrolysis chemistry of furan resins are provided by thermogravimetric analysis (TGA), thermovolumetric analysis (TVA), differential thermal analysis (DTA), analysis of gaseous products, and by arresting the pyrolysis at various stages to analyse the residue. The techniques and results of thermal degradation of organic polymers are described by Madorsky [36].

Rather variable data on carbon yields resulting from the pyrolysis of furan resins are reported in the literature. Fitzer and Schafer [37], employing heating rates of both 12 and 30 °C h^{-1} for temperatures up to 1000 °C, obtained a 60% carbon yield, whereas Ozanne [38] using a heating rate of 150 °C h^{-1} to 900 °C obtained only 50%. On the other hand, O'Neill *et al.* [39], using heating rates between 420 and 540 °C h^{-1} to a final temperature of 950 °C, found a high carbon yield of 59%. These differences are considered to result from a difference in the amount of cross-linkage in the resins tested [37].

On the basis of all the data available, the mechanism for the pyrolysis of a furan resin (I) is shown in Fig. 4.9 [37]. The primary pyrolysis product is water, being formed from the free hydroxyl groups. Subsequently the methylene bridges rupture, forming CH_4 as a by-product, and the furan ring is opened (II). The latter process is recognized by the release of carbon monoxide, carbon dioxide and again water. It is not possible to give a definite formulation of the intermediate stages. According to the observed CO formation, the furan ring is stable up to approximately 275 °C.

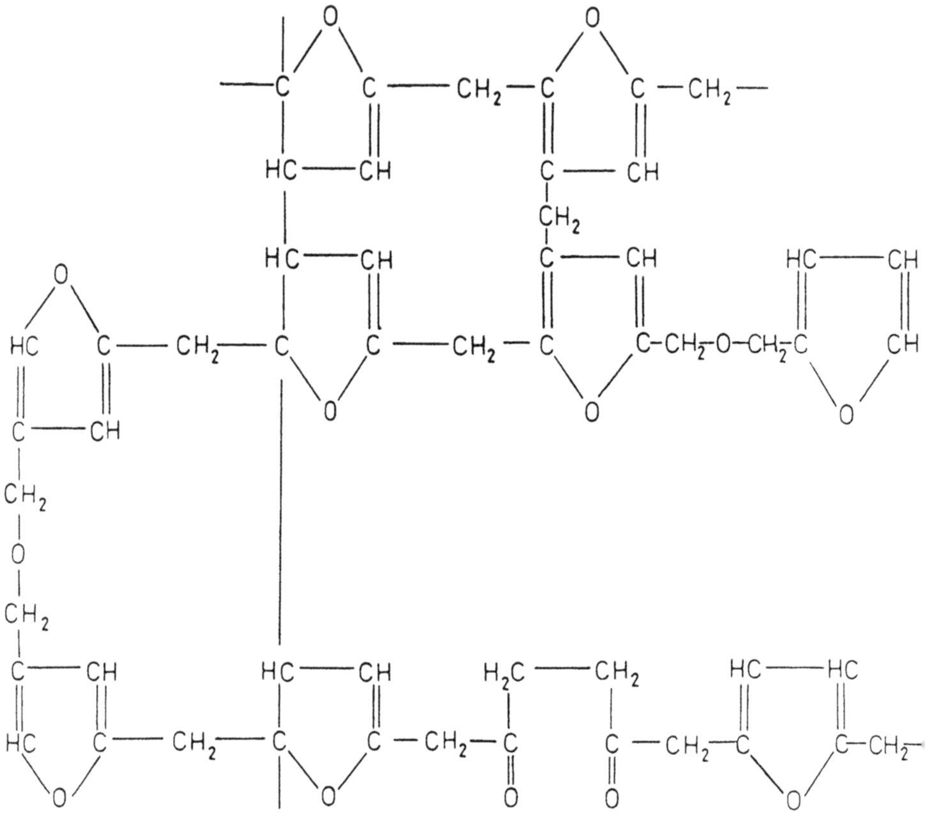

Fig. 4.8 Cross-linking in furan resins.

Between 300 and 400 °C the IR band for the C–O–C bond of the furan ring, at a wavelength of 5.85 μm, loses its intensity, due to the rupture of the furan ring. In the gas analysis, this temperature is characterized by a maximum in the first carbon monoxide formation as well as by release of water. A quantitative calculation reveals that, taking into account the hydroxyl group decomposition to water, half the furan ring oxygen must be released as water. The other half is liberated as carbon monoxide.

At 400 °C, new IR bands begin to appear which indicate the beginning of a build-up of aromatic systems. These aromatic systems are formed spontaneously from the fragments of the furan rings. They have also been identified by Braun [40] using Diamond's X-ray profile analysis.

At temperatures above 450 °C a reaction of pyrolysis water with remaining methylene bridges occurs, (III) in Fig. 4.9. This secondary reaction to form alkanone groups is readily recognized from the decrease in the formation of water above 450 °C. In the same temperature range, the formation of methane ceases because of the oxidation of the methylene bridges and significant amounts of hydrogen are released. Above 460 °C

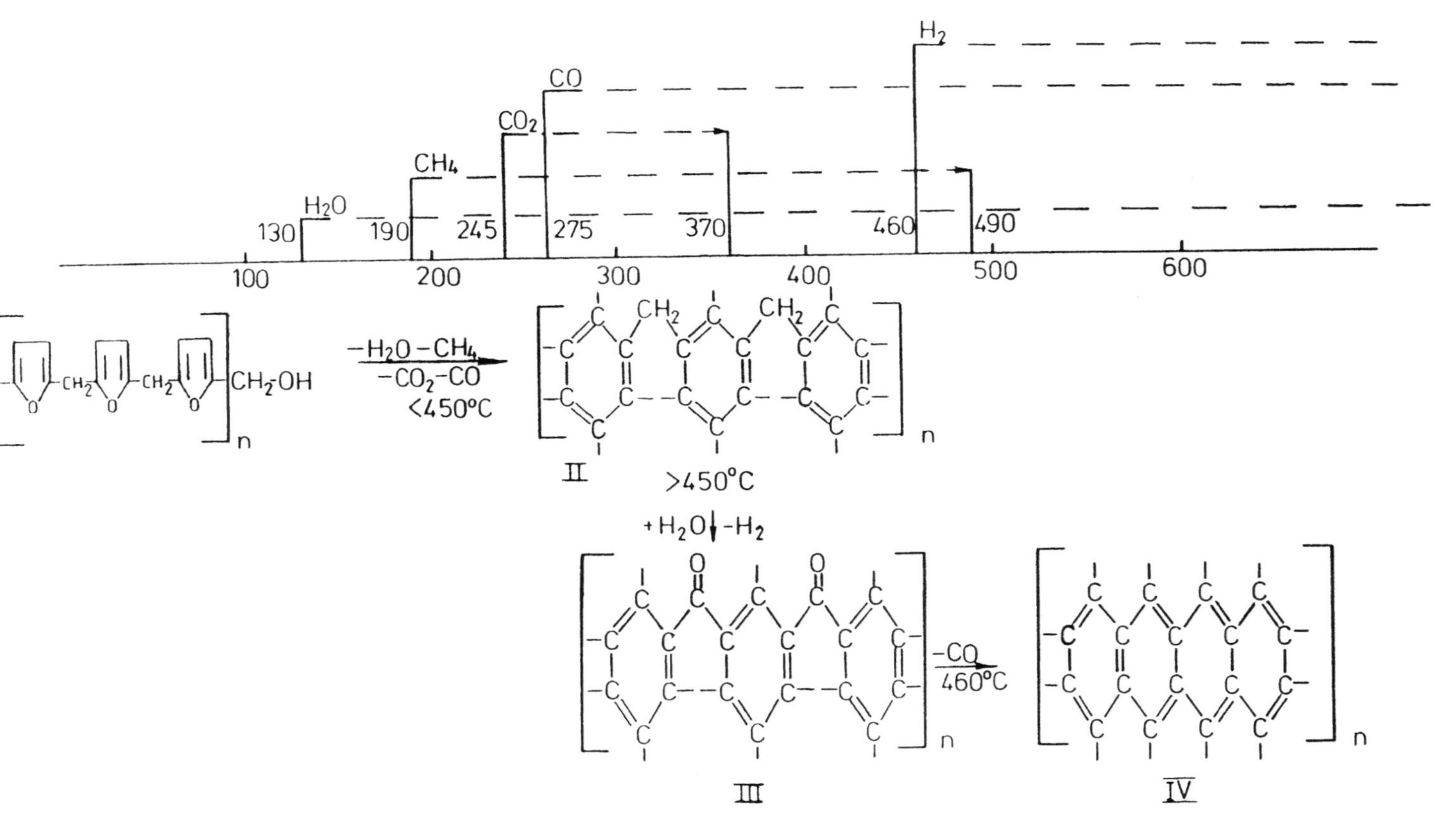

Fig. 4.9 The pyrolysis chemistry of furan resins [36].

the carbon groups liberate carbon monoxide, leading to a highly unsaturated aromatic C–H residue (IV) capable of forming cross-linked aromatic systems.

During pyrolysis polyfurfuryl alcohol exhibits around 20% linear shrinkage [37]. This shrinkage corresponds to the known behaviour observed during the production of commercial isotropic carbon products [41].

The density of the carbon produced is around 1.55 g cm^{-3}. When comparing this bulk density with that of graphite (2.267 g cm^{-3}) a difference of 30% appears which is attributed to the different structural arrangement.

It has also been shown [37] that a variation in the degree of cross-linking in the starting materials will affect the consecutive chemical steps of the decomposition and the formation of gaseous by-products. A decrease in the amount of cross-linking will diminish the carbon yield in the residue by promoting the formation of low molecular weight compounds.

4.5 CARBONIZATION OF COMPOSITES

In a composite laminate or structure, the matrix acts as a medium for the effective transfer of load on to the reinforcing fibres. The properties of the matrix and the fibre/matrix interface thus make important contributions to the mechanical properties of the composite. Failure in polymer composites will most likely initiate at the fibre/matrix interface so that bonding in this region is of prime concern. Strong fibre/matrix adhesion is generally achieved by some form of surface treatment of the fibres [42,43]. The matrix in carbon–carbon composites is very brittle in nature. The microstructure of the carbon matrix will therefore play a major role in the initiation of failure, whereas fibre/matrix bonding is responsible for the transmission of cracks. Pyrolysis of the polymer composites strongly influences both the microstructure of the carbon matrix and the nature and strength of the fibre/matrix interface [44,45].

The shrinkage of the resin matrix as a result of carbonization is observed to be the most critical factor governing the mechanical properties of the carbon–carbon product. The surface properties of the carbon fibres have been demonstrated to control the pyrolysis shrinkage and subsequent microcrack density in the carbon matrix [44]. It is the formation of such cracks which results in composites of low strength. Bradshaw and Vidoz correlated the strength of different carbon–carbon composites to the strain to failure of the matrix [46], which they believed to be controlled by the differential shrinkage between fibre and matrix and the consequential crack formation. The difference in fracture behaviour of carbonized polymers has been attributed to differences in fibre matrix bonding both before and after pyrolysis [47]. Manocha [48] observed the changes occurring during the carbonization of polymer composites using dilatometer measurements of shrinkage in tandem with microstructural observations [49].

The dimensional changes observed during the pyrolysis of a carbon fibre reinforced polymer occur as a direct result of thermal degradation and shrinkage of the polymer matrix [44,45]. In a unidirectionally reinforced composite the changes in the fibre axis direction are controlled by the longitudinal thermal expansions of the carbon fibres, which are very small. As a result, there is very little measurable dimensional change in that direction. In the direction perpendicular to the fibres, however, appreciable changes are observed. Very little bulk dimensional change is observed in woven or multi-directional fibre-reinforced composites because of the fibres. The matrix on the other hand shrinks significantly, resulting in the production of a system of cracks.

Figure 4.10 shows the general trends in the dimensional changes (change in thickness of unidirectional (UD) composites) observed in an unreinforced polymer (furan resin) and the same material reinforced with non-surface-treated (relatively weak fibre/matrix bonding) and surface-treated (strong interface) carbon fibres on pyrolysis. Each of the materials undergoes a reversible thermal expansion up to around 200 °C. Further heat treatment results in the onset of shrinkage in the range 300–450 °C. Shrinkage begins at around 300 °C for the unreinforced polymer, reaching a maximum between 350 and 700 °C. In the composites, the onset of shrinkage is suppressed by 50–100 °C. The fibre reinforcement is found to influence both the onset of shrinkage as well as the rate. In composites made with surface-treated carbon fibres shrinkage starts at 350 °C, while in those made with non-surface-treated fibres it begins at around 420 °C. The maximum shrinkage in both composite systems occurs between 450 and 650 °C. Between 650 and 800 °C the rate of shrinkage is reduced to about half that measured between 450 and 650 °C. Shrinkage continues up to about 1100 °C but is very slow above 800 °C.

Microscopical observations show the fibres to be well bonded to the matrix. At around 500 °C gaps begin to appear between the fibres and matrix in composites with non-surface-treated fibres. Composites with surface-treated fibres display a mixed type of microstructure with microcracks formed in the bulk matrix as well as at the fibre matrix interface. As the HTT is increased towards 1000 °C, the interface debonding increases in the non-surface-treated fibre composites, while in those composites using surface-treated fibres the interface appears to become progressively stronger.

The shrinkage of the matrix on pyrolysis is due to the degradation of the polymer chains in the 300–600 °C range and coalescence of the C–C chains in the 600–800 °C temperature range [21]. In the case of composites using non-surface-treated fibres, the fibre/matrix interface is of the weak 'mechanical' type [44,45], much weaker than the C–C bonding within the

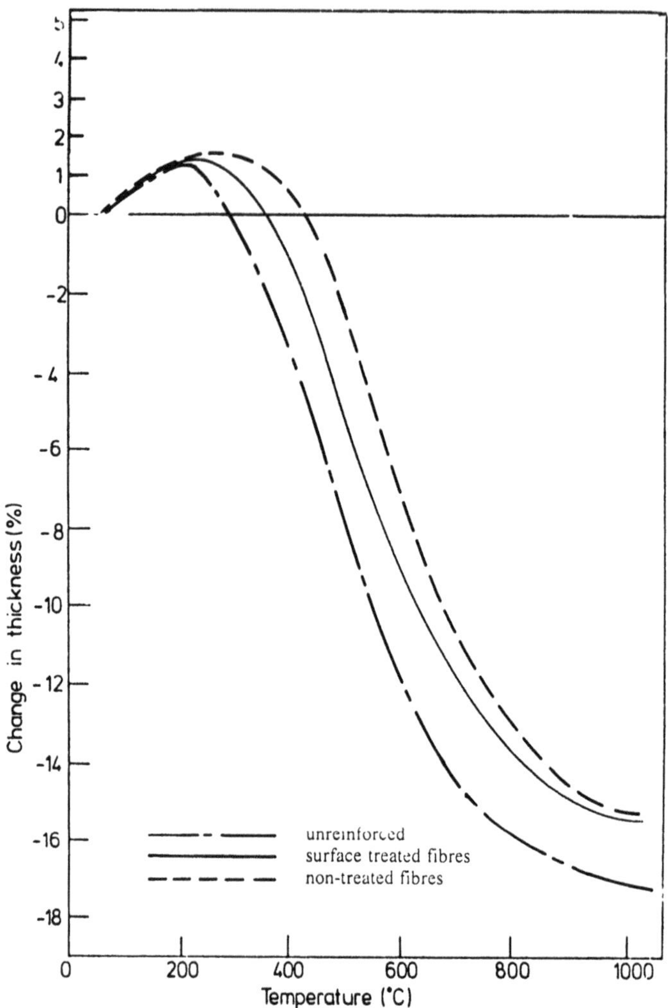

Fig. 4.10 Dimensional changes in composites during pyrolysis [45].

isotropic carbon matrix. When carbonized the matrix shrinks away from the fibres with little displacement of the fibres. The result is a gap between matrix and fibres and a low bulk shrinkage. By contrast, the bond between surface treated fibres and matrix is of the strong 'chemical' variety [44,45]. The strong interface developed in the polymer system is carried over into the carbonized structure with the result that the matrix does not shrink away from the fibre. During pyrolysis, the fibres become displaced by the shrinkage of the matrix. The consequence is a large bulk shrinkage of the composite and a large degree of matrix cracking.

The extent of the chemical bond between fibres and matrix will vary with the type and degree of surface treatment, be it oxidation to produce reactive groups or sizing with some form of coupling agent. The strength of the interface may therefore be tailored to suit the required properties. Mechanical testing shows there to be an optimum fibre/matrix bond strength when developing carbon–carbon composites with good mechanical properties. Too strong an interface produces brittle composites exhibiting catastrophic failure and poor strength, as matrix cracks pass straight through any fibre they intersect absorbing very little work of fracture. If the bond strength is too low, composites will tend to fail in shear with poor translation of fibre strength. Treating the fibres so that an intermediate bond strength is achieved will result in a mixed mode fracture and a resulting higher strength.

4.6 GRAPHITIZATION

In carbon–carbon composites with a matrix derived from a thermosetting resin, the microstructure is very much dependent on the HTT. When an unreinforced thermosetting resin is pyrolysed it will form an isotropic carbon [13]. No evidence of a graphite-like structure is apparent from X-ray analysis. When the resin is heat treated in the composite form with carbon fibres present, a graphitic anistropy can be detected by density measurements, X-ray diffraction and optical microscopy (Fig. 4.11) [49]. With furan resins this phenomenon is very marked and anisotropic regions begin to develop in the matrix at temperatures as low as 1000 °C. Above 2200 °C these anisotropic regions gradually adopt a graphite crystal structure, such that at 2800 °C the matrix is essentially all graphite.

The observations cannot be explained on the basis of the known structures and properties of the respective components. It is believed, therefore, that there must be some interfacial effect occurring between the carbon fibres and the isotropic carbon matrix during heat treatment [50]. During carbonization the resin shrinks by up to 50% by volume, while the fibres change very little in dimension. It has been postulated that the driving force for the graphitization of the 'non-graphitizing' matrix is the stress accumulation caused by the difference in the coefficients of thermal expansion between the matrix and fibres. Kamiya and Inagaki [51] reported the existence of such a stress accumulation when heat treating natural graphite/isotropic carbon composites. They estimate these stresses to be of the order of 300 MN m^{-2}.

An abrupt graphitization of isotropic carbon has been observed at pressures above 3 kbar [52,53]. Glassy or isotropic carbon has also been reported to graphitize at high temperatures by the addition of the natural graphite powder [29,54]. On the basis of the evidence observed to date, it would appear that the graphitization of isotropic carbons at high temperatures is a direct result of the accumulation or application of stress. For this reason the phenomenon is referred to as **stress graphitization** [55].

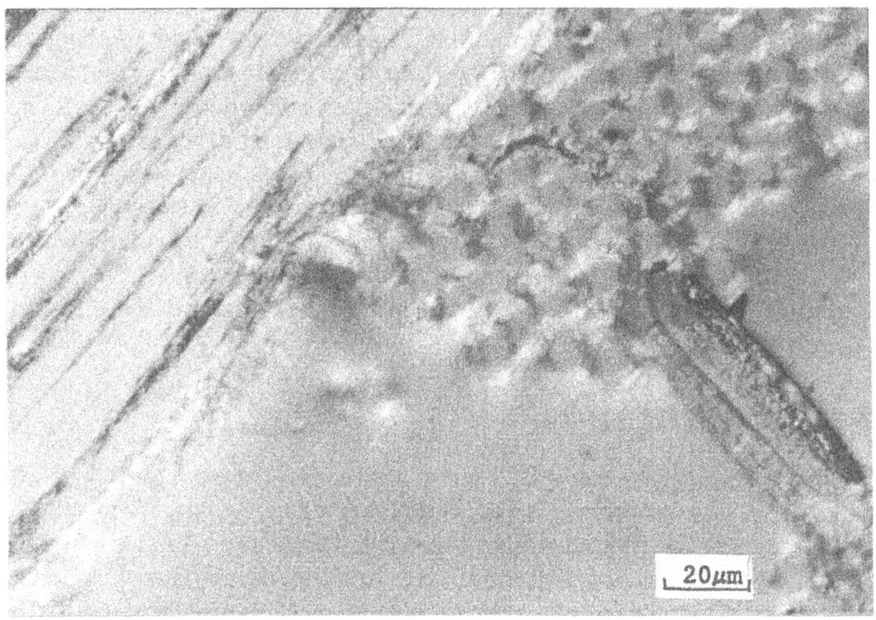

Fig. 4.11 Evidence of graphitization of the isotropic carbon matrix in polymer-derived carbon–carbon systems after heat treatment to 2500 °C as shown by polarized light microscopy.

It has been suggested, using optical microscopy techniques, that the graphite microstructure above an HTT of 2200 °C is orientated in such a way that the graphite planes in the matrix surround the fibre. The graphite layers are divided into a number of domains and, as a whole, are observed to envelope the fibres [56]. The temperature for the manifestation of such a graphitic microstructure varies according to the fibre type; the higher the modulus of the fibre the lower this temperature will be. Based on their observations, many workers suggest that the graphitization of the matrix begins at the fibre/matrix interface and progresses with time and temperature into the bulk. One must, however, question the validity of the microscopical evidence presented in this case. Optical anisotropy observations from polarized light microscopy are used to explain the detailed molecular reorientation occurring in regions often only 1 or 2 μm in size between 7 μm diameter fibres. It is the opinion of the author that a certain degree of 'artistic (or scientific!) licence' has been employed such that microscopical observations beyond the limit of their iresolution are being used to support what appears to be the most likely explanation. The development of a graphitic structure in the matrix of carbon–carbon composites is crucial to the understanding of their mechanical properties. While the hypothesis that graphitization beings at the fibre/matrix interface and progresses into the 'bulk' appears the most plausible, it cannot be conclusively proved without the use of higher resolution techniques.

Recently, studies on the origins of graphitization in resin-derived isotropic carbon matrices have been carried out using high-resolution TEM and selected area X-ray diffraction (SAD) techniques [57]. The material investigated was produced from high-modulus mesophase pitch-based fibres (Thornel P 75) and a polyaromatic thermoplastic resin (PEEK) matrix (Chapter 5). Graphitic domains were observed to form at a relatively low temperature in this system (1600 °C). A sympathetic alignment of the graphene structure in the matrix with that in the fibre was evident but this could not be linked to crystal nucleation at the fibre surface. Stress graphitization was believed to be responsible for the observation made but, as yet, it has not been possible to obtain detailed knowledge of the mechanism for the observed phenomena.

Carbon–carbon composites made by the thermoset polymer route are very often given a 'graphitization' treatment to temperatures up to 2500 °C. This results in a higher density, improved ablation resistance and the 'opening up' of porosity to aid reimpregnation cycles. The mechanism of the process and the microstructures obtained are, at present, not fully understood. The production of composite artefacts thus often occurs without a full understanding of the structural and physical changes which occur during manufacture.

4.7 IMPREGNATION TECHNOLOGY

Carbon–carbon composites formed by the thermoset resin technique are made by pyrolysing a laminated structure in an inert atmosphere to around 1000 °C. The resulting carbon fibre reinforced isotropic carbon matrix composites may be subsequently heat treated to higher temperatures of approximately 2500 °C in order to graphitize the matrix. The process is extremely inefficient, with up to half of the matrix being lost during carbonization, requiring several steps to produce the final artefact. It is, however, considerably quicker and cheaper than CVD processing, and the equipment required, in what is an ambient pressure fabrication route, is not limited by size or the requirement of a large initial capital investment. Furthermore, high-quality precursors, often exhibiting extremely complex geometries, may be produced by exploiting the technology developed for the 'conventional' composites industry.

4.7.1 Hand lay-up

The simplest method of making a composite is to lay the fibres on to a mould by hand, paint on the resin, build up the required number of layers

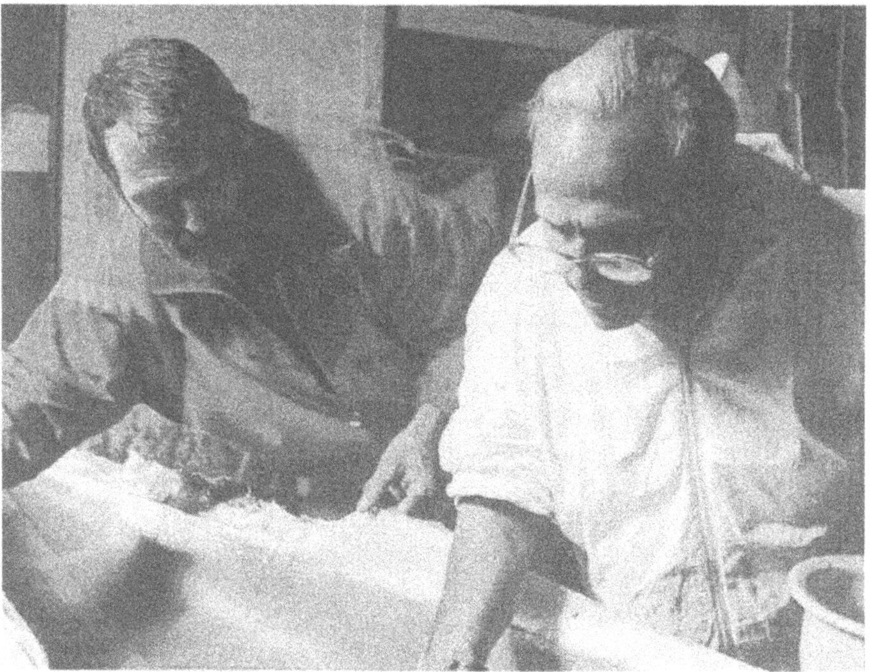

Fig. 4.12 Hand lay-up of a Kevlar/carbon fabric reinforced racing kayak using GRP tooling (courtesy Gaybo International Ltd.).

and allow to cure. Very little capital investment is required and, although very labour intensive, hand lay-up is widely used because of its great versatility. The fibres can be in the form of random mats or woven fabrics. Very large structures may be produced and the resin can be sprayed rather than brushed. Consolidation may be aided by the use of a vacuum bag or the application of pressure. Additionally, tooling can be made relatively inexpensively (Fig. 4.12).

4.7.2 Pultrusion

The fibres are impregnated with a liquid resin and pulled through a die shaped to produce the desired cross-section in the product. A great number of shapes are available by changing the shape of the die which is heated to promote curing of the resin. Once hardened, the material is cut into suitable lengths (Fig. 4.13).

4.7.3 Prepreg

Composite structures are designed to have a precisely defined quantity of fibres in the correct location and orientation with a minimum of polymer

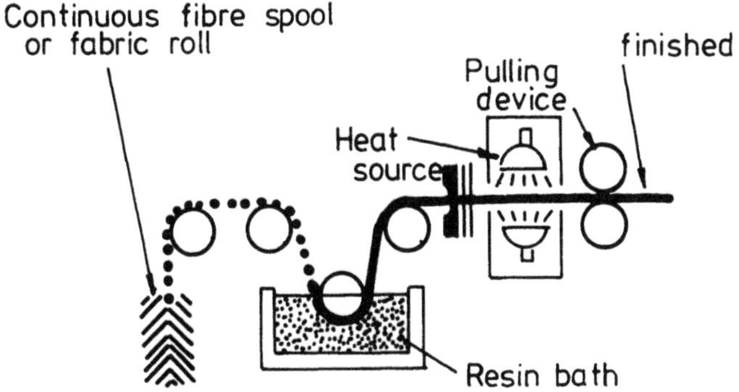

Fig. 4.13 Continuous pultrusion process.

to provide the support. The composites industry achieves this precision using **prepreg** as an intermediate product. Prepreg is a broad tape of aligned or woven fibres, impregnated with polymer resin. A laminated plaque or component is fabricated by stacking successive layers of prepreg followed by some form of curing cycle. Prepregs designed as carbon–carbon precursors are generally produced with a higher resin content to account for weight loss on carbonization.

Fibres are converted into a prepreg either by bringing a number of spooled tows into a collimated tape (Fig. 4.14) or by impregnating a woven fabric. The fibre contructions may be impregnated using either a hot-melt or solvent-coating process. The hot-melt operation consists of heating a matrix resin to obtain a low viscosity to produce a well-dispersed fibre–resin mass. The volume fraction of fibre is controlled by the number of tows brought into the web or by the construction of the weave. The resin is cast on to the substrate paper either on the prepreg line or in a separate filming operation in order to obtain the desired fibre : resin ratio. The prepreg is calendered to obtain a uniform thickness and to close fibre gaps before being wound on a core. Substrate paper is generally left between layers of tape. The paper is coated with a release film.

Solvent coating is achieved by immersing the fibre web or fabric into a bath containing 20–50% of a solvent/resin mixture followed by some form of drying process (Fig. 4.14).

Resin content is controlled either by the use of metering rolls after impregnation or by adjustment of solvent : resin ratio. The drying process reduces volatiles and advances the resin molecular weight.

The finished product from either route is a thin sheet of fibre-reinforced resin which is generally wound on a cardboard core and interleaved with release paper. Table 4.3 shows typical ranges of production dimensions.

A general rule of thumb is that unidirectional prepregs tend to be made by the hot-melt process, whereas fabrics are usually impregnated by

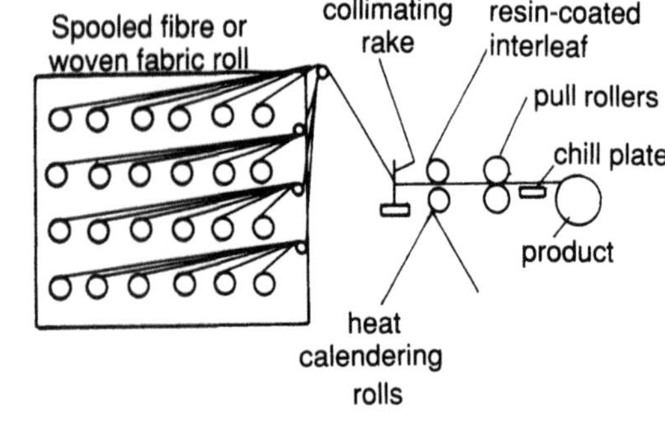

(a)

Drying oven
(solvent removal)
product
wind-up spool
metering
rollers
Dry fibre creel
or
woven fabric
solvent-resin
bath
chilled roller

(b)

Fig. 4.14 Typical prepreg operations: (a) hot-melt process; (b) solvent process.

Table 4.3 Prepreg product dimensions

Parameter	*Product range*
Thickness	0.08–2.5 mm
Resin content	28–45% (normally ±2%)
Areal weight of dry fibre	30–300 g m^{-2}
Width	25–1500 mm
Package size	5–250 kg

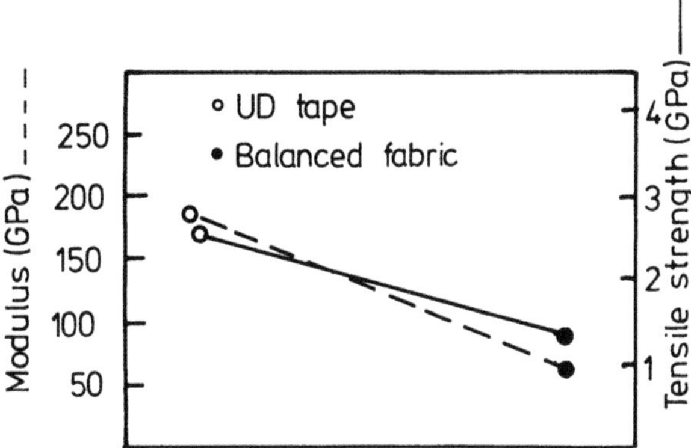

Fig. 4.15 Comparison of mechanical properties between tape and fabric prepregs.

the solvent route. Thermoplastic resin prepregs (Chapter 5) tend to be manufactured by a combination of pultrusion and solvent impregnation.

Carbon reinforcing fibres are, by their nature, extremely anisotropic. Consequently, unidirectional prepreg tapes reinforce primarily in the 0° direction (fibre axis). Tapes offer the best translation of fibre properties because the fibres are not crimped or distorted as in fabric prepregs. Significant differences in mechanical properties exist between tape and fabric. Figure 4.15 shows typical tensile property differences between tape and fabric prepregs. The differences arise due to the fabrics having a significant proportion of fibres perpendicular to the applied load and as a result of fibre crimping. The weave pattern of a fabric controls the handling characteristics and, to a degree, influences the mechanical properties of a product which uses it as reinforcement. The plain weave, interlacing one warp yarn over and under one fill yarn, exhibits the greatest degree of stability with respect to yarn slippage and fabric distortion (Fig. 4.16). The eight-harness satin weave has one warp yarn interlacing over seven and under one fill yarn in an irregular pattern. The result is a pliable fabric which readily conforms to compound contours. A fabric in this form will possess improved mechanical properties over a plain weave, as a consequence of these being fewer fibre distortions at cross-overpoints. Furthermore, since the movement of fibres is less restricted, an eight-harness satin weave will conform more readily to the shape of a mould, but at the expense of reduced energy absorption and fabric stability. This is because the greater the number of cross-over points, the greater will be the energy transference to adjoining fibres. A number of fabric forms exist, each attempting to exploit a different compromise of stability, conformity and energy absorption, etc. (Fig. 4.17).

The measure of the adhesion of the prepreg to the surface of tooling or other prepreg sheets is known as **tack**. The level of tack is carefully

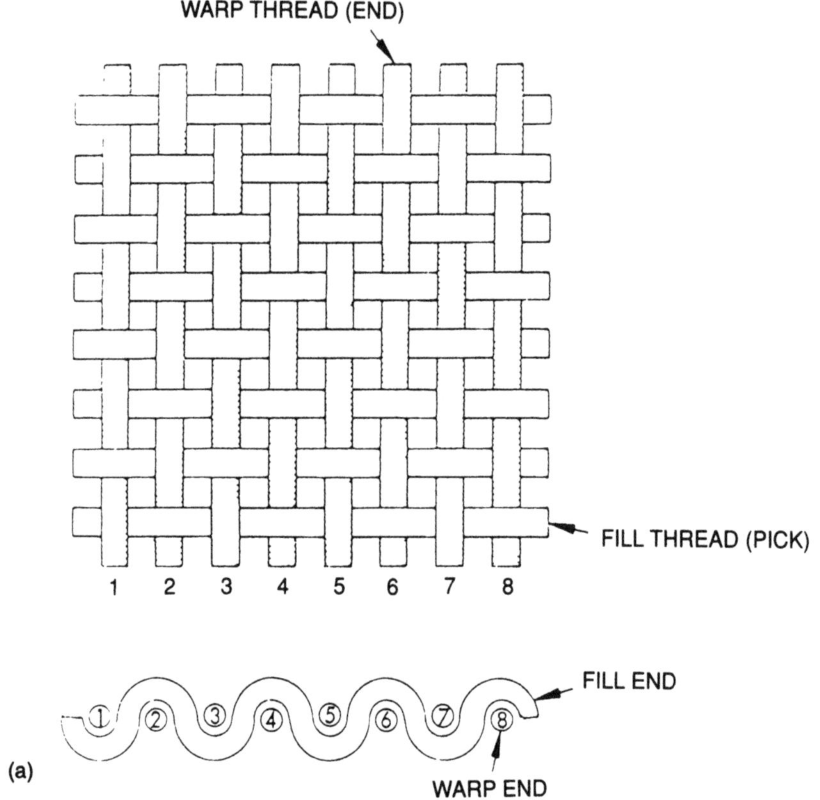

controlled in order to optimize lay-up operations and, as such, may vary considerably from process to process. It is affected by the apparent viscosity of the resin. Tack may be increased by increasing resin and volatile contents, by retarding polymerization or by increasing lay-up room temperature or humidity. Resin formulations may thus be reapportioned or additives blended in in order to control tack. Thermoplastic prepregs, by nature of their being prepolymerized, do not possess tack, adding difficulty to processing. Successive layers may, however, be spot welded together to alleviate this problem. Ideally the tack qualities should be adequate to allow the prepreg to adhere to the prepared tooling surface or preceding plies of a laminate, but light enough to part from the backing film without loss of resin.

Flow is the measure of the amount of resin squeezed from a specimen as it cures (under heat and pressure) between press platens. Flow measurement indicates the capability of the resin to fuse successive plies in a laminate and to bleed out volatiles and reaction gases. Flow can be an indicator of prepreg age or advancement of polymerization. It is often desirable to optimize resin content and viscosity to attain adequate flows.

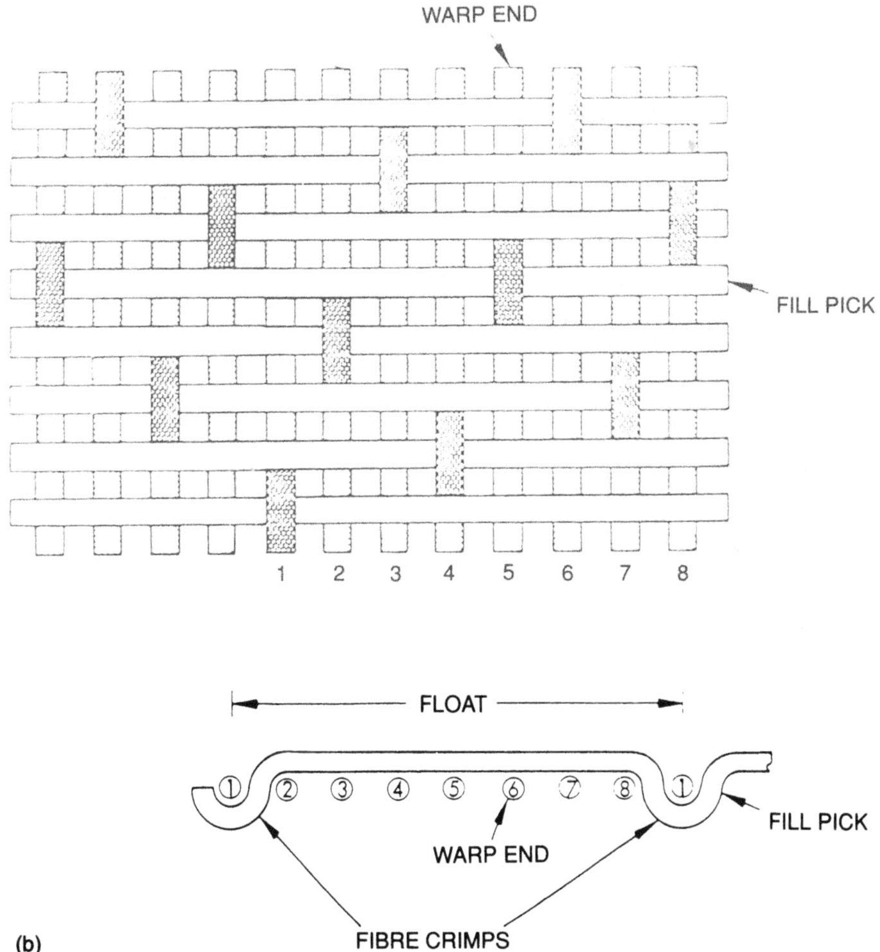

Fig. 4.16 Fabric construction forms showing degree of fibre crimp, of (a) plain weave, (b) eight-harness satin weave.

Prepreg flow may also be controlled by adding thickening or thixotropic agents to the resin. The length of time a specimen remains between heated platens until the resin gels, or reaches a very high viscosity, is known as the **gel time** of the prepreg [58]. Gel time measurements may be used to estimate the degree of advancement and 'shelf life' of the material. Most prepregs are formulated to attain a useful life of 10 days or more under standard laboratory conditions. Cold storage is often used to lengthen shelf life. The formability of a material around contours is denoted **drape** and is critical to fabrication costs. Tapes are typically less drapable than fabric forms of prepreg. It is paramount that prepregs be staged to desirable tack and drape qualities.

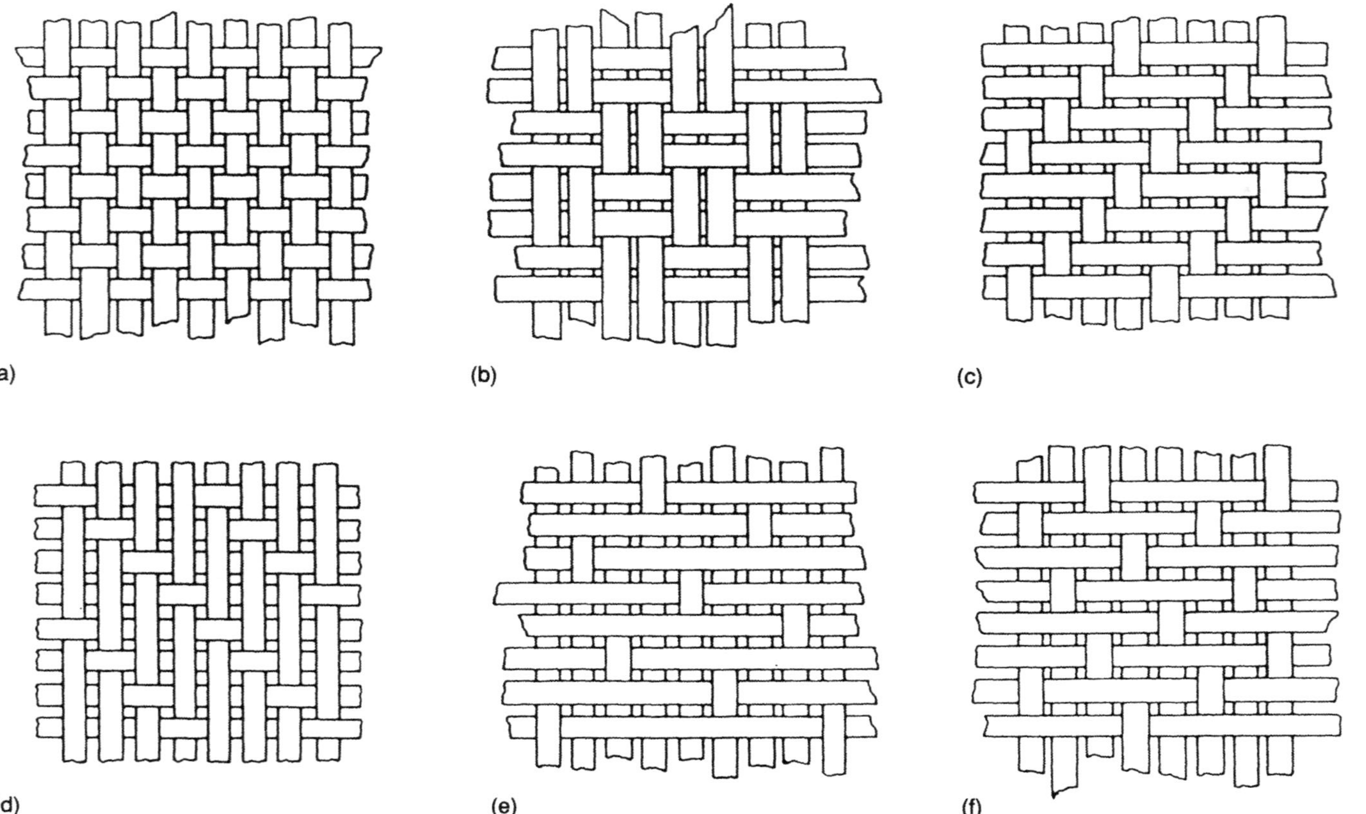

Fig. 4.17 Woven fabric constructions: (a) plain weave; (b) basket weave; (c) twill weave; (d) four-harness satin weave; (e) eight-harness satin weave; (f) five-harness satin weave.

Historically, prepregs have been laid up by hand with an operator cutting lengths of material and laying them in a mould prior to autoclave cure (Chapter 7). As previously stated, unidirectional tapes are extremely anisotropic with properties maximized in the fibre direction. The mechanical response of a unidirectional laminate is characterized by high strength and modulus parallel to the fibres and poor transverse properties; 0°/90° laminates are produced to improve transverse properties (at the expense of axial) while balanced 0°/90° ± 45° laminates approach isotropic in plane strength asymptotically with increasing thickness – for this reason they are referred to as **quasi-isotropic** (Fig. 4.18). The laminator cuts lengths of tape and places them on the mould surface in the desired ply orientation. The process uses one of the cheaper forms of prepreg and requires low initial capital investments in both equipment and facilities. Unfortunately, this method results in a high wastage of material, fabrication time, manpower costs and operator-to-operator variability. Indeed the scrap factor in such an operation may exceed 50% depending on complexity and size of components. Lay-up times may be reduced by using templates to aid the cutting process and a number of companies are in the process of developing computer numerically controlled (CNC) tape-laying machines.

Woven fabric prepregs can be made to conform to the complex geometries of low-stressed artefacts by a process known as **darting**. The technique involves slitting the prepregs at locations where folds would normally occur in a lay-up. The excess material is removed and the edges butted together. Alternatively, the prepreg may be slit at points where a crease would normally form and the excess material allowed to overlap. When the former method is employed an additional ply may be required to compensate for the weak butt joints in the laminate. In the case of highly stressed constructions the fabric prepregs should be cut in predetermined patterns so that joints in successive plies do not coincide. When fabric reinforcements are laid up on complex double curvature shapes, the weave patterns become distorted and the fibre directions changed. As a result ply alignments in heavily draped lay-ups are difficult to control. Off-axis angle ply fabrics of, for example ±60° ±45°, are used to compensate for undetermined deficiencies.

The high shrinkage occurring during carbonization and the brittle nature of the isotropic carbon matrix have dictated that commercially exploited carbon–carbon composites formed by the thermoset route are generally produced from woven prepregs. Laboratory specimens, on the other hand, are often produced in unidirectional forms to facilitate easier analysis by the reduction of experimental variables.

Once laid up, the composite structures require to be cured under temperature and pressure prior to carbonization. This is carried out by compression moulding in a heated press, or by use of an autoclave (Chapter 7).

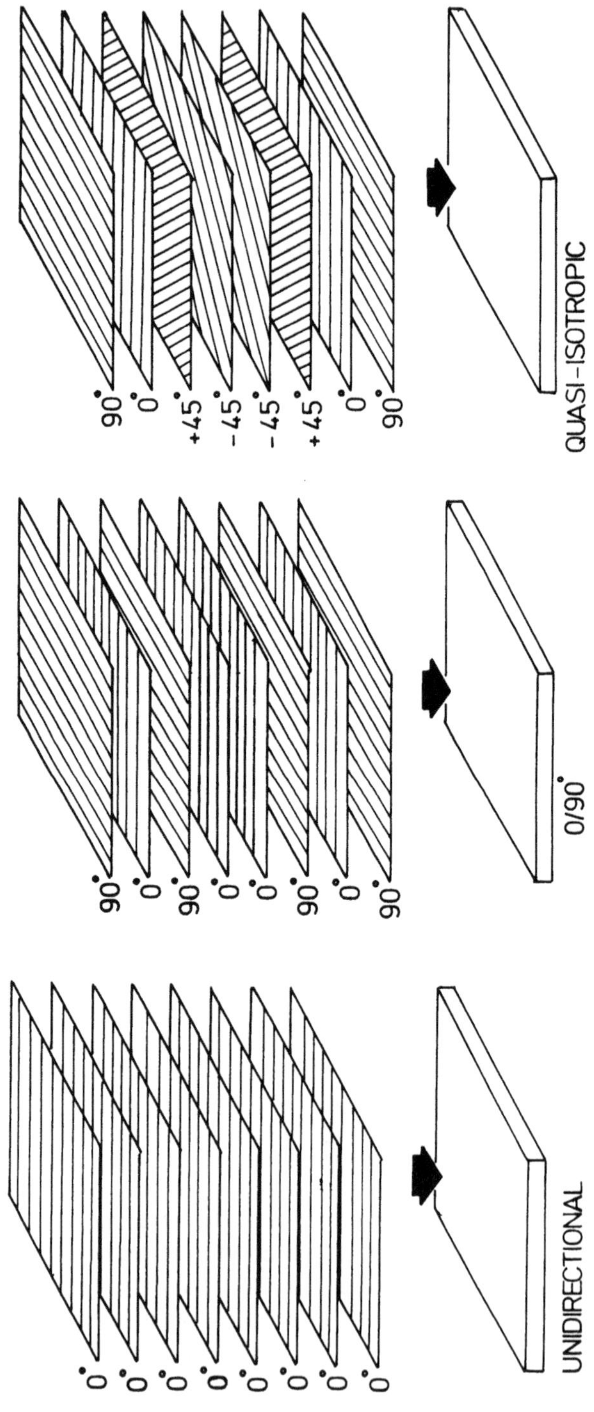

Fig. 4.18 The most commonly used lay-ups of UD prepregs.

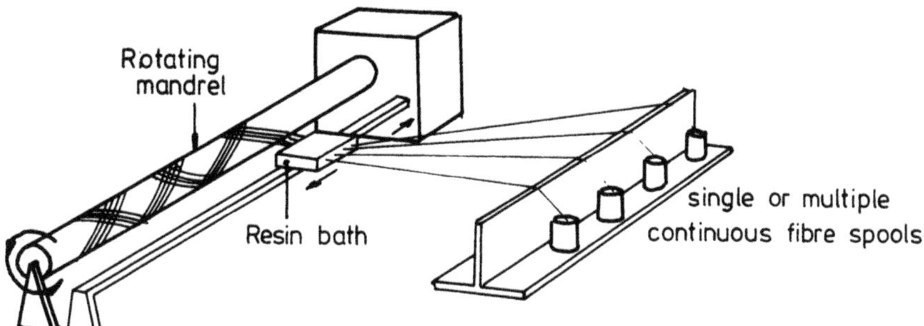

Fig. 4.19 Filament-winding process.

4.7.4 Filament winding

The filament-winding technique allows the high speed, precise lay-down of continuous fibre reinforcement in predescribed patterns. Resin-impregnated fibre tows are wound over a rotating male mandrel (Fig. 4.19). In the **wet winding** process the fibres are solution impregnated just prior to being wound on the mandrel. The **dry winding** method, on the other hand, uses single-tow prepreg as its feedstock. Two types of winding machine exist; rotating mandrel and stationary mandrel. There are, in addition, two methods of winding available: **polar** (or planar) in which each layer of fibres is wound without spaces or cross-overs, and **helical** winding, in which both spaces and cross-overs occur. In both cases the fibres are laid on to the mandrel in a helical pattern with the helix angle being chosen to suit the application. The construction of the mandrel requires considerable skill. It must not collapse under the pressure resulting from the fibre-winding tension and must be easily removed once the process is complete. Segmented metal forms (usually steel faced with plaster) are most commonly used.

The mandrel may be cylindrical, spherical or any shape that does not exhibit re-entrant curvature. Filament-winding operations have the capacity to vary the winding tension, wind angle or resin content in each layer of reinforcement until the desired thickness and resin content of the composite are achieved along with the required direction of strength. The great advantage of the process is that the cost is generally lower than the prepreg cost for most materials [59]. The lower costs arise because a relatively expensive fibre can be combined with an inexpensive resin to produce a relatively inexpensive composite. Coupled with this is a high-speed lay-down of fibre. Table 4.4 illustrates the major advantages and disadvantages of filament winding.

The fabrication of the composite winding is completed by the curing operation. The structure can be cured easily in an oven without the need of pressure. Recently certain manufacturers have used vacuum bags and

Table 4.4 Advantages and disadvantages of filament winding

Advantages	*Disadvantages*
The highly repetitive nature of fibre placement (from layer to layer and from part to part).	Shape of component must be such that the mandrel can be removed.
The capacity to use continuous fibres over a whole component area (without joints) and to orientate fibres easily in load direction.	Inability to wind reverse curvature.
Elimination of capital expense of autoclave.	Inability to change fibre path easily (in one ply).
Large structures can be built (larger than any autoclave).	Need for mandrel, which can be complex and expensive.
High fibre volume is obtainable.	Poor external surface, which may hamper aerodynamics.
Lower cost for large numbers of components.	
Relatively low material costs because fibre and resin can be used in their lowest cost form rather than as prepreg.	

autoclaves to produce more compact void-free laminates. The majority of carbon–carbon precursor windings are carbonized to produce tubes for various applications (Fig. 4.20), although some are slit to produce channels in a variety of cross-sections.

4.7.5 Compression moulding compounds

A relatively cheap precursor moulding compound may be produced by impregnating a single-fibre tow with a thermoset resin. The impregnated tow is pultruded through a circular die and chopped into short lengths of up to 50 mm in length. The chopped 'matchsticks' are packed into a metal tool and compression moulded between the platens of a heated press. The cured composite is carbonized and subsequently densified by reimpregnation or CVD for use as a brake material. Generally, mesophase pitch-based fibres are used in what is only a lightly loaded composite.

4.8 HIGH CARBON YIELD MATRIX PRECURSORS

In addition to the commonly used phenol/formaldehyde and furan resins, a wide range of 'exotic' polymers have been used as carbon–carbon matrix

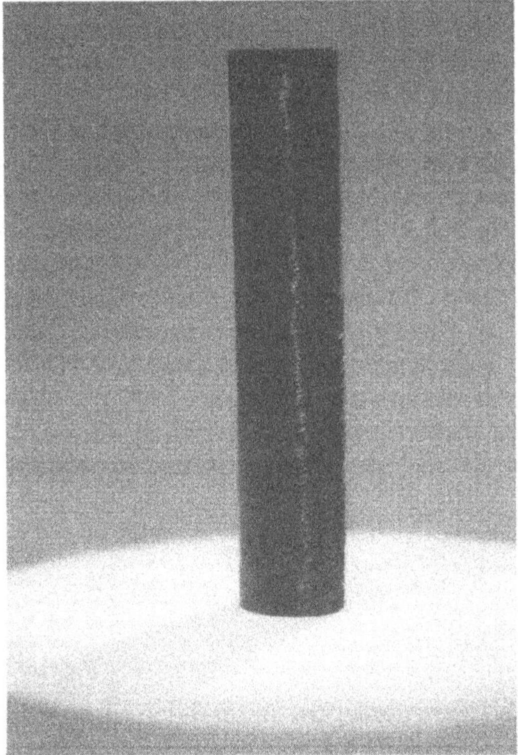

Fig. 4.20 Filament-wound carbon–carbon tube.

precursors. While some attention has been given to the use of thermoplastics (such as PEEK) the majority of work has been carried out with thermosetting systems.

For the fabrication of carbon–carbon the resin should give a high carbon yield on pyrolysis, exhibit low shrinkage and cure rapidly at low temperature (without the evolution of volatiles). It should also be possible to prepare prepreg that exhibits good tack and drape properties from the resin.

Among the commercial high-temperature polymer systems available polyimide has received the most attention as a possible carbon source. As with phenol/formaldehyde and furan resins, extensive data on the properties and processing of polyimides exist. Burger *et al.* [20] prepared carbon–carbon using a series of polyimides as matrix precursors. The polymers had carbon yields of approximately 60% but exhibited high shrinkage and had poor plastic flow on carbonization. The mechanical properties of the polyimide-derived composites were only slightly superior to those prepared from a pitch binder. Generally, polyimides have not been widely used for production of carbon–carbon because of a combination of high costs and problems associated with their processing such as the use of non-volatile aggressive solvents and evolution of volatiles during polymerization.

Poly (*p*-phenylene) has long been recognized as one of the most stable of organic structures, while it possesses an extremely high carbon content (95%). Numerous attempts to prepare composites from polyphenylene have been made but little success has been achieved as the material is an infusible powder that is insoluble in known solvents. Even if a composite could be prepared, the absence of a means of cross-linking the polymer chains results in the loss of oligomeric phenylene units on carbonization. This leads to a much reduced carbon yield (*c*. 72%) over the theoretical when pyrolysed at atmospheric pressure [60].

Recently studies have been concentrated on the use of soluble precursors that can be easily processed, then converted to polyphenylene *in situ* [61].

$$\xrightarrow{\Delta} \quad + \ 2\ \text{ROM}$$

RO OR

While this approach allows the preparation of prepreg it suffers from a number of drawbacks. The prepolymer is only soluble in high boiling solvents that are difficult to remove. The decomposition to give polyphenylene proceeds at low temperature but generates volatile by-products. In common with previous polyphenylenes the lack of a cross-linking mechanism results in carbon yields of *c*. 70% while the precursor polymer is at present only available on a small scale at a high cost.

Oligomers containing the acetylenic (ethynyl) group (–C≡C–) have recently been the subject of considerable research [62]. These materials can be cured thermally to give cross-linked products having good physical and mechanical properties, coupled with good solvent and moisture resistance. A key feature of acetylene-containing systems is that curing proceeds without the evolution of volatiles, thus making the fabrication of thick void-free laminates considerably easier. A number of efforts to utilize acetylene containing reins as carbon–carbon matrix precursors have been made. These have usually consisted of acetylene-substituted benzenes or oligomeric materials derived from them. Some of the more common routes used to prepare acetylene-containing materials are as follows:

1. from an aryl acetyl compound by treatment with the Vilsmeyer reagent followed by basic cleavage;
2. from olefins by a halogenation–dehydrohalogenation procedure;
3. from an aryl acetyl compound by a halogenation–dehydrohalogenation strategy;

4. from an aryl halide via a palladium–copper catalysed coupling reaction with acetylene.

All of these methods have one or more limitations associated with them. Vilsmeyer chemistry is extremely sensitive to reaction conditions and has often proved difficult to scale up. Both this route and the reactions in (3) frequently give poor yields. Method (2) not only assumes the aryl olefin is readily available but also involves an elimination step where the choice of base and reaction conditions can be critical. The first three methods all tend to produce tar-like materials along with the desired acetylenic compound. These are usually separated by distillation, although aryl acetylenes can decompose explosively at elevated temperatures.

The coupling reaction in (4) is usually carried out with a mono-protected acetylene, with the protecting group being subsequently removed. The main drawbacks to this route are the expense of the catalyst and the starting aryl halides, difficulties in the removal of the protecting groups and problems in the removal of final traces of catalyst. A residual catalyst has a detrimental effect on the long-term thermo-oxidative stability of the resin.

Of the resins used as carbon precursors most work was carried out with so-called H-Resins produced by Hercules. These consisted of acetylene-terminated oligomers derived from diethynyl benzoene such as (X) [63].

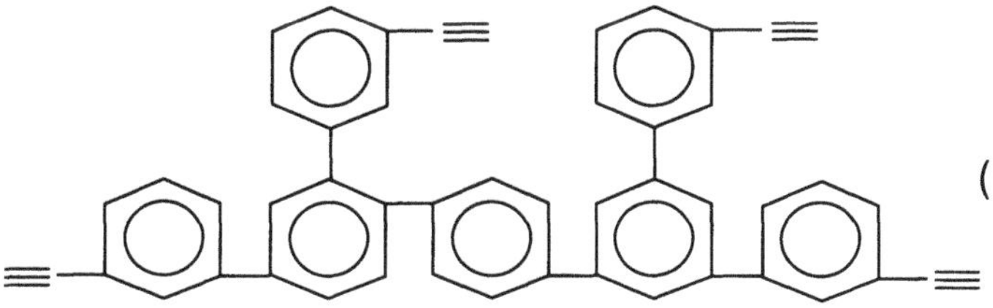

(X)

These resins have very high carbon contents which on curing give a highly cross-linked solid with a carbon yield of *c.* 80%. Despite these favourable properties H-Resins were withdrawn due to a combination of processing difficulties (H-Resins consisted of a fine powder that proved difficult to prepreg) and high cost.

Subsequent to the withdrawal of H-Resins sporadic attempts to develop a carbon precursor based on acetylene resins have been reported.

The use of low-melting diethynyl aromatic compounds of general formula (Y) was described by workers at Hughes Aircraft [64].

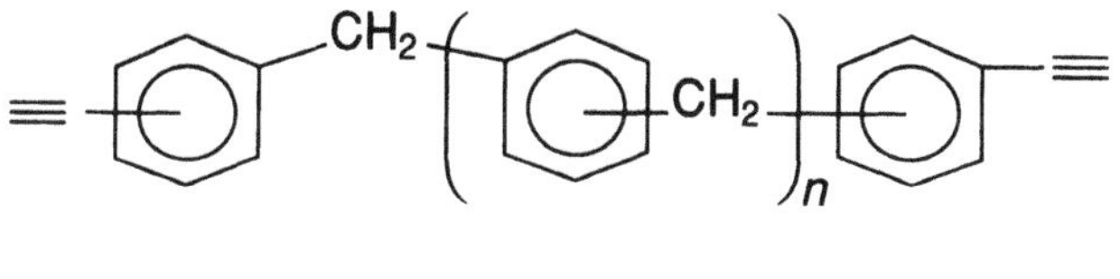

CH_2 CH_2 n (Y)

These are low-viscosity melts that can easily impregnate fibrous preforms and then undergo rapid polymerization on curing. In the case of 4,4′-diethynyldiphenylmethane (Y, n = 0) the carbon yield was reported as 91%.

As the carbon obtained from acetylene resins is derived from a thermoset it usually exhibits only short-range order (glassy, isotropic carbon). The preparation of 'graphitizable' carbon from acetylenic materials has been described by Bilow [65]. This used a low-melting mixture of diethynyl benzene and ethynyl pyrene which could be polymerized in a controlled manner (without risk of a violent exotherm) at moderate pressure, while still retaining a high carbon yield.

Most recently, Hyperion Catalysis International have introduced a carbon precursor resin. Little detailed information is available but it appears to be an acetylene-containing polyphenylene.

Despite the wide range of resins available, no acetylenic materials have gained wide acceptance as carbon precursors. Problems such as lack of prepreggability, highly exothermic cure reactions, porosity and type of carbon microstructure generated are all potentially soluble, but as yet no cost-competitive system has emerged. The methods of synthesis of acetylene groups described above (along with other more esoteric routes) [66] all suffer from the use of either expensive starting materials or reagents, poor yields or multi-step procedures. Many of the routes are difficult to scale up for bulk resin production.

In conclusion, while none of the 'exotic' resins described above can yet fulfil all the requirements of a carbon matrix precursor the highly desirable prospect of reducing carbon–carbon fabrication costs via a high char precursor will ensure continued research in this area.

REFERENCES

1. Kobayashi, K., Sugawara, S., Toyoda, S. and Honda, N. (1968) *Carbon*, **6**, 359.
2. Mackay, H. A. (1970) *Carbon*, **8**, 517.
3. McAllister, L. E. and Taverna, A. R. (1971) *Proc. 10th Biennial Conf. on Carbon*, Bethlehem, Pa, Paper no. FC-40.
4. Lewis, J. C., Redfern, B. and Cowlard, F. C. (1963) *Solid State Electronics*, **6**, 251.
5. Cowlard, F. C. (1970) *Design Eng.*, March, 49.
6. Bokros, J. C., Atkins, R. J., Shim, H.S., Hanbold, A. D. and Agarwal, N.K. (1976) Carbon in prosthetic devices, in *Petroleum Derived Carbons* (eds M. L. Deviney and T. M. O'Grady) Chem. Soc., Washington DC, p. 237.
7. Kaae, J. L. (1975) *Carbon*, **13**, 246.
8. Kaae, J. L. (1971) *J. Nucl. Mat.*, **38**, 42.
9. Yamada, S. (1968) *A Review of Glass-like Carbons*, Battelle Memorial Institute, Columbus, Ohio.

10. Kawarmura, K. and Jenkins, G. M. (1972) *J. Mat. Sci.*, **7**, 1099.
11. Kaae, J. L. (1972) *J. Biomed. Mat. Res.*, **6**, 279.
12. Stanitski, C. L. and Mooney, V. (1973) *J. Biomed. Mat. Res.*, **7**, 97.
13. Savage, G. M. (1985) PhD thesis, Univ. London.
14. Cowlard, F. C. and Lewis, J. C. (1967) *J. Mat. Sci.*, **2**, 507.
15. Madorsky, S. L. J. (1953) *Res. Nat. Bur. Stds*, **51**, 327.
16. Mackay, H. A. (1969) *Sandia Labs Report, no. SC-RR-68-651*.
17. Kotlensky, W. V. (1973) *Chem. Phys. Carbon*, **9**, 173.
18. Fitzer, E. and Gkogkdis, A. (1986) in *Petroleum-derived Carbons ACS Symp Se no. 303* (eds J. C. Bacha, J. W. Newman and J. L. White, Am. Chem. Soc., Washington, DC. p. 347.
19. Fitzer, E., Heym, M. and Karlisch, K. (1974) *Proc. 4th Lond. Int. Conf. on Carbon*, p. 172.
20. Burger, A., Fitzer, E., Heym, M. and Terwiesch, B. (1975) *Carbon*, **13**, 149.
21. Jenkins, G. M. and Kawamura, K. (1976) *Polymeric Carbons, Carbon Fibre, Glass and Char*, Cambridge Univ. Press.
22. Lum, R., Wilkins, C. W., Robbins, M., Lyons, A. M. and Jones, R. P. (1983) *Carbon*, **21**(2), 111.
23. Conley, R. T. (1967) *J. Macromol. Sci.*, **A1**, 81.
24. Kumar, A. Kulshershtha, A. K. and Gupta, S. K. (1980) *Polymer*, **21**, 317.
25. Zinke, A. (1951) *J. App. Chem.*, **1**, 257.
26. Ouchi, K. and Honda, H. (1956) *J. Chem. Soc. Jap.*, **77**, 147.
27. Ouchi, K. and Honda, H. (1959) *Fuel (London)*, **38**, 429.
28. Ouchi, K. (1966) *Carbon*, **4**, 59.
29. Fitzer, E., Schafer, W. and Yamada, S. (1969) *Carbon*, **7**, 643.
30. Billmeyer, F. W. Jr (1970) *Text Book of Polymer Science*, 2nd edn, Wiley-Interscience, New York, p. 486.
31. The Quaker Oats Company, *Technical Bulletin 131-A*.
32. Delmonts, J. (1969) Furan resins in *Modern Plastics Encyclopedia*, McGraw-Hill, New York, p. 138.
33. Downing, P. A. (1978) *The Chemical Engineer*, April, 272.
34. Dunlop, A. P. and Reineck, E. A. (1947) *Proc. Tech. Conf. Am. Chem. Soc., Chicago Sect.*, North Western Univ., 24 Jan.
35. The Quaker Oats Company, *Technical Bulletin 205-B*.
36. Madorsky, S. L. J. (1964) *Thermal Degradation of Organic Polymers*, Interscience, New York.
37. Fitzer, E. and Schafer, W. (1970) *Carbon*, **8**, 597.
38. Ozanne, N. (1968) PhD thesis, Univ. Grenoble.
39. O'Neill, H. J., Boquist, G. W., Putcher, R. E. and Griffith, J. S. (1964) *Wadd Tr. 81–72*, Vol. XL.
40. Braun, W. (1970) PhD thesis, Univ. Karlsruhe.
41. Yamada, S., Saks, H. and Ishi, T. (1965) *Carbon*, **3**, 253.
42. Delmonte, J. (1981) *Technology of Carbon and Graphite Fibre Composites*, Van Nostrand Reinhold, New York.
43. Manocha, L. M. (1982) *J. Mat. Sci.*, **17**, 3039.
44. Fitzer, E., Geigl, K. H. and Huettner, W. (1980) *Carbon*, **18**, 265.
45. Manocha, L. M., Yasuda, E., Tanabe, Y. and Kimura, S. (1988) *Carbon*, **26**, 333.
46. Bradshaw, W. G. and Vidoz, A. E. (1978) *Am. Ceram. Soc. Bull.*, **57**, 193.

47. Boyne, L. (1974) *Proc. 4th Lond. Int. Carbon and Graphite Conf.*, p. 215.
48. Manocha, L. M. (1988) *Composites*, 19 (4), 311.
49. Kimura, S., Yasuda, E., Tanaka, H. and Yamada, S. (1975) *J. Ceram. Soc. Jap.*, **83**, 122.
50. Hishiyama, Y., Inagaki, M., Kimura, S. and Yamada, S. (1974) *Carbon*, **12**, 249.
51. Kamiya, K. and Inagaki, M. (1973) *Carbon*, **11**, 429.
52. Noda, T. and Kato, H. (1965) *Carbon*, **3**, 289.
53. Kamya, K., Inagaki, M., Mizutani, M. and Noda, T. (1968) *Bull. Chem. Soc. Jap.*, **41**, 2169.
54. Nakamura, S., Ishii, T. and Yamada, S. (1964) *Proc. Symp. on Carbon*, Tokyo.
55. Saxena, R. R. and Bragg, R. H. (1978) *Carbon*, **16**, 373.
56. Yasuda, E., Kimura, S. and Shibusa, Y. (1980) *Trans. Jap. Soc. Comp. Mat.*, **6**, 14.
57. Payne, R. S., von Bradsky, G., Gray, G. and Savage, G. M. (1990) *Proc. Int. Conf. Adv. in Elec. Mic.*, Seattle., p. 1024.
58. Agarwol, B. D. and Broutman, L. J. (1980) *Analysis and Performance of Fibre Composites*, Wiley, New York.
59. Freeman, W. T. and Stein, B. A. (1985) *Aerospace America*, **44**, Oct. p. 44.
60. Fewell, L. (1976) *Chromatog. Sci.*, **14**, 564.
61. Ballard, D. G. H., Courtis, A., Shirley, I. M. and Taylor, S. C. (1983) *J. Chem. Soc. Chem. Commun.*, 954.
62. Hergenrother, P. (1985) in *Encyclopedia of Polymer Science and Engineering*, **1**, 2nd edn, Wiley, New York p. 61.
63. Fitzer, E. (1987) *Carbon*, **25**, 163.
64. Austin, W. B. and Bilow, N. US patent, 4,284,834.
65. Bilow, N. US patent, 4,369,297.
66. Anderson, C. US patent, 4,665,246.

Thermoplastic Matrix Precursors

5

5.1 INTRODUCTION

It has been discussed, thus far, how the fabrication of carbon–carbon composites is achieved by the impregnation of fibre tows, weaves or skeletons (3-D structures or felts) with thermosetting resins or by chemical vapour infiltration with gaseous hydrocarbons. All of the processes presently practised are slow and expensive, and fail to exploit fully the strength of the reinforcing fibres [1]. Vapour infiltration methods require low reaction rates to maintain a uniform deposition throughout a porous body.

The resin-based process, on the other hand, requires a number of cycles of impregnation and carbonization to attain useful levels of density because the carbon yield is limited and/or the matrix may be exuded, in part, by pyrolysis. Carbon–carbon components are used in critical applications where no other material can serve, but serious constraints of cost, processing time and reliability of manufacture limit their use in other systems for which their properties are also well suited.

A great deal of work has been carried out to investigate the suitability of pitches, derived from coal tar or petroleum, and polyaromatic thermoplastic resins as matrix precursors. The aim is to realize the advantages of high carbon yield and ease of graphitizability characteristic of these materials to achieve lower processing costs. Upon pyrolysis, aromatic polymerization drives the majority of pitches through a liquid crystalline state known as the 'mesophase' in which the graphitizability of the coke is established by parallel alignment of the large aromatic molecules [2]. Pitch-based carbons are thus almost unique among structural materials in forming their microstructures by liquid crystal mechanisms [3–5]. The only other notable structural material formed via a liquid crystal precursor are 'Kevlar' polyaramid fibres manufactured by Du Pont [6].

Table 5.1 Typical coal-tar distillation data

Product	*Boiling range* (°C)	*Weight* (%)
Light oil	Below 200	1
Naphthalene oil	200–230	12
Creosote oil	230–300	6
Anthracene oil	Above 300	20
Pitch	Residue	61

5.2 PITCH

The use of pitches as matrix precursors in carbon–carbon composites is an extension of the technology used in the graphite electrode processing industry. A large data base, covering impregnation, carbonization and graphitization of both coal-tar and petroleum pitches, is therefore readily available. Pitches have a low softening point, low melt viscosity, high carbon yield and tend to form graphitic carbon structures. Pitches used as carbon–carbon precursors are a refined oligomeric mixture of polynuclear aromatic hydrocarbons which are thermoplastic in behaviour.

5.2.1 Coal-tar pitch

Coal-tar is a by-product of the coking of bituminous coals to produce cokes. Metallurgical cokes are produced at high temperatures (900–1100 °C) while low temperatures (≈600 °C) yield domestic smokeless fuels. The low-temperature process gives a smaller amount of tar than the high-temperature process. Pitch is obtained from the coal tar by distillation and heat treatment processes. Typical coal-tar distillation data are given in Table 5.1.

Pitch is the residue which follows the removal of the heavy (creosote or anthracene) oil fractions. Pitches are complex mixtures containing many different individual organic compounds, and the precise compositions and properties vary according to the source of tar and the method of removal of low molecular weight fractions. Smith *et al.* [7] report that roughly two-thirds of the compounds so far isolated from coal-tar pitch are aromatic, the remainder being heterocyclic. Many of the compounds are substituted, with the methyl group being the most prevalent. The great majority of the coal-tar pitch components contain between three and six rings, with boiling-points in the range 340–550 °C. Coal-tar pitch can therefore be considered to consist predominantly of carbon and hydrogen with small amounts of nitrogen, oxygen and sulphur. An estimate of the degree of aromacity within the pitch may be made from its C/H ratio.

Table 5.2 Characteristics of typical carbon matrix precursor pitches from three different feedstocks

		Petroleum Pitch		*Coal-tar pitch*	
Property	*Steam cracker tar*	*1*	*2*	*1*	*2*
Softening point (°C)	110	117	110	101	113
Coking value (wt% at 550 °C)	52	54	56	57	60
Aromatic C (%)	78	82	80	89	88
C/H ratio	1.37	1.44	1.57	1.77	1.76

5.2.2 Petroleum pitch

Petroleum pitch is a readily available product and may be obtained from the bottom of catalytic crackers. It is the heavy residue obtained from a catalytic cracking process, from steam cracker tar, a by-product of the steam cracking of naphtha or gas oils to produce ethylene or any residues from crude oil distillation or refining. A number of processes may be used to manufacture pitch from the aforementioned feedstocks, including thermal treatment, vacuum or steam stripping, oxidation, simple distillation or a combination of these [8]. In common with coal-tar pitch, the chemical and physical characteristics of petroleum pitch are very dependent on the process and conditions employed, especially the process temperature and heat treatment time. Generally, longer times and higher temperatures produce pitches with increased aromacity and higher anisotropic contents. Petroleum pitches are usually less aromatic than coal-tar pitch. Typical properties of the various types of pitch are shown in Table 5.2.

5.3 CHARACTERIZATION OF PITCHES

Fundamental to the understanding of the properties of carbon–carbon composites derived from pitch is an accurate chemical characterization of the precursor. Solvent fractionation and extraction are used to divide the pitch into a broad range of constituents which can be further subdivided using more sophisticated techniques as the need arises. As is normally the case in carbon science, a system of nomenclature has existed for many years describing the various fractions obtainable. Unfortunately, the coal-tar and petroleum industries use their own naming routines which do not correspond. The petroleum industry defines the following species:

Carboids are insoluble in CS_2.
Carbenes are soluble in CS_2 but insoluble in CCl_4.
Asphaltenes are insoluble in light paraffinic hydrocarbons (n-pentane, for example) but usually soluble in CS_2, CCl_4 and C_6H_6 – they are condensed, highly aromatic molecules.

Pre-asphaltenes is the term used to describe the fraction which is insoluble in solvents such as benzene, but soluble in pyridine.

Despite a degree of variation in the nomenclature used by different workers in the field [9], carbon–carbon matrix precursor pitches are most often fractionated using the following solvents:

1. The fractions which are insoluble in *quinoline* or *pyridine* are very high molecular weight aromatic compounds and solid impurities known as the 'C' component QI or 'α resins'.
2. *Benzene* or *toluene* insoluble material, which is soluble in quinoline or pyridine, is known as the 'C_2' component or 'β resins'.
3. Material which is insoluble in *petroleum ether* or *n-hexane* but soluble in benzene (or toluene) is referred to as the 'resinoid' fraction and is identified with asphaltenes, while the soluble material is called the 'crystalloid' fraction.

The solubility spectrum for a pitch can be very useful in its characterization especially in understanding the fundamentals of carbonization. Separation procedures, such as distillation [10], solvent fractionation [11,12] and column chromatography [13], have been coupled with various constitutional methods including elemental [14] and molecular weight analysis [15], IR [16] and UV [17] spectroscopy to show that pitches are highly complex mixtures of predominantly aromatic compounds. Figure 5.1 illustrates the type and percentage of structures identified in a typical coal-tar pitch [18]. Despite the fact that over the last two decades substantial progress has been made in the application of new methods for separation and characterization of carbonaceous materials [19], impregnating pitches remain a poorly understood 'hotch-potch' of compounds.

Gas chromatography (GC) is a useful tool with which to study the low molecular weight and volatile components of pitch. Greinke and Lewis [20] used GC to identify aromatic hydrocarbon components containing up to six condensed rings in the distillates collected from both coal-tar and petroleum pitches. Larger molecular components may be separated using high pressure liquid chromatography (HPLC). The separation is based on the presence of functional groups as well as on molecular size [21]. The method most applicable to the analysis of the larger molecular components of pitch is gel permeation chromatography (GPC). The technique involves separation of a solution of the sample on a porous polymer gel and is used to measure size and molecular weight distribution of high polymers. Edstrom and Petro [22] used GPC to measure the molecular size distribution within a number of pitches, using tetrahydro-furan as a solvent. The work was augmented by Tillmans *et al.* who employed quinoline as a solvent, allowing a larger fraction of the pitch to be measured [23].

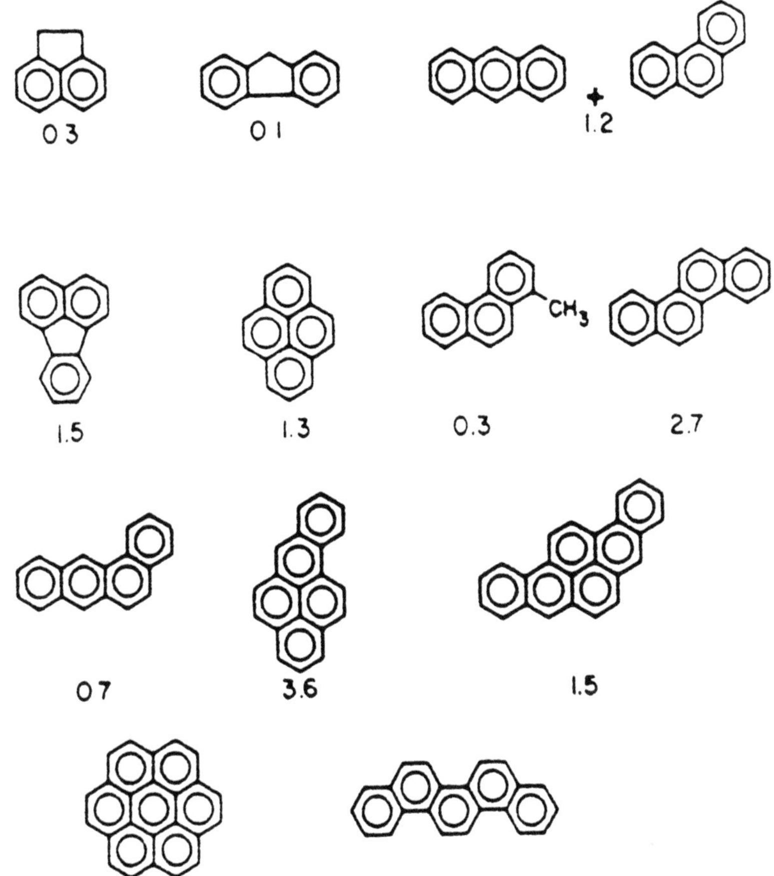

Fig. 5.1 Polynuclear aromatic components of pitch [18].

Quantitative information concerning the distribution of hydrogen and carbon in pitch molecules can be obtained using nuclear magnetic resonance (NMR). Dickinson [24] combined H and C^{13} NMR with other molecular constitutional data to produce average structures for various petroleum pitches. A number of other structural analysis techniques such as mass spectroscopy, X-ray diffraction, neutron inelastic scattering and electron spin resonance (ESR) have been used to study the reaction mechanisms during the carbonization of pitch.

5.4 PYROLYSIS OF PITCH

Pitch is converted to carbon by a process of pyrolysis. The mechanism of carbonization is one of aromatic growth and polymerization [25]. Figure 5.2 shows a schematic representation of the carbonization process whereby a small aromatic structure may ultimately attain the three-dimensional

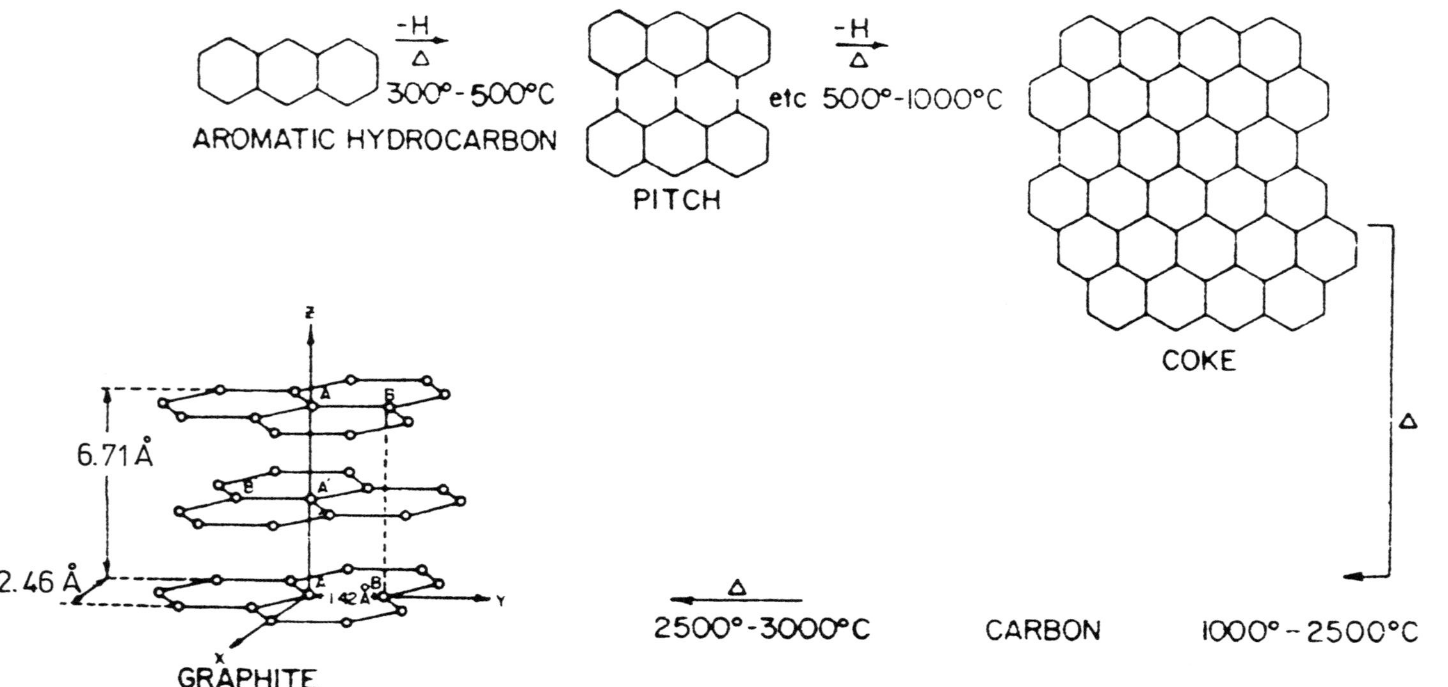

Fig. 5.2 Schematic diagram for the pyrolysis of pitch.

order of graphite via an aromatic polymeric structure. Before the carbonization process commences, non-aromatic structures are converted to aromatics, the key building blocks for carbon. Some polymers also degrade into tar-like liquids in the early stages of pyrolysis, PVC being a prime example. The overall process of carbonization is extremely complex. One may, however, consider individually a number of processes representative of the major reactions involved:

1. C–H and C–C bond cleavage to form reactive free radicals;
2. molecular rearrangement;
3. thermal polymerization;
4. aromatic condensation;
5. elimination of side chains and hydrogen.

All of the above processes can occur in parallel during the heat treatment but may be studied separately from a mechanistic viewpoint.

The initial thermal reaction is the carbonization process; although poorly understood, it is believed to involve the formation of a free radical intermediate. Two types of radical may be formed by bond cleavage of the aromatic molecule. The σ radical is produced by the breaking of an aromatic C–H bond. The reaction is a high-energy process since a bond dissociation energy of $\approx$420 kJ mol^{-1} must be overcome [26]. The σ radical intermediate is very unstable and the free electron localized. The second, π, radical is rather more stable, with much less energy ($\approx$325 kJ mol^{-1}) being required to break the methyl C–H bond. The unpaired electron is resonance stabilized and simple π radicals such as the benzyl radical have been detected by ESR [27]. Stein *et al.* [28,29] predicted the initial stages in the pyrolysis of polynuclear aromatic hydrocarbons using the 'radicals approach'. The addition polymerization of anthracene was proposed to proceed by the reactions delineated in Fig. 5.3 [30]. The π radical (I) could be generated by a small amount of the anthryl σ radical. More detailed observations of unstable radicals are required, however, before the initial reactions of carbonization are clarified.

Thermal rearrangement is another important step in the early stages of carbonization. Figure 5.4 shows examples of thermal rearrangement for (a) acenaphthylene and (b) bifluorene, which can transform unstable five-membered rings into more stable six-membered ring systems without the loss of carbon atoms. Example(c) methylene–phenanthrene, illustrates a similar rearrangement which involves the loss of carbon atoms. Either the rearranged or starting molecule serves as the building block in carbonization. One of the factors making carbonization so complex is the presence of so many possible polymerization sites in an aromatic molecule. Anthracene, for example, may form 11 different reaction products from the simple dimerization reaction [31].

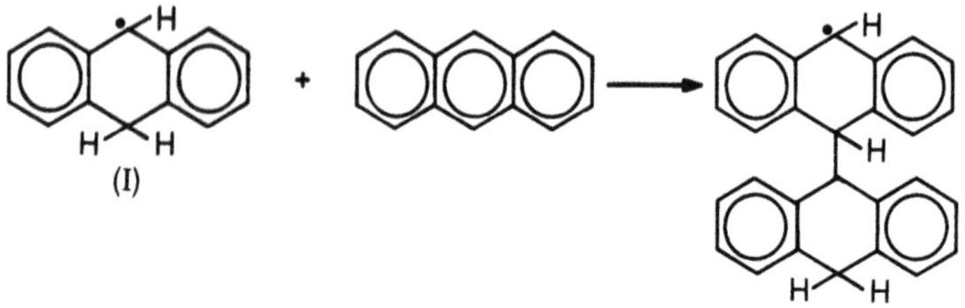

Fig. 5.3 Initial reactions in the pyrolysis of anthracene.

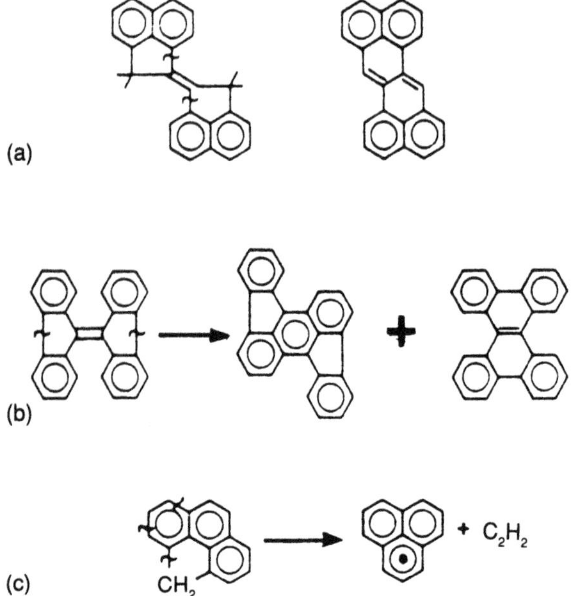

Fig. 5.4 Thermal rearrangement reactions during pyrolysis of polyaromatics.

Once formed, the reacting intermediate can undergo direct polymerization as shown in Fig. 5.5, illustrating the thermal polymerization of naphthalene to form naphthalene polymer. These initial polymerization reactions involve the loss or redistribution of hydrogen which can often be accomplished through internal hydrogen transfer. Figure 5.6 lists the major condensable volatile materials identified during the pyrolysis of several polynuclear aromatic hydrocarbons and illustrates the importance of hydrogen transfer. The products all consist of hydrogenated derivatives of the original compound with hydrogen added at the most reactive position in the molecule.

It can be seen in Fig. 5.7 that the polymerization process may occur in two stages, giving rise to either non-condensed or condensed polymers. In the case of naphthalene, the loss of two hydrogen atoms between two

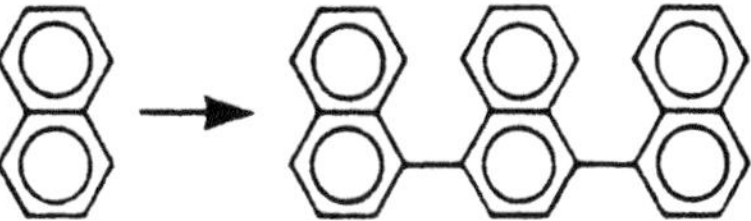

Fig. 5.5 Thermal polymerization of naphthalene.

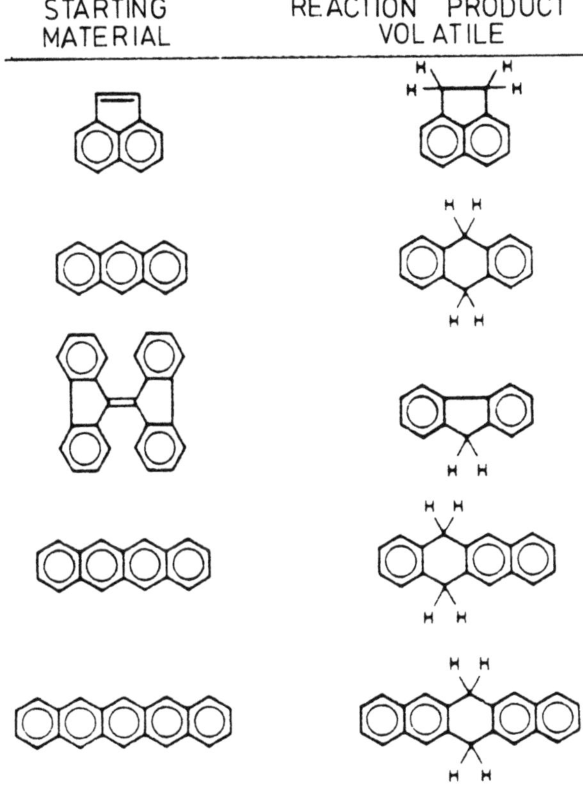

Fig. 5.6 Hydrogenated reaction products from the pyrolysis of aromatic hydrocarbons.

reacting molecules results in polymers in which the units are linked by single bonds. An additional loss of two hydrogen atoms produces a fully condensed polymer. Despite the non-condensed and condensed polymers having virtually the same molecular weights, they vary considerably in structure and properties. The non-condensed polymers are non-planar and their reactivities and ionization potentials change very slowly with increasing degree of polymerization. The condensed polymers, on the other hand, are fully planar and show marked changes in reactivity and ionization potential with increasing size. The relative role of these two processes is critical in the carbonization of pitch.

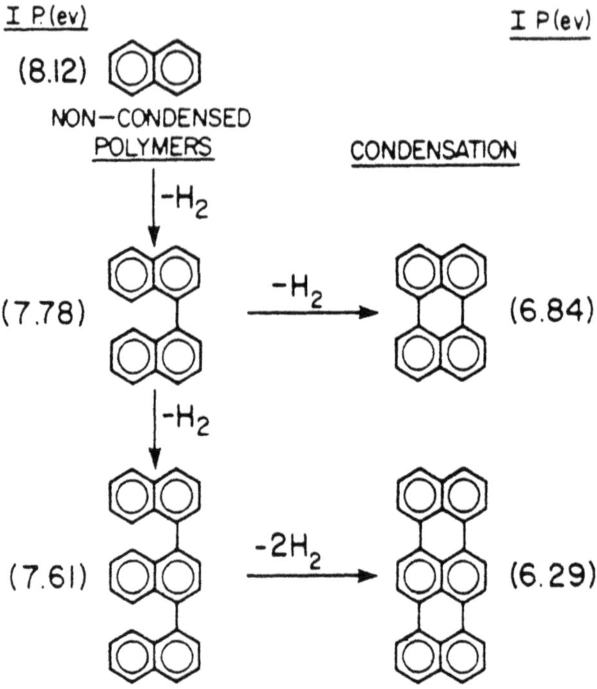

Fig. 5.7 Polymerization–condensation process in carbonization.

Lewis and Singer used the reaction scheme in Fig. 5.8 to show how odd-alternate radicals are involved in the polymerization condensation process in carbonization [32]. The polynuclear aromatic radicals produced are relatively reactive, as is illustrated for naphthalene. A non-condensed polymer (III) is formed by rapid polymerization. This polymer contains 3 non-condensed naphthalene units and a total of 30 carbon atoms. The loss of a single hydrogen atom from the polymer results in the formation of a free radical (IV) containing 30 carbon atoms with a single sp^3 tetrahedral carbon. The loss of an additional hydrogen from the π radical compound would create a fully condensed molecule.

The carbonization process is further complicated because the eventual polymerization to carbon occurs in two dimensions. Of paramount importance is how well the aromatic building blocks polymerize to develop a perfect graphitic network. The molecule zethrene (Fig. 5.9(a)) has the perfect shape and reactivity to polymerize in two dimensions to a planar graphitic structure whereas the compound tetrabenzonaphthalene shown in Fig. 5.9(b) has a non-planar structure, cannot polymerize without creating vacancies and is therefore poorly graphitizing.

The most significant feature of liquid phase pyrolysis is the development of a liquid crystal phase in the pyrolysing liquid prior to the formation of carbon [2]. Above around 400 °C spheres, initially around 0.1 μm

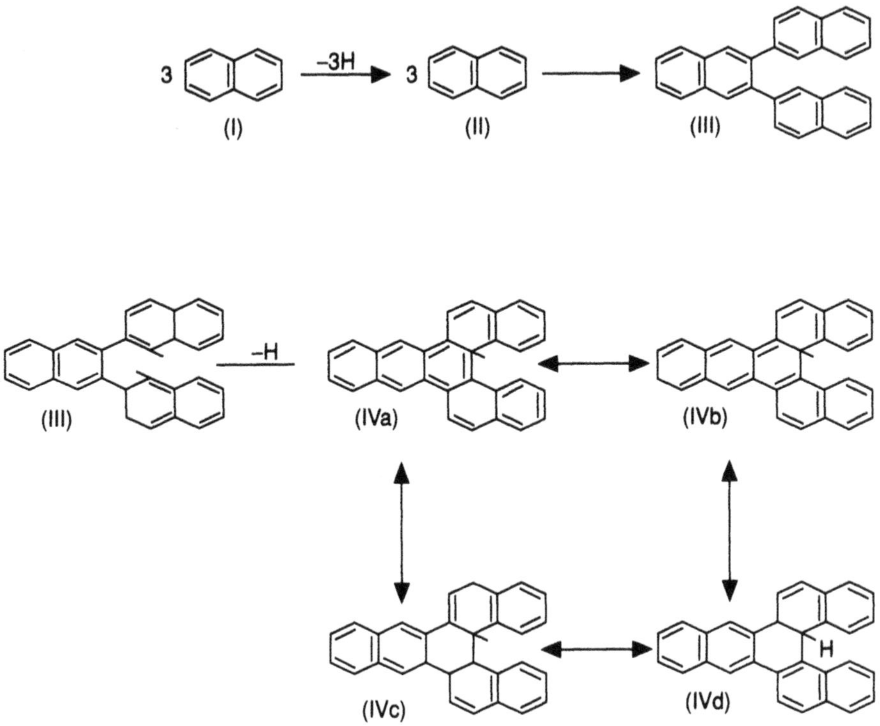

Fig. 5.8 Formation of stable odd-alternate free radical structures in carbonization.

in diameter, are observed in the isotropic liquid pitch. The spheres exhibit a highly oriented structure and are known as 'mesophase' [33]. The mesophase comprises polynuclear aromatic species which are stacked in approximate parallel array to form a discotic nematic liquid crystal system. The lamellar molecular structure formed within this temperature region (400–450 °C) may be considered as the 'embryonic graphite'. Figure 5.10 shows a schematic diagram of the type of molecular arrangement one would expect to find within the mesophase [4]. Under prolonged heating the spheres coalesce to form larger regions of extended order. The volume fraction of the mesophase in the two-phase emulsion gradually increases as pyrolysis proceeds, until the whole liquid is transformed to the anisotropic phase (Fig. 5.11) which subsequently solidifies to form carbon at around 500–600 °C. The structure of the domains of preferred orientation within the mesophase is determined by a number of factors. The reactivity of the precursors and the viscosity of the isotropic parent phase from which the mesophase grows control the nucleation and growth phenomena. The mesophase may be oriented by shear (Fig. 5.11). As a consequence, the shear history during pyrolysis, especially in the latter stages when the system is highly viscous and relaxation times are long, is of great significance in

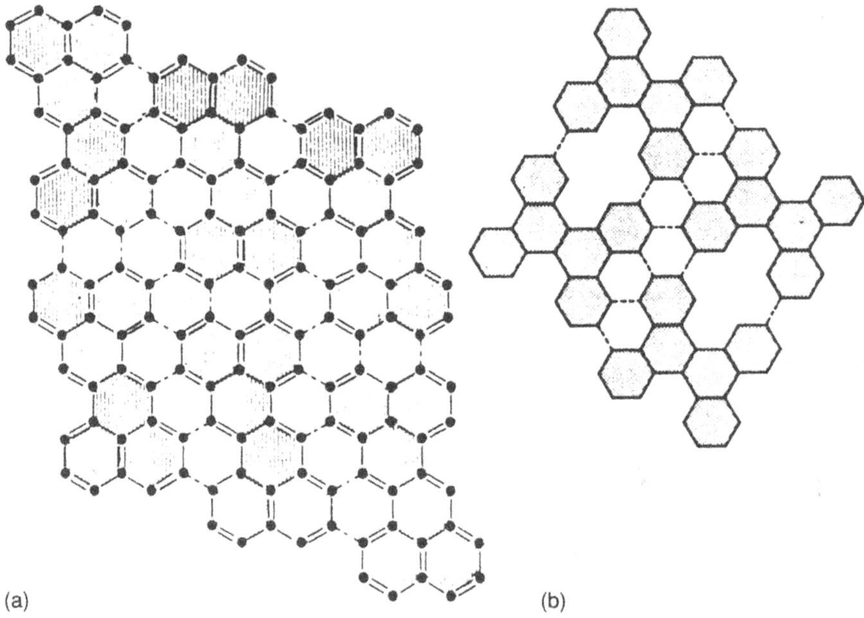

Fig. 5.9 Two-dimensional polymerization scheme for (a) zethrene and (b) tetrabenzonaphthalene.

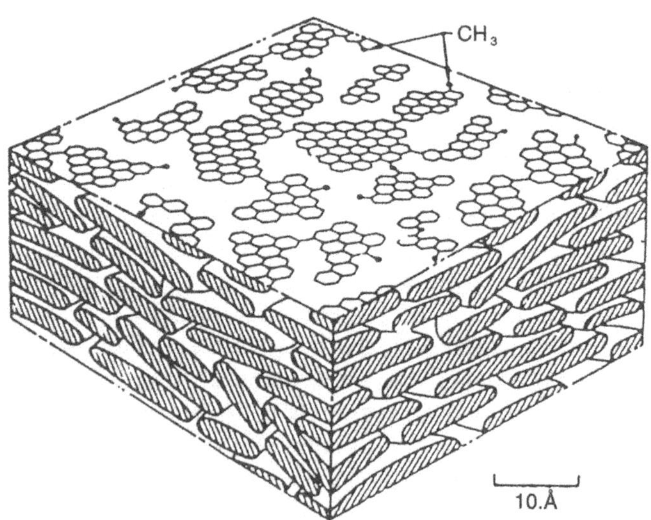

Fig. 5.10 Schematic representation of the molecular arrangement within the mesophase [4].

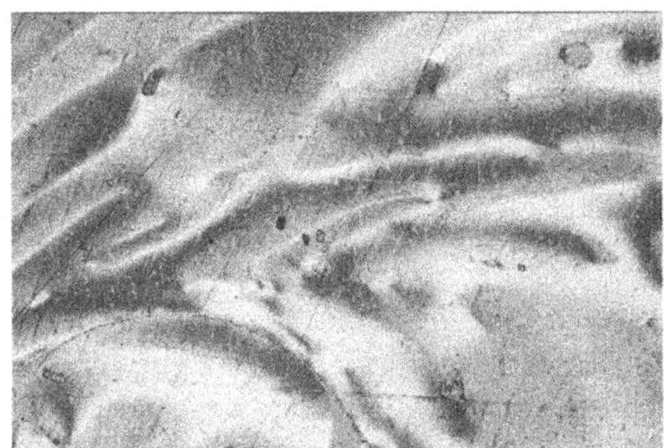

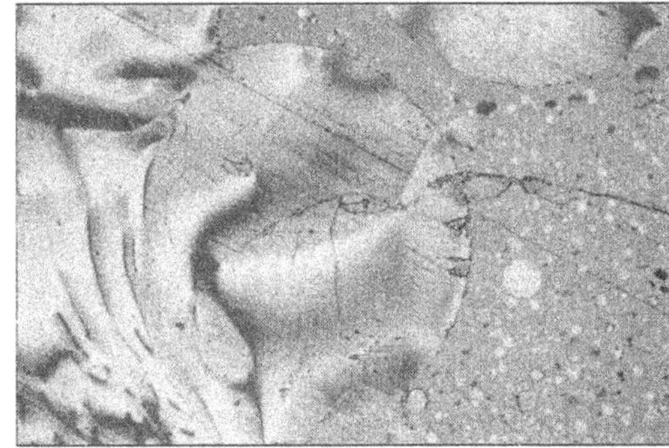

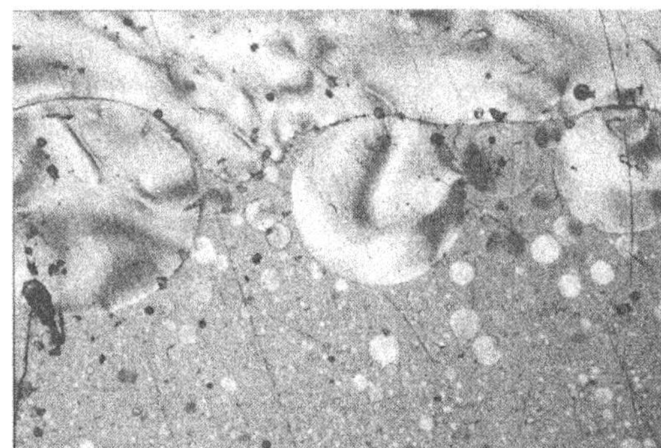

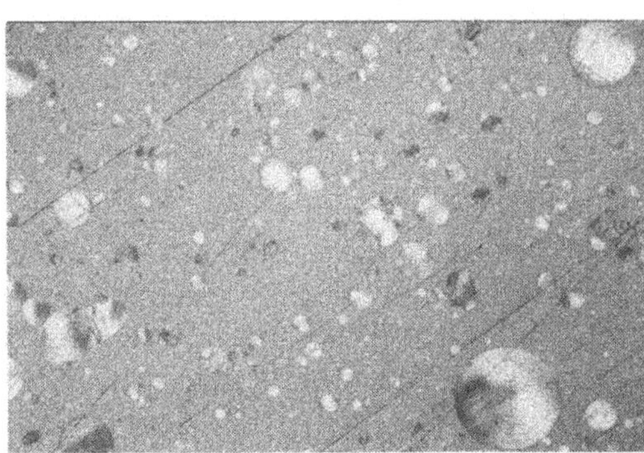

Fig. 5.11 Polarized light micrographs showing the development of mesophase during liquid phase pyrolysis of pitch between 400 and 500 °C (Courtesy Tonen Ltd).

the development of the microstructure of the carbon–carbon composite. Recent work [34] has shown how the mesophase may be aligned using a magnetic field. The polynuclear aromatic molecules of the mesophase have a disc-like shape which tends to align parallel to an applied magnetic field. It is suggested that this technique could be used to improve the mechanical properties of the matrix but, as yet, no conclusive results have been presented.

At high temperatures (>2300 °C) the lamellar arrangement within the pitch-derived carbon favours the formation of a graphitic crystal structure. The high density and small layer spacing measured in pitch-derived matrices of carbon–carbon (Fig. 5.12) is indicative of such a graphitic structure [35].

5.5 CARBON YIELD FROM PITCH

Pitch is a suitable matrix precursor for carbon–carbon composites only under the precondition that high carbon yields are obtainable. The yield depends very much on the composition of the precursor pitch and the conditions of pyrolysis. Decreasing the heating rate, the application of pressure or the use of chemical additives prior to the thermal decomposition of the pitch will all increase its carbon yield. Each of the aforementioned pyrolysis variables serves to improve the yield by restricting the evolution of volatile molecules present in the original pitch. The retention of these species in the carbonizing liquid as the temperature is raised allows them to participate in the aromatic growth and polymerization processes. For a fixed set of pyrolysis conditions, the carbon yield is influenced very strongly by the proportion of β resins and quinoline insolubles. The carbon yield thus generally increases with increasing amounts of high molecular weight compounds. Impregnating pitches are often fractionated to increase their content of those components which provide the highest carbon yield. It should be remembered, however, that this will result in higher viscosities which will affect the final microstructure (Section 5.7) and make processing more difficult. A balance often requires to be struck between the adventages and disadvantages of increasing the content of high molecular weight species in the pitch.

Bhatia and co-workers [36,37] have been able to show that a coal-tar pitch of higher softening point (SP) and β resins content exhibited a superior carbon yield to that of a conventional binder pitch.

High values of the molecular weight related properties of SP, benzene or toluene insolubles (BI), quinoline insolubles (QI) and β-resins (BI-QI) were found to be of significance in the pitch's suitability as a carbon matrix precursor since they are known to influence strongly properties such as viscosity and C/H atomic ratio. They were further able to show a remarkable

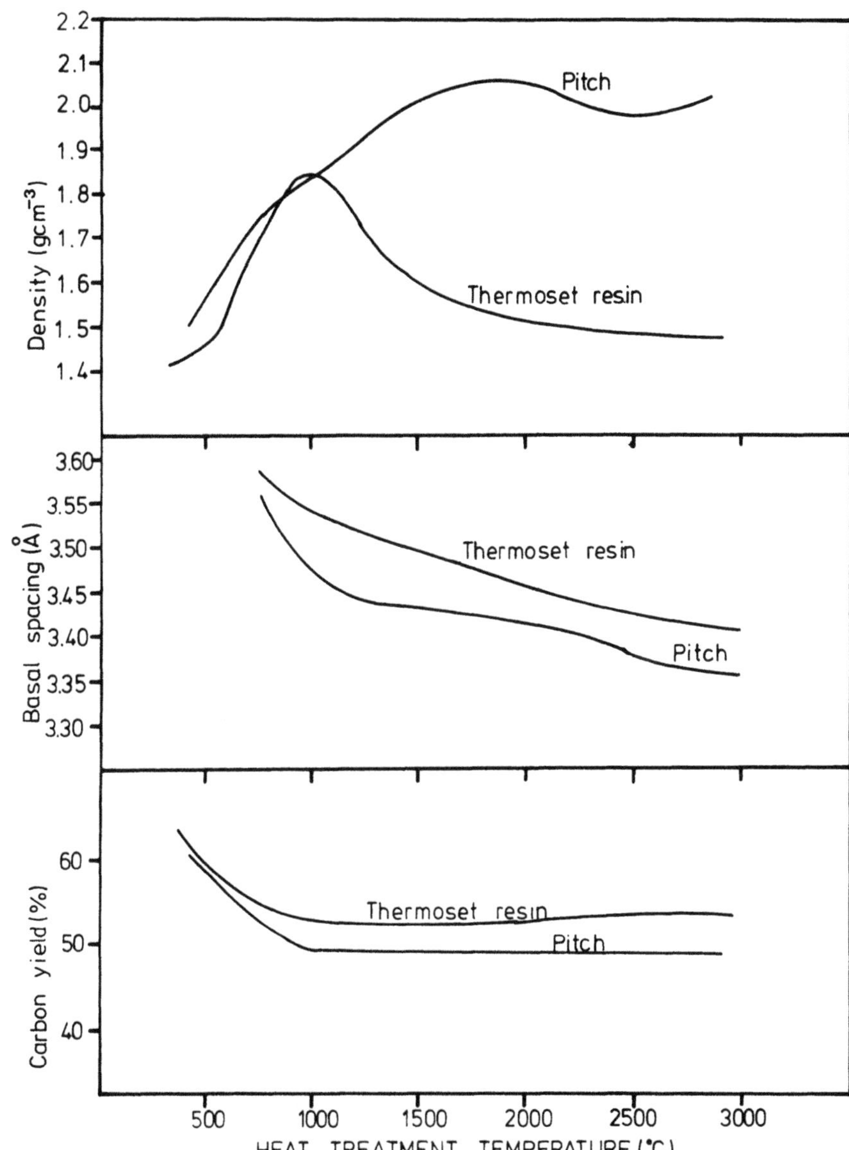

Fig. 5.12 Characteristics of resin and pitch as a function of HTT [35].

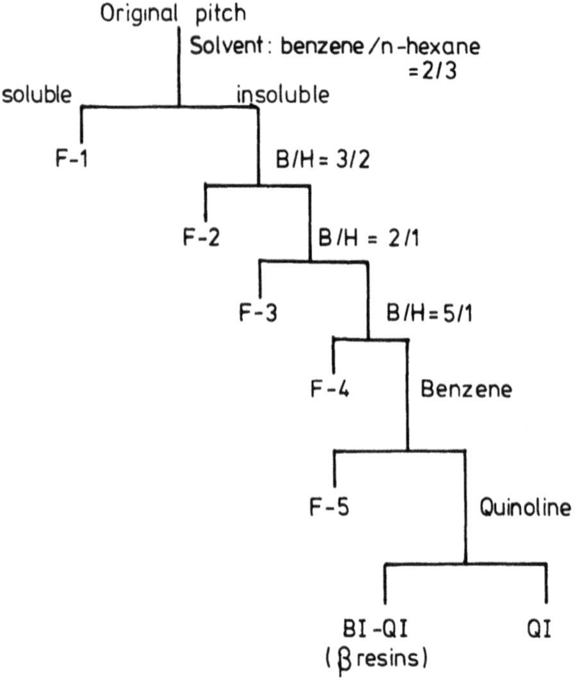

Fig. 5.13 Solvent fractionation of a coal-tar pitch.

agreement between the experimentally measured carbon yields of a number of pitches and those calculated theoretically [38].

The theoretical evaluation of the carbon yield of a pitch is based on the law of mixtures. That is to say, the carbon yield of a pitch is the sum total of contributions of its three components, namely, QI, β resins and benzene solubles (BS) obtained by solvent fractionation as shown in Fig. 5.13. The carbon yields of QI and β-resin fractions are believed to be almost constant and independent of the origin of the pitch [39,40] being ≈95 and 85% respectively [41]. The carbon yield of the BS fraction is known to vary from 30 to 55% depending on the SP of the pitch. The higher the SP of the pitch the higher will be the average molecular weight of the BS fraction and, as a consequence, the higher will be its contribution towards the carbon yield of the pitch. The relationship for the carbon yield of a pitch can be written as [38]

$$\mathrm{CY} = \mathrm{QI} \times 0.95 + (\mathrm{BI\text{-}QI}) \times 0.85 + \mathrm{BS} \times K, \tag{5.1}$$

where CY is the carbon yield, QI and BI are quinoline insoluble and benzene insoluble fractions respectively, and K a constant such that $0.3 < K < 0.55$, which is dependent on the SP of the pitch.

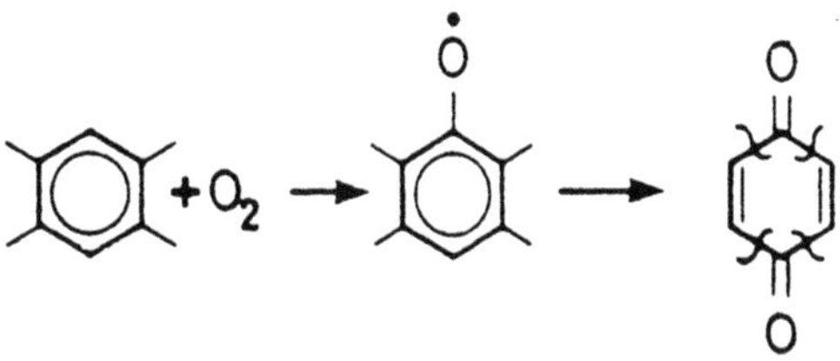

Fig. 5.14 Reaction of oxgen with aromatic hydrocarbons [46].

5.6 THE INFLUENCE OF ADDITIVES ON CARBONIZATION

The *carbonization reactions* of pitches can be considerably altered by the use of reactive additives, the most widely used of which is oxygen [42]. Oxidation treatment is used to 'thermoset' the pitch in order to prevent bloating during carbonization and to increase its carbon yield by retarding the volatilization of low molecular weight compounds. Severe oxidation has been found to reduce significantly the graphitizability of the pitch [43,44] and inhibit mesophase development [45]. Mild oxidation has been demonstrated to induce dehydrogenative polymerization without necessarily adversely affecting graphitizability [45]. Electron spin resonance studies of the oxidation of aromatic hydrocarbons in air identified aryloxy radicals (Fig. 5.14) as polymerization intermediates [46,47]. The decrease in graphitizability symptomatic of extensive oxidation, was believed due to the formation of quinones (Fig. 5.14) which pyrolyse with the loss of CO to produce non-planar radicals.

Sulphur has also been extensively used as an additive to improve carbon yield [48,49]. The final structure of the carbon may be related to the amount of sulphur added. High sulphur contents produce non-graphitizing carbons whereas for lower levels the reverse is true. Sulphur-containing radicals have been proposed as intermediates in these reactions [50]. The reaction of sulphur with aromatic hydrocarbons is thought to produce sulphur-containing polymers in which the sulphur is stable to high temperatures [51]. A number of other additives, such as $AlCl_3$ [52], alkali metals [53], anhydrides [54] and $FeCl_3$ [55], have also been used to convert low molecular weight aromatic hydrocarbons to carbon and improve the carbon yield of pitch, with varying degrees of success.

5.7 CONTROL OF MICROSTRUCTURE IN PITCH-DERIVED CARBON–CARBON COMPOSITES

There are three basic methods for the production of carbon–carbon composites, namely infiltration by CVD (Chapter 3) , impregnation with a thermosetting resin (Chapter 4) and impregnation with pitch. The microstructure of the matrix carbon produced by CVD may be controlled by the deposition temperature and partial pressure of the hydrocarbon gas to obtain isotropic, rough laminar and smooth laminar textures. The microstructure of the matrix derived from a thermosetting resin is mainly isotropic at HTTs below 1500 °C and coaxially aligned above 2300 °C.

Control of the microstructure is necessary to optimize thermomechanical properties [56]. The microstructure of a pitch-derived carbon matrix is generally dominated by the character of the pitch [57]. The complex polycylic aromatic nature of pitch results in detailed information with respect to carbonization being difficult to obtain. It is possible, however, to subdivide a pitch into a number of fractions of similar molecular weight distributions as defined by their solubility in certain solvents. It is apposite to break down the pitch into three groups: quinoline insolubles, β resins and benzene solubles. Kimura *et al.* [56] report approximate molecular weight ranges for the QI fraction of 1500 and above, 500–1500 for the β resins and between 300 and 460 for the benzene solubles. Carbonization of the benzene solubles produces a strongly anisotropic coarse-flow microstructure texture (Chapter 1), β resins carbonize to a 'coarse mosaic' texture with an average anisotropic domain size of ≈100 μm, whereas the QI fraction produces an isotropic carbon [58]. The BS, β resin and QI fractions are observed to be carbonized in a liquid phase, a more viscous liquid phase and a solid phase respectively. The extent of the orientation of the basal plane of the resultant carbon is proposed to be strongly affected by the difference in the mobility of polycyclic aromatic molecules during the carbonization, suggesting that the microstructure is governed by the viscosity of the pitch. A fractionation and mixing technique can be used to control the viscosity of the pitch by variation of the molecular weights and by the mixing-in of non-fused fine particles. Both techiques are therefore useful methods in the microstructure control of carbon–carbon composites formed from pitch.

A difference is observed between carbons derived from BS fractions themselves and those in fibre-reinforced composites [56]. A longer range of preferred orientation is observed in the composites compared with pitch alone. This is believed to result from a combination of preferred growth of mesophase spherules on carbon fibres, the frictional force near the fibres during mesophase movement and the shear force due to the shrinkage of the matrix. The carbon produced from the β resins, on the other hand, displays a decrease in the size of anisotropic domains when combined with carbon fibres, perhaps due to the fibres increasing the fraction's viscosity by acting as an obstacle to molecular movement. The microstructure of QI-derived carbon shows no changes in the presence of fibres.

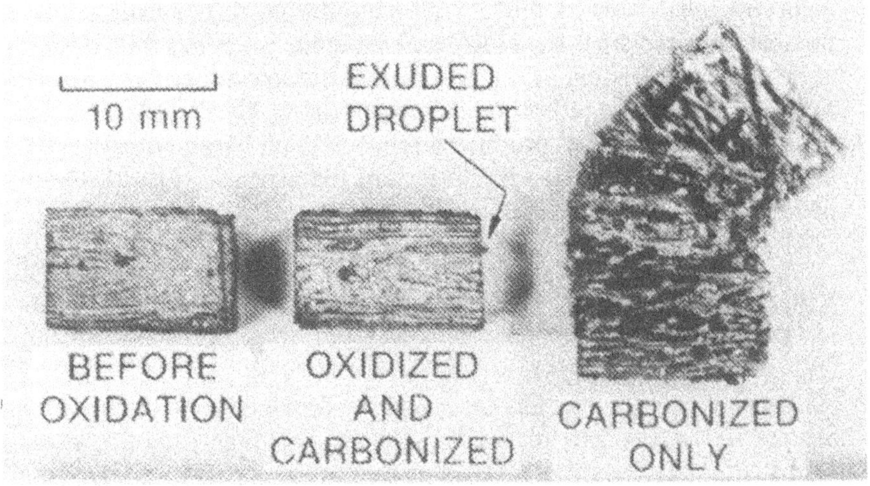

Fig. 5.15 The effect of prior oxidation on the carbonization of 2-D pitch-based carbon–carbon [61] (reproduced from *Carbon, an International Journal*, by kind permission of Pergamon Press).

5.8 LOW-PRESSURE COMPOSITES PROCESSING

Mesophase pitches are known to bloat seriously upon carbonization even under substantial applied pressure. Studies of petroleum coking have shown foaming to commence with the coalescence of bulk mesophase [59, 60]. If carbon–carbon composites are to be efficiently processed from pitch matrix precursors it is paramount that the matrix be stabilized in place prior to carbonization [61]. White and Sheaffer [61] carried out an investigation into the feasibility of stabilizing the pitch matrix by an oxidation process analogous to that used to stabilize pitch fibres prior to carbonization [62]. They postulated that it would be possible to stabilize the pitch within a fibre preform provided the shrinkage cracks, resulting from the severe mismatch in thermal expansivities between fibre and unhardened mesophase, allowed sufficient connected access for the entry of oxygen and evolution of gaseous oxidation products. Their initial work utilized orthogonal 3-D fibre preforms impregnated with Ashland A240 petroleum pitch. Specimens were oxidized in oxygen for up to 100 h at 222 °C followed by carbonization to 1100 °C. Subsequent microscopic investigation revealed that oxidation not only prevented bloating of the samples but also retained the mesophase spherules. Further experiments were carried out using 2-D woven fabric preforms, such materials being highly vulnerable to bloating due to lack of reinforcement in one direction. Matrix bloating destroyed unoxidized samples whereas those treated as before remained intact (Fig. 5.15).

White and Sheaffer concluded that oxidation processing of pitch-based composites was able to prevent bloating during carbonization, preserve the matrix microstructure established while the pitch was fluid or deformable and, by restriction of the volatilization of low molecular weight compounds, increase the carbon yield. An obvious application of this technique would be the ability to produce large 2-D pitch-based structures hereto impossible due to the size limitations of the HIP machines (Section 5.9) presently used in their fabrication.

5.9 HIGH-PRESSURE PROCESSING OF PITCH-DERIVED CARBON–CARBON

At atmospheric pressure the carbon yields obtained from pitches are only around 50% by weight, i.e. similar to those from high-yield thermosetting resins, although oxygen-stabilized pitches are reported to produce carbon yields in excess of 70% [61]. Yields as high as 90% can be obtained, however, by carbonizing the pitch under high pressure [63]. The effect of carbonization pressure on carbon yield is illustrated in Fig. 5.16 [64].

Pressure applied during pyrolysis also affects the matrix microstructure formed. Figure 5.17 shows micrographs of a petroleum pitch carbonized under an applied pressure of 6.89 and 68.9 MPa respectively followed by heat treatment to 2200 °C [35]. The coke formed at the lower pressure is needle-like, possibly due to deformation of the mesophase by gas bubble percolation. The higher the pressure the more coarse and isotropic will be the microstructure. This is thought to be due to the suppression of gas formation and escape during the carbonization reaction. High pressures are also observed to lower the temperature at which mesophase formation occurs. At very high pressures (≈200 MPa) coalescence of mesophase does not occur, so that an optimum of around 100 MPa is generally chosen for high-pressure processing of carbon–carbon [65].

The characteristics of pitches when chosen as matrix precursor for carbon–carbon processing may be summarized as follows:

1. Carbon yields at atmospheric pressure are about 50% by weight. Carbonization under high pressure (≈100 MPa) can result in yields in excess of 90% for some pitches.
2. Carbon microstructures are graphitic.
3. Matrix density is high (≈2 g cm^{-3}).
4. Applied pressure affects matrix microstructure.

5.9.1 Hot isostatic pressure impregnation carbonization

As a result of the low carbon yields generally obtained, all liquid impregnation techniques essentially follow three steps to completion:

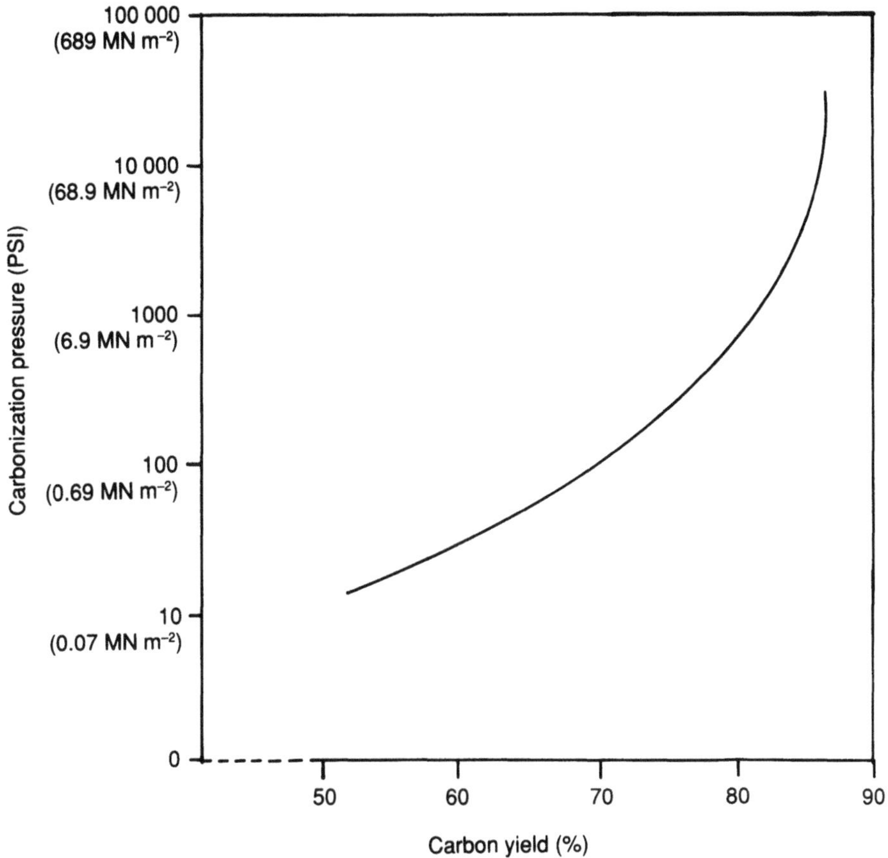

Fig. 5.16 Effect of carbonization pressure on carbon yield from petroleum pitch.

1. Impregnate fibres and form basic shape.
2. Carbonize to form porous carbon structure.
3. Reimpregnate/carbonize to improve density and hence properties, this can involve multiple steps before the component is complete.

It is thus fairly easy to conclude that the cost of carbon–carbon is not controlled by material costs as in carbon-fibre reinforced plastics, but that the greatest cost lies in fabrication. Carbon yields of 50–60% achieved with thermosetting resins, although low, represent a conversion efficiency of around 95% of the carbon actually available.

The only practical route available to lowering the cost of carbon–carbon fabrication is to attempt to improve the carbon yield and processability of pitches or other high-carbon-content compounds.

As shown in Fig. 5.16, the application of pressure during the carbonization of pitch can increase the carbon yield from 50% at ambient pressure to over 90% at 100 MPa. This pressure effect on the carbon yield of

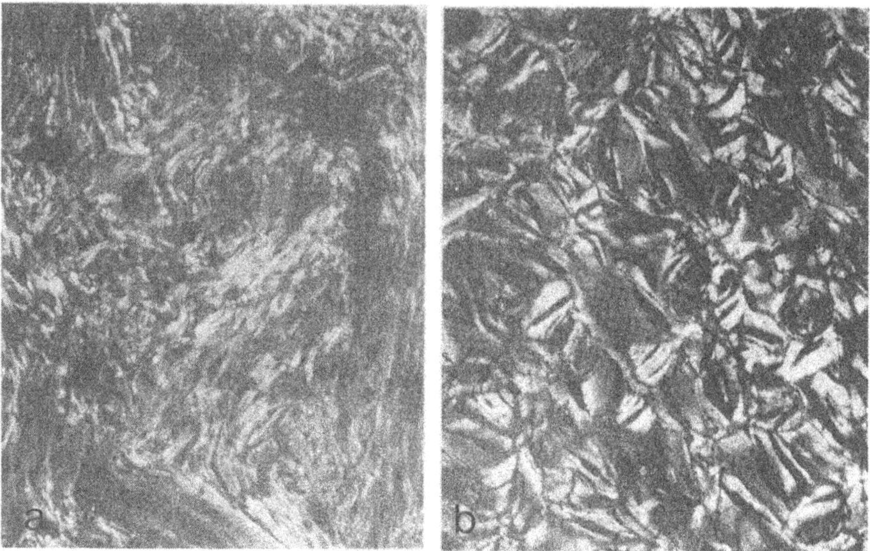

Fig. 5.17 Microstructure formed by petroleum pitch carbonized under (a) 6.89 and (b) 68.9 MPa applied pressure – subsequently graphitized to 2700 °C [35] (reproduced by kind permission of Elsevier Science Publishers).

pitches has become the basis of a process for the densification of carbon–carbon composites known as hot isostatic pressure impregnation carbonization, or HIPIC for short. The technique uses high isostatic inert gas pressure effectively to impregnate and densify carbon–carbon composites during the melting and carbonization stages of the pyrolysis cycle [66,67]. A schematic diagram of a complete HIPIC densification cycle is shown in Fig. 5.18.

5.9.2 Equipment and process requirements

The traditional hot isostatic press (HIP) design consists of a large externally water-cooled pressure vessel within which is situated a furnace surrounded by thermally insulating material referred to as a thermal barrier (see Fig. 5.19). In this design, pressurizing hot zone gases and reaction products have free access to all areas within the pressure vessel.

During the carbonization stage of the HIPIC process, reaction products which fall into one or more of the following categories are produced:

1. carbonaceous materials forming 'sooty' deposits which are otherwise chemically inert and present no threat to the mechanical integrity of the pressure vessel, e.g. condensed carbon black and soot particles;
2. chemically active materials which may produce a corrosive environment within the pressure vessel, e.g. ammonia;

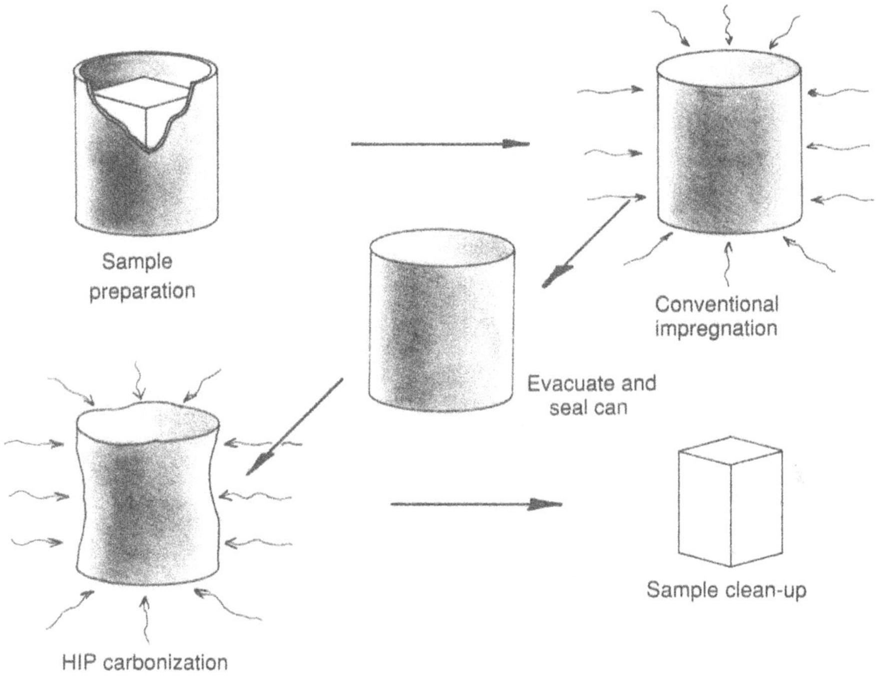

Fig. 5.18 Schematic diagram of HIPIC process.

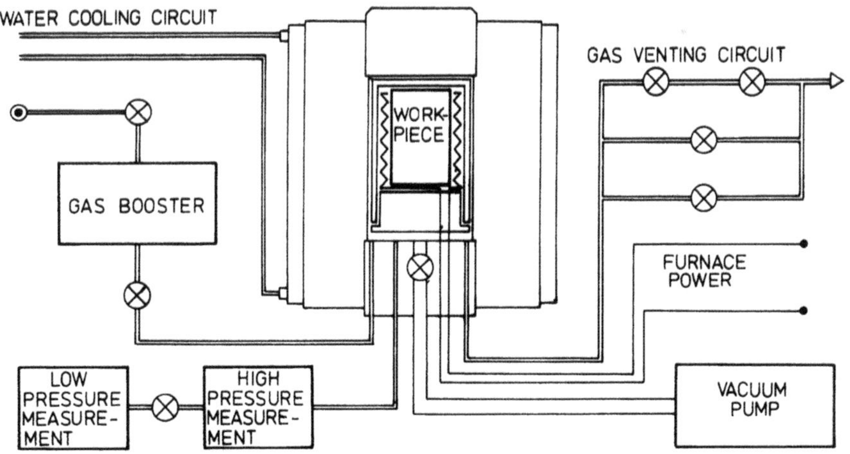

Fig. 5.19 Schematic diagram of standard HIP unit.

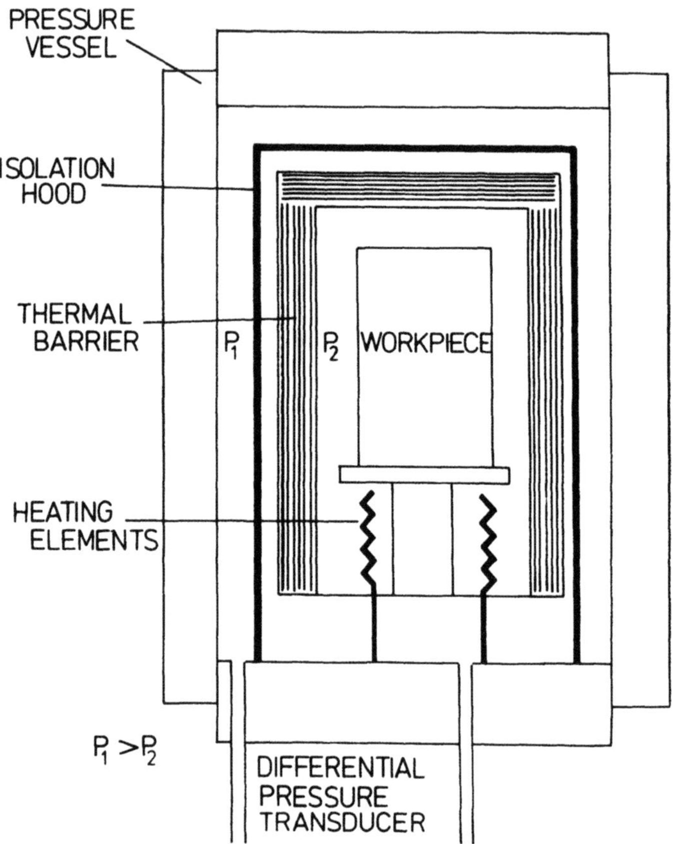

Fig. 5.20 Schematic diagram of the two gas isolation hood and differential pressure transducer gas management system developed for HIPIC processing.

3. atomic and/or molecular hydrogen which under certain conditions can result in embrittlement of the mild steel pressure vessel [68].

The risk of hydrogen embrittlement is small but nevertheless deserves attention. The effects of the reaction products in groups 1 and 2 present more of a cleanliness and good operating procedure problem. In order to combat these problems, the HIPIC process must be carried out in a specially designed HIP fitted with a novel gas management system [69], a schematic of which is shown in Fig. 5.20. Gas control depends upon the establishment of a differential pressure between the work zone and the pressure vessel inner wall. A differential pressure transducer operating at a system pressure of up to 2000 bar establishes and maintains a slight overpressure of 1.5 bar on the outside of the isolation hood. This has the effect of isolating the hot zone gases, and causing all gas movement to be from the outside in. Careful differential pressure control by a microprocessor leads to a clean, reliable and inherently safer impregnation and carbonization cycle. A dry fibre preform or porous carbon–carbon laminate is vacuum-impregnated with molten pitch, placed inside a metal container, or can, and surrounded by an excess of pitch. The can is then evacuated and sealed (preferably using an electron beam weld).

The sealed can is placed within the work zone of the HIP unit and the temperature raised, at a programmed rate, above the melt point of the pitch, but not so high as to result in weight loss due to the onset of carbonization. The pressure is increased and maintained at around 100 MPa (1.5 kbar).

The pitch initially melts, expands within the can and is forced by isostatic pressure into the pores in the sample. A sealed container will act like a 'rubber bag' and aid the pressure transference to the workpiece. As the pitch begins to carbonize the high isostatic pressure will maintain the more volatile fractions of the pitch impregnant in a condensed phase. The pressure not only increases the carbon yield but also prevents liquid from being forced out of the pores by pyrolysis products [64]. After processing the preforms are removed from their container and cleaned up by removing any excess carbonized liquid from the surface. The matrix may be graphitized by controlled heating to temperatures above 2300 °C [69]. The complete cycle as shown in Fig. 5.18 must be repeated until the required composite density is achieved.

In order to fabricate high-quality composites with good mechanical properties it is essential that the carbonization process be carried out slowly (over 1–3 days) and under strict process control. The control subsystem of an HIP unit used in carbon–carbon technology must integrally link, monitor and maintain all of the time temperature and pressure conditions intrinsic to the process. Because of the long time periods involved it is requisite that the entirety of the operation be controlled by a micro- or minicomputer from a detailed program input by the operator. In the interests of product quality, it is preferable to have the capability to log all of the data with respect to each carbonization cycle. Small irregularities introduced into the composites during the various fabrication subcycles may be very detrimental to the properties of the final component.

5.10 THERMOPLASTIC POLYMER MATRIX PRECURSORS

Recent work [70] has demonstrated the capability of fabricating carbon–carbon composites from a variety of exotic polyaromatic thermoplastic resin precursors such as polyetherether ketone (PEEK) and polyetherimide (PEI) using HIP technology. Over the last 5 or 6 years a novel enabling technology has evolved for the processing of composites made from the aforementioned resins reinforced with carbon fibres [71]. A family of carbon

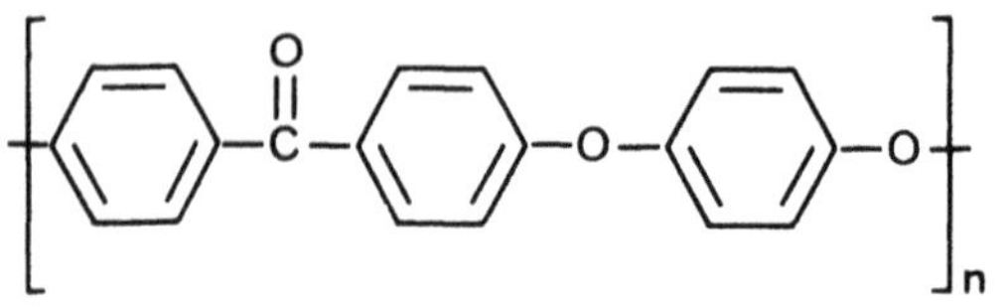

(a)

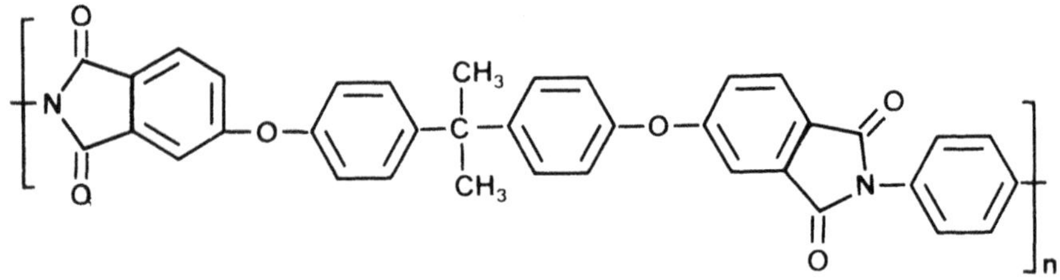

(b)

Fig. 5.21 'Exotic' thermoplastic carbon matrix precursors: (a) PEEK; (b) PEI.

fibre/thermoplastic prepregs is produced commercially ICI under the trade name 'APC', which stands for Aromatic Polymer Composite [72], as an intermediate to component manufacture in applications such as the aerospace industry, where weight saving is of paramount concern. Such materials are of interest as carbon–carbon precursors because of the attractive C/H ratios of the resin matrices resulting from a high degree of aromacity (Fig. 5.21). There is thus the potential of high carbon yields (i.e. >70%).

The thermoplastic nature of the precursors, whilst aiding component fabrication [71], requires the application of pressure during carbonization to maintain the integrity of the sample [70]. It has been found that good quality carbon–carbon materials can be made by hot isostatic pressing of these precursors without the need of the encapsulation essential when pyrolysing pitch precursors. It is, as yet, not fully clear what is the precise mechanism behind the achievement of such results. It is believed, however, that the resin precuror, being thermoformable, flows sufficiently during the initial part of the process to form an effective seal around and/or throughout the workpiece whereby the isostatic pressure is applied to the whole of the workpiece, yet enabling volatiles generated in the pyrolysis to diffuse out without significant distortion.

Although expensive (the prepreg costs from £120 kg^{-1} upwards) the properties observed in carbon–carbon from the 'APC' family of materials is generally superior to those derived from phenolic matrix composites even after multiple reimpregnation/carbonization cycles (Table 5.3).

The microstructures observed in these composites differ considerably from their more 'conventional' counterparts. The matrix component of the composites, heat treated to 1000 °C, although isotropic in nature, retains many of the characteristics of the polymer from which it was derived (Fig. 5.22) as indicated by the relatively 'strong' bonding observed between

Table 5.3 Mechanical properties of unidirectional carbon–carbon composites

Precursor	*Density* (g cm^{-3})	*Flexure strength* (MN m^{-2})	*Flexure modulus* (GN m^{-2})	*Transflexure strength* (MN m^{-2})
Carbon/PEEK	1.55	900	190	10
Carbon/PEI	1.47	563	135	4
Carbon/phenolic	1.40	630	130	10
Carbon/phenolic after 4 impregnation cycles	1.46	950	178	10

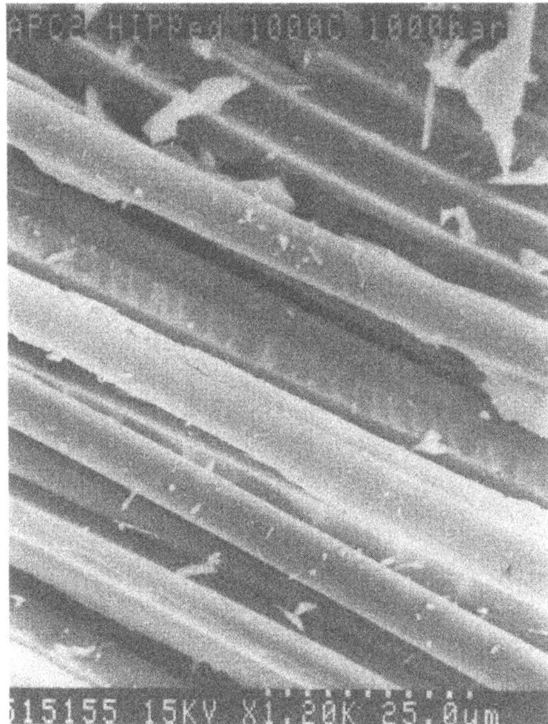

Fig. 5.22 Fracture face of carbon–carbon made from APC-2 precursor.

fibre and matrix. A number of techniques may be employed to improve the bonding between fibre and matrix (Table 5.4) such as pre-oxidation of fibres, heat treatment to 'pseudo-cross-link' the polymer matrix and/or the addition of heterogeneous cross-linking agents to the materials such as sodium. It is possible to obtain a very strong, almost 'polymeric' type of bonding at the fibre/matrix interface (Fig. 5.23). Unfortunately, when inducing such bonds, one is confronted by the classic brittle fibre/ brittle matrix problem; cracks propagating through a brittle matrix under

Table 5.4 Effect of precursor–pretreatment on the fibre/matrix interface in UD carbon–carbon produced from CF/PEEK

Pre-treatment to precursor	*Transflexure strength* (MPa)
Unoxidized carbon fibres	4
Standard fibre oxidation treatment (APC 2)	10
4 × standard fibre oxidation treatment	12
Resin cross-linked by heat treatment	17
Resin cross-linked by chemical additions	25
Resin cross-linked by chemical addition and heat treatment	32

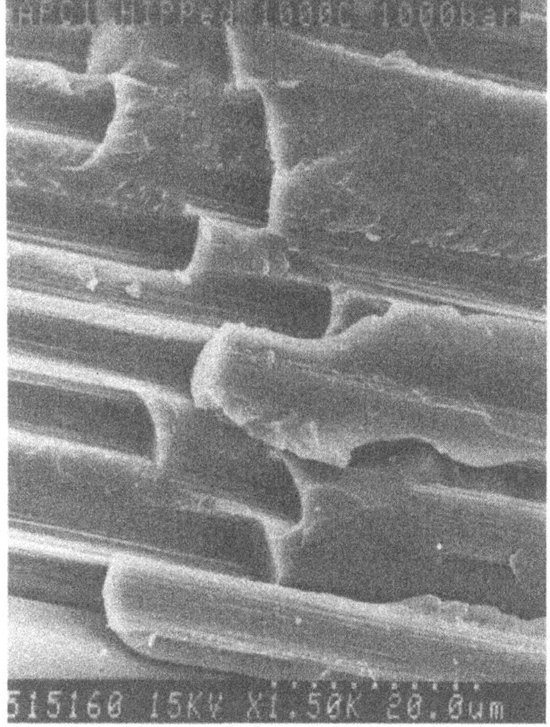

Fig. 5.23 Transflexure fracture face of cross-linked carbon/PEEK derived carbon–carbon showing strong 'polymeric' fibre/matrix interface.

Table 5.5 Comparison of ultra-high modulus carbon–carbon with that made from other precursors

Precursor	*Density* (g cm^{-3})	*Transflexural strength* (MN m^{-2})	*Flexural strength* (MN m^{-2})	*Flexural modulus* (GPa)
APC-2	1.55	10	900	190
P75/PEEK	1.83	16	600	325
Phenolic	1.40	10	630	130

mechanical stress are not deflected at a strong interface and carry straight on through the brittle fibre resulting in catastrophic failure at relatively low levels of loading. A weak interface, on the other hand, diminishes the efficiency of the transfer of load from matrix to fibres. Carbon–carbon composites will thus always exhibit lower strengths than the materials from which they were derived. A balance must therefore be struck between effective load transfer and tendency to brittle failure. The interface in thermoplastic-derived materials may be 'engineered' with a good degree of accuracy and reproducibility, thus allowing the required optimization.

In space applications, such as satellite structures, the most important design parameter is specific stiffness, i.e. modulus per unit weight. To attack that market niche, several products have been developed which are composites reinforced with ultra-high modulus carbon fibres made from mesophase pitch precursors [73–75]. In that area, ICI have developed a prepreg material similar to APC-2 but which employs high modulus (517.5 GPa) Thornel P75 fibres [76]. Since APC-2 had proved a very useful carbon–carbon precursor, it was a logical step to investigate the possibility of forming similar materials using the ultra-high modulus product.

Table 5.5 shows a comparison of the properties of carbon–carbon from P75/PEEK with those from other precursors. The long flexure fracture faces in Figs. 5.24 and 5.25 show a transition between pull-out and transverse cracking modes of failure. What is most apparent when viewing the micrographs is the very low level of voidage as confirmed by the high density obtained (although some of the increase is due to the higher fibre density). The matrix, although glassy and isotropic in nature, again exhibits a 'polymeric' type fracture behaviour in that it is not catastrophically cracked away from the fibres during testing. The flexure strength of the P75-based material represents a drop of only 25% over its polymeric 'parent', whereas the APC-2 derived material is more than 50% lower than that of its precursor, indicating a more efficient stress transfer at the fibre/matrix interface.

It is known that graphitization can be induced in a thermoset-based matrix for carbon–carbon and that thermoplastic polymers and pitches

Fig. 5.24 Axial flexure fracture face of P75/PEEK carbon fibre reinforced carbon.

form a graphitizable carbon. The transition between amorphous and crystalline carbon occurs slowly up to 2000 °C. Above this temperature defects become highly mobile, resulting in a dramatic increase in the rate of ordering. It has been suggested that graphitization, in sheaths oriented along the fibre axis, results from a combination of thermally activated processes and stress induced by the difference in thermal expansion coefficients between the fibre and matrix. This process is usually thought to commence at around 2000 °C. Observations made during TEM studies indicate that the process begins at temperatures below 1600 °C when PEEK is HIPed [77]. It is

Fig. 5.25 Texturing of carbon microstructure in P75/PEEK derived carbon–carbon composite indicative of ordering of matrix microstructure at higher temperatures (HTT 1600 °C).

possible that the application of external pressure during pyrolysis aids crystallographic alignment. One can detect a gradual texturing of the matrix structure in the electron micrograph in Fig. 5.25.

5.11 SUMMARY

The fabrication of carbon–carbon composites from thermosetting resin precursors is severely limited due to the relatively low carbon yields obtainable from resins such as phenolics and furans. Acetylene-terminated resins, although possessing high potential carbon yields, are beset by prohibitively high raw materials costs. The use of pitches as matrix precursors is an extension of the technology used in the graphite electrode processing industry. A large data base covering impregnation, carbonization and graphitization of both coal-tar and petroleum pitches is therefore readily available. Pitches have a low SP, low melt viscosity, high carbon yield and tend to form graphitic carbon structures. The complex oligomeric nature of pitches results in their being difficult to categorize with any great degree of accuracy, thus presenting a severe problem of quality control in the final product. It is, however, possible to divide the pitch into three subgroups: QI, β-resins and benzene solubles, dependent on their molecular weight as 'defined' by solubility in certain solvents. As a general rule of thumb, the greater the proportion of the high molecular weight species the higher will be the carbon yield. The SP of the pitch is another indication of carbon yield; the higher the SP the greater the average molecular weight of the pitch and thus the higher will be its carbon yield. As an example, one may consider the Ashland series of refined petroleum impregnation pitches A240, A400 and A420 where the figures represent the pitches' SPs in °F, A420 having the highest carbon yield.

Pitch-based matrices thus offer higher yields than do the phenolic resins conventionally used in 2-D composites, and are generally much lower in cost than polymer systems of equivalent yield. Furthermore, pitch-based systems exhibit the potential for a wider variety of matrix microstructures, ranging from glassy carbon to fully graphitizable mesophase microstructures. The thermoplastic nature of the pitches results in severe bloating and loss of low molecular weight compounds when carbonized under ambient conditions. Oxidation treatment to cross-link the pitch can be used to eliminate the bloating problem and increase the carbon yield. The true potential yields of pitches can, unfortunately, only be achieved by the application of high pressure in specially designed HIP apparatus. Exploitation of the HIP process is limited by the physical size of the equipment and the high initial capital investment requirement to set up a production facility, but it is probably the most efficient route to the manufacture of good quality, thick sectioned, high density artefacts such as rocket nozzles and re-entry vehicle nose cones. Pitches are also useful as reimpregnants in thermoset-based processing when they are often blended with the resin to improve carbon yield.

Recently it has been shown that an HIP is capable of converting thermoplastic resin matrix composites to carbon–carbon. Furthermore the materials so formed have generally exhibited superior properties (density, strength, stiffness, etc.) to those derived from the more 'traditional' phenolic precursors. These observations are particularly true when the precursor material is a combination of fibres derived from the mesophase pitch (Thornel P75) and a PEEK matrix. Microstructural observations have shown that the carbon–carbon composites formed by HIPing this relatively novel precursor retain a good deal of their 'polymeric' characteristics (e.g. 'well wetted' fibre/matrix interfaces and low porosity), while acquiring the high-temperature properties unique to 'all-carbon' systems.

The properties measured for the HIPed materials compare favourably and often exceed those of competitive materials (even after those competitors have been subject to a series of densification cycles). It is fair to say then that the composite produced, and particularly those from the P75/PEEK precursor, may be considered as near-finished materials and therefore ready for exploitation as prototype components. Although the properties of carbon–carbon from the APC genre of composites have good properties they will be prohibitively expensive for many applications because of the high cost of raw materials. It is therefore important to target applications requiring 'high-added-value' artefacts. Applications where further research effort could/should be placed are: fasteners, biomedical implants and, most especially, space structures where high costs can be justified by the reduction of launching costs.

REFERENCES

1. Fitzer, E. (1987) *Carbon*, **25**, 163.
2. Brooks, J. D. and Taylor, G. H. (1968) *Chem. Phys. Carbon*, **4**, 243.
3. White, J. L. (1975) *Prog. Solid-State Chem.*, **9**, 59.
4. Zimmer, J. E. and White, J. L. (1982) *Adv. in Liq. Cryst.*, **5**, 157.
5. White, J. L. and Buechler, M. (1986) in *Petroleum-derived Carbons*, **303**, Am. Chem. Soc. Symposium Series, Washington DC, p. 62.
6. Blades, H. (1972) US Patent 3,767,756.
7. Smith, F. A., Eckle, T. F., Osterholm, R. J. and Stichel, R. M. (1966) in *Bituminous materials*, Vol. **3** (ed. A. J. Hoiberg), Interscience, New York, p. 57.
8. Newmann, J. (1976) in *Petroleum Derived Carbons* (eds M. L. Deviney and T. M. O'Grady,) AC5 Sym. Ser. 21. p. 52.
9. McNeil, D. (1966) in '*Bituminous Materials*, **3** (ed. A. J. Hoiberg), Interscience, New York, p. 139.
10. Franck, H. G. (1963) in *Chemistry and Coal Utilization*, supplementary vol. (ed. H. H. Lowry), Wiley, New York, p. 592.
11. Smith, J. W. (1966) *Fuel*, **45**, 233.
12. Halleux, A. and de Greef, H. (1963) *Fuel*, **42**, 185.
13. Vahrman, M. (1950) in *Progress in Coal Science*, (ed. D. H. Bangham), Interscience, New York, p. 60.
14. Van Krevelen, D. W. (1950) *Fuel*, **29**, 269.
15. Phillips, G. and Wood, L. J. (1955) *J. App. Chem. (Lond.)*, **5**, 326.
16. Brooks, J. D. and Steven, J. R. (1964) *Fuel*, **43**, 87.
17. Friedel, R. A. (1966) in *Applied Infrared Spectroscopy*, (ed. D. N. Kendall), Wiley, New York, p. 312.

18. Lewis, I. C. and Singer, L. S. (1969) Preprints of fuel division, *Am. Chem. Soc.*, **13**, 86.
19. Friedel, R. A. (ed.) (1970) *Spectrometry of Fuels*, Plenum Press, New York.
20. Greinke, R. A. and Lewis, I. C. (1975) *Anal. Chem.*, **47**, 2151.
21. Bartle, K. D., Collin, G., Stadelhofer, J. W. and Zander, M. (1979) *J. Chem. Tech. Biotechnol.*, **29**, 531.
22. Edstrom, T. and Petro, B. A. (1968) *J. Poly. Sci. Part C*, **21**, 171.
23. Tillmanns, H., Ulsamer, W. and Pietzka, G. (1976) *Carbon 76*, Int. Carbon Conf. Preprints, 2nd p. 557.
24. Dickinson, E. M. (1980) *Fuel*, **59**, 290.
25. Lewis, I. C. (1982) *Carbon*, **20**(6), 519.
26. Badger, G. M. (1965) *Prog. Phys. Organ. Chem.*, **3**, 1.
27. Carrington, A. and Smith, I. C. P. (1965) *Mol. Phys.*, **9**, 137.
28. Stein, S. E., Golden, D. M. and Benson, S. W. (1977) *J. Phys. Chem.*, **81**, 314.
29. Stein, S. E. (1981) *Carbon*, **19**, 421.
30. Livingstone, R., Zeldes, H. and Conradi, M. S. (1979) *J. Am. Chem. Soc.*, **101**, 4312.
31. Lewis, I. C. (1980) *Carbon*, **13**, 191.
32. Lewis, I. C. and Singer, L. S. (1981) *Chem. Phys. Carbon*, **17**, 1.
33. Brooks, J. D. and Taylor, G. M. (1967) *Carbon*, **3**, 185.
34. Zimmer, J. E. and Weitz, R. L. (1988) *Carbon*, **26**(4), 579.
35. Burns, R. L. and McAllister, L. E. (1976) *Proc. 12th Propulsion Conf.*, Palo Alto.
36. Manocha, L. M., Bhatia, G. and Bahl, O. P. (1982) *Proc. 1st Indian Carbon Conference*, New Delhi, December, p. 325.
37. Bahl, O. P., Manocha, L. M., Singh, Y. K., Bhatia, G., Aggarwal, R. K. and Dhami, T. L. (1986) *Proc. 4th Int. Carbon Conf.*, Baden-Baden, June, p. 38.
38. Bhatia, G., Aggarwal, R. K. and Bahl, O. P. (1987) *J. Mat. Sci.*, **22**, 3847.
39. Blackley, T. H. and Earp, F. K. (1958) *Industrial Carbon and Graphite*, Soc. of Chem. Ind., London.
40. Chari, S. S., Bhatia, G. and Aggarwal, R. K. (1978) *J. Sci. Ind. Res.*, **37**, 502.
41. King, L. F. (1978) in *Analytical Methods for Coal and Coal Products*, **2** (ed. C. Karr), Academic Press, New York.
42. Fitzer, E., Meuller, K. and Schaffer, W. (1971) in *Chemistry and Physics of Carbon*, **7** (ed. P. L. Walker jr), Marcel Dekker, New York, p. 237.
43. Kipling, J. J., Sherwood, J. N., Shooter, P. V. and Thompson, N. R. (1964) *Carbon*, **1**, 315.
44. Otani, S. (1965) *Carbon*, **3**, 31.
45. Barr, J. B. and Lewis, I. C. (1978) *Carbon*, **16**, 439.
46. Lewis, I. C. and Singer, L. S. (1981) *J. Phys. Chem.*, **85**, 354.
47. Lewis, I. C. and Singer, L. S. (1969) *Proc. 9th Biennal Conf. on Carbon*, Boston, USA, June, p. 120.
48. Christie, N., Fitzer, E., Kalka, J. and Schafer, W. J. (1969) *Chem. Phys. Physicochem. Biol.*, p. 50.
49. Fitzer, E., Huttner, W. and Manocha, L. M. (1980) *Carbon*, **18**, 291.
50. Kipling, J. J., Shooter, P. V. and Yound, R. N. (1966) *Carbon*, **4**, 333.
51. Lewis, I. C. and Greinke, R. A. (1982) *J. Polym. Sci. Chem. Ed.*, **20**, 1119.
52. Mochida, I., Inoue, S., Maeda, K. and Takeshita, K. (1977) *Carbon*, **15**, 9.

53. Mochida, I., Nakamura, E., Maeda, K. and Takeshita, K. (1976) *Carbon*, **14**, 123.
54. Evans, S. and Marsh, H. (1971) *Carbon*, **9**, 747.
55. Otani, S. and Oya, A. (1972) *Bull. Jap. Chem. Soc.*, **45**, 623.
56. Kimura, S., Yasinda, K., Inagaki, M. and Yasuda, E. (1984) *Rpt. Res. Lab. Eng. Mats. Tokyo Inst. Tech. no. 9.*
57. Kamramura, K. Kimura, S., Yasuda, E. and Inagaki, M. (1982) *Tanso*, **109**, 46.
58. Sanda, Y. (1978) *Nenryokokaishi*, **57**, 117.
59. White, J. L. (1976) in *Petroleum-derived Carbons*, **21**, Am. Chem. Soc. Symp. Series, Washington DC, p. 282.
60. Weinberg, V. A., White J. L. and Yen T. F. (1983) *Fuel*, **62**, 1503.
61. White, J. L. and Sheaffer, P. M. (1989) *Carbon*, **27**(5), 697.
62. Singer, L. S. (1977) US patent 4,005,183.
63. Fitzer, E. and Terwiesch, B. (1973) *Carbon*, **11**, 570.
64. Lachmann, W. L., Crawford, S. A. and McAllister, L. E. (1978) *Proc. Int. Conf. on Composite Mats*, Met. Soc. of AIME, New York.
65. Gray, G., Hunter, A., Payne, R. S. and Savage, G. M. (1990) *Powder Met. Rept*, **45**(4), 290.
66. Burns, R. L. and Cook, J. L. (1974) *Petroleum-derived Carbons*, **21**, ACS Symp. Series, Am. Chem. Soc.,Washington DC, p. 139.
67. Chard, W., Conaway, M. and Neizz, D. (1974) *Petroleum-derived Carbons*, Vol. 21, ACS Symp. Series, Am. Chem. Soc., Washington DC, p. 155.
68. Gangloff, R. P. and Wei, R. (1977) *Metallurgical Transactions, A, USA*, 1043.
69. Dietrich, H. and McAllister, L. E. (1978) *Am. Ceram. Soc. 80th Ann. Meeting, Detroit.*
70. Norton-Berry, P., Savage, G. M. and Steel, (1990) M. L. Eur. Patent App. H35016/EP.
71. Gray, G. and Savage, G. M. (1989) *Metals and Materials*, Sept; 513.
72. ICI Fiberite Corp. (1986) *APC 2 Data Sheet no. 2*, (1986) Orange.
73. Silverman, E. M. and Jones, R. J. (1988) *Proc. SAMPE. Int. Symp.*, **33**, 1418.
74. Silverman, E. M. and Jones, R. J. (1988) *SAMPE J.*, **24**(4), 33.
75. Silverman, E. M., Sathoff, J. E. and Forbes, W. C. (1989) *SAMPE J.*, **25**(5), 39.
76. Barnes, J. A. and Cogswell, F. N. (1989) *SAMPE Quart.*, **20**(3), 22.
77. Payne, R. S., von Bradsky, G. Gray, G. and Savage, G. M. (1990) *Proc. Conf. Adv. in Elec. Mic.*, Seattle, Sept. p. 1024.

Oxidation and oxidation protection

6

6.1 INTRODUCTION

Over the past decade the development of structural carbon–carbon materials has received a great deal of attention. Potential uses have been cited in future generation military aircraft, missile systems and a number of proposed hypersonic aerospace vehicles. All of the possible applications take advantage of the excellent high-temperature properties of carbon and the benefits of fibre reinforcement, most especially high strength and strength retention at temperatures in excess of 2000 °C. It is important to note that all such applications involve operation for extended periods of time in oxidizing environments. Since the composites have already demonstrated the mechanical requirements, it is generally concluded that the development of reliable oxidation protection is crucial to carbon–carbon attaining its full potential. The method accepted as the most feasible way to protect carbon–carbon composites involves **coating** of the outer surfaces of the material with appropriate refractory materials in order to prevent oxygen attacking the substrate. Additionally, protective compounds, known as **inhibitors**, may be placed within the composite.

The provision of oxidation protection for carbon materials has been the subject of intense research over the last 50 years. The first noteworthy patent to be granted in this field was issued to the National Carbon Company in the USA in 1934 [1]. Up until roughly 25 years ago, the major driving force for oxidation protection was for polycrystalline graphite used as electrodes in the steel industry or for furnace heating elements.

In the 1960s, the 'space race' demanded ablation and oxidation-resistant graphites for rocket propulsion and as re-entry components. The emphasis in present-day studies is firmly on carbon–carbon composites. The first efforts in oxidation protection of carbon–carbon were conducted in the early 1970s and the stimulus was provided by the requirements of the space shuttle programme. The components involved, leading edges and nose-cap, used low-performance rayon-based fibre composites [2]. The investigation of methods for providing oxidation protection for high-performance structural carbon–carbon has only been ongoing over the last 10 or so years. A 'potted history' of the development of carbon oxidation protection systems is given in Fig. 6.1.

The earlier mentioned National Carbon Company patent claimed a dual protection system based on an inner silicon carbide layer and an outer glaze of B_2O_3. It has since been recognized that glassy materials are an important constituent of oxidation protection systems for carbon, a whole host of patents having been granted using variations of this theme. The original patent also exemplifies the possible use of P_2O_5 and SiO_2 glasses. P_2O_5 glasses can be effective but are limited in operation by volatilization to fairly low temperatures, while silica glasses do not generally possess the required wetting characteristics to afford protection. B_2O_3 glasses, on the other hand, have the ability to provide protection over a wide range of temperatures both in the form of surface coatings and inhibitors within the body of the composite. This ability arises from a combination of thermal stability and appropriate viscosity and wetting properties.

A prominent milestone in the ongoing 'saga' of oxidation protection of carbon was the development of JTA graphite by Union Carbide and a number of related materials [3–5]. Developed for the production of re-entry components for the US space programme, these materials contained non-oxide ceramics of boron, silicon, zirconium and hafnium. Oxidation produced a borate glass coating capable of providing protection for periods of several hours at temperatures of up to 1700 °C. A pure borate glass is only able to give primary protection for limited times above around 1100 °C due to the volatility of B_2O_3. Although limited in application, the work clearly demonstrated the effectiveness of boron additions (preferably in unoxidized form – to ease processing) to the body of a carbon in laying the foundations for further development.

A great many coatings, other than glasses and glass formers, have been added to carbon–carbon for high-temperature operation in oxidizing environments. Chown and his team investigated the suitability of coatings of SiC, produced both by CVD and direct reaction with molten silicon [6]. A number of other refractory carbide and boride coatings have also been investigated by Chown [6] and Fitzer *et al.* [7]. The results obtained indicated that silicon carbide was capable of providing reliable protection for long time periods, provided the temperature remained below 1700 °C and flaws in the coatings could be eliminated. SiC coatings are found to decompose rapidly above 1700 °C, whereas a sintered coating of zirconium carbide and boride (ZrC and ZrB_2) will allow short-term protection up to temperatures of roughly 2200 °C.

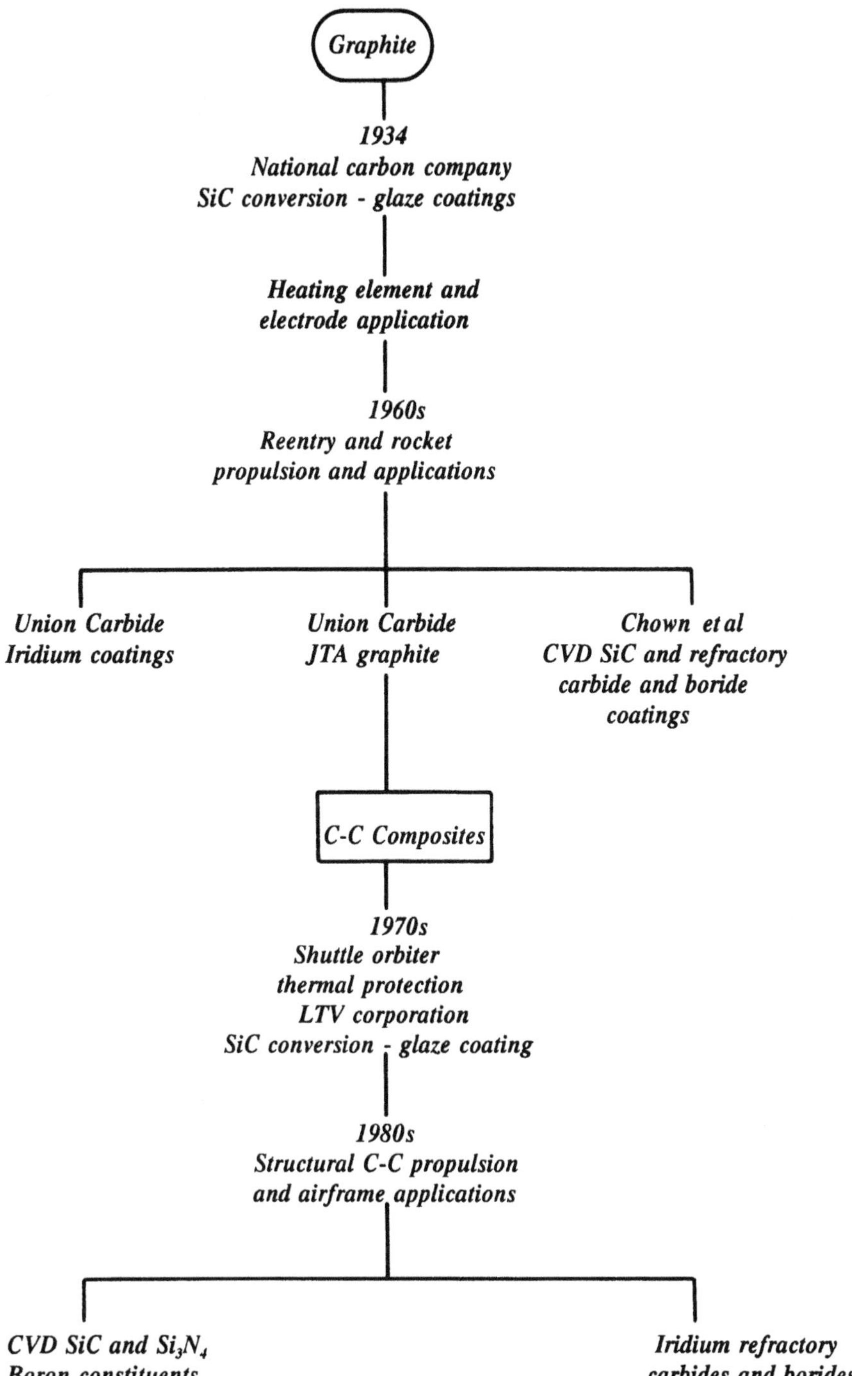

Fig. 6.1 Development of oxidation protection for carbon materials.

The temperature limit observed for SiC appears to apply to all silicon-based materials used in oxidation protection. Molybdenum disilicide ($MoSi_2$) and silicon nitride (Si_3N_4) also show this limitation which is probably endemic to the silicon family of materials. Higher-temperature refractories based on zirconium and hafnium are able to operate at temperatures well above silicon, but only for limited periods of time as a result of the rapid rates of diffusion of oxygen through these oxides.

In the middle of the 1960s iridium coatings were used to protect graphite for short periods of time in the 2000–2100 °C temperature range [8, 9]. Iridium melts at 2440 °C, has very low permeability to oxygen up to 2100 °C, does not react with carbon below 2280 °C, and forms an effective barrier to the diffusion of carbon. It is thus an attractive material for the basis of a high-temperature protection system. The problems associated with iridium include erosion due to the formation of a volatile oxide, a relatively high coefficient of thermal expansion and poor adhesion to carbon. At 2000 °C, in flowing air, an iridium coating has been observed to erode at a rate of ≈0.13 mm h^{-1} [10]. The erosion problem could, perhaps, be averted by combining the iridium with a refractory oxide coating, with an inner carbide layer to improve adhesion. Despite showing promise, iridium has never been deployed to any great extent due to high costs and limited availability.

The driving force for research into the protection of carbon–carbon arose in the 1970s with the development of leading edges and nose-caps for the space shuttle programme (Fig. 6.2). The carbon–carbon parts for the shuttle are made by the low-pressure resin technique from rayon-based fabric and a phenolic resin. Following three reimpregnation cycles with a phenolic resin, a ceramic coating of silicon and aluminium is added by pack cementation. The part is packed both inside and out with ceramic powder and fired to 1650 °C. A silicon carbide coating is formed on the top two layers of the laminate. To protect the coating from spallation, due to differential thermal cracking, the surface is impregnated with tetraethylorthosilicate (TEOS). The TEOS is cured to leave a silicon residue throughout the coating, further reducing the area of exposed carbon and providing a glass sealant for cracks in the carbide layer.

The coefficient of thermal expansion (CTE) of the rayon-based fibres used in the shuttle's re-entry parts is roughly twice that of the PAN-based fibres used in today's structural or 'advanced' carbon–carbon (ACC). Rayon-based carbon–carbon has a CTE closer to that of SiC or Si_3N_4 and thus has less mismatch than PAN systems. New structural applications require PAN-based materials which tend to amplify the problems of thermal shrinkage mismatch. Applications in jet engines demand longer and longer operating times at higher and higher temperatures. Exhaustive research is ongoing, developing newer materials and combinations of materials in an attempt to meet these new demands. The oxidation

Fig. 6.2 Oxidation-protected carbon–carbon re-entry components of US space shuttle.

protection systems presently in use are generally modifications of the earlier silicon-based systems, using boron [11–13].

All of the contemporary protection systems successfully operating at temperatures of 1500 °C or below, for extended time periods, use boron in one form or another. The benefits obtained from the inclusion of boron and its compounds in carbon–carbon composites are disclosed in a number of recent patents and papers [14–18]. Oxidation protection at temperatures above 1500 °C requires higher refractories as a result of the volatilization of B_2O_3. Short-term protection has been demonstrated at temperatures in excess of 2000 °C, but long-term protection above 1500 °C requires materials systems which have yet to be identified, let alone introduced into service. It is possible, however, to define some of the parameters and problems needing to be addressed in order to achieve the goal of a truly reusable high-temperature (>1500 °C) protection system.

6.2 OXIDATION BEHAVIOUR OF CARBON–CARBON

The oxidation of carbon has been studied, historically, in terms of the combustion of coal. 'Bouduard's reaction' and the steam gasification reaction have been used to describe carbon's oxidation by CO_2 and H_2O gases. At low to moderate temperatures, the rate-controlling steps of the following two reaction schemes are known to be chemical in nature [19,20]:

$$C(s) + CO_2(g) \rightarrow 2CO\ (g) \tag{6.1}$$

$$C(s) + H_2O(g) \rightarrow CO(g) + H_2(g). \tag{6.2}$$

Catalysts are thus effective in the acceleration of these oxidation reactions.

The rates of oxidation are found to increase exponentially with temperature. Above 700 °C the rate-controlling step generally changes to the diffusion of gaseous species through the boundary layer close to the solid carbon [21]. By contrast, carbon–carbon composites are usually used in air. The oxidation reaction therefore proceeds according to

$$2C(s) + O_2(g) \rightarrow 2CO(g). \tag{6.3}$$

The above reaction has a large driving force, even at very low oxygen partial pressures, as a result of the large negative value of Gibbs free energy change. The rate of oxidation is thus controlled not by the chemical reaction itself, but rather by the transport of the gaseous species to and from the reaction front.

Over a number of years, the Japanese workers Kimura and Yasuda have studied the oxidation behaviour of carbon–carbon between 650 and 850 °C using a thermobalance [22]. The relationship between percentage weight loss and time at a number of different temperatures for the oxidation of a 67% by weight of fibres composite (HTT 2800 °C) is shown in Fig. 6.3. The rate of oxidation increases with temperature but decreases with time. Yasuda *et al.* [22] found the matrix to be more reactive, with a slower oxidation of the fibres. Figure 6.4 shows the effect of composite HTT on the rate of oxidation. The rate clearly decreases with increase in HTT, despite a higher oxidation temperature. A number of other workers have observed the same phenomena independently. In Fig. 6.4, the rate of oxidation, as defined by the slope of the curves, is increasing with time.

An Arrhenius plot (log gasification rates versus $1/T$) for several carbon–carbon materials shows the rate of oxidation to decrease with increase in HTT (Fig. 6.5). At the lower temperatures (below around 600–800 °C) oxidation is controlled by the reaction of oxygen with active sites on the carbon surface [23]. At higher temperatures the rate-limiting step is the diffusion of oxygen through the boundary layer at the surface of the composite [24]. The reaction-controlling steps of the oxidation of carbon–carbon are thus very similar to that of graphite [23].

The general rate of reaction dw/dt, in the zones of the chemical and diffusion-controlled steps respectively, may be expressed as follows [23]:

$$\frac{dw}{dt} = \frac{R}{n} K_s S_v C_r^m \phi + C_r^m K_s f \tag{6.4}$$

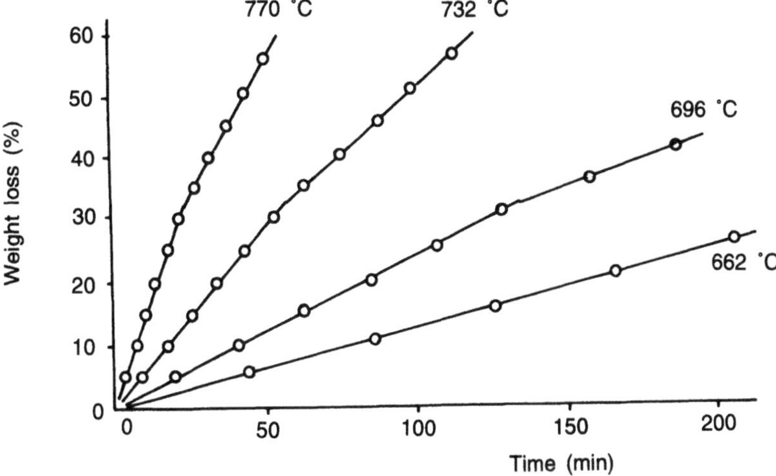

Fig. 6.3 Oxidation of carbon–carbon composite in air [22].

$$\frac{dw}{dt} = \frac{(C_g - C_r)}{\delta} D_f, \tag{6.5}$$

where R is the thickness for a planar sample and radius for a cylinder or sphere, n is 1, 2 or 3 for a plane, cylinder or sphere respectively, K_s the rate constant per unit of reacting surface, S_V the specific internal surface area expressed per unit volume, C_r the concentration of reactant gas at the exterior surface of the reacting specimen, m the true order of reaction, ϕ Thiele modulus, defined as the ratio of the actual rate of reaction to that which would occur if the reacting gas concentration were uniform throughout the material, f a 'roughness' factor of the exterior surface, C_g the concentration of reactant gas in the gas stream, and D_f the diffusion coefficient of the reactant through the boundary layer of thickness S.

The first term on the right-hand side of equation (6.4) represents the reaction occurring **within** the solid, while the second term describes the reaction occurring on the exposed exterior surface. The change in surface area (ΔS_v) is generally proportional to the square of the change in length $(\Delta l)^2$ [25], i.e.

$$\Delta S_v = \alpha(\Delta l)^2 \tag{6.6}$$

where α is a constant. Neglecting the surface oxidation terms in equation (6.4) it becomes

$$\frac{dw}{dt} = \frac{R}{n} K_s S_v C_r^m \phi. \tag{6.7}$$

Inserting equation (6.6) into (6.7)

$$\frac{dw}{dt} = \frac{R}{n} K_s C_r^m \phi\alpha(\Delta l)^2. \tag{6.8}$$

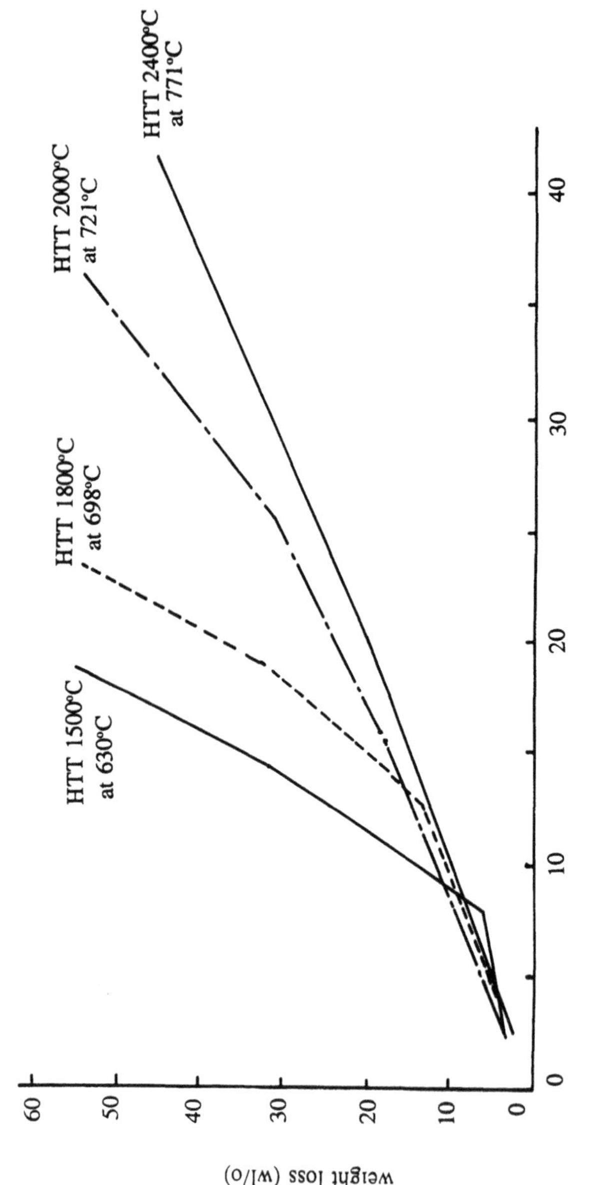

Fig. 6.4 Oxidation of carbon–carbon composites heat treated to different final process temperatures.

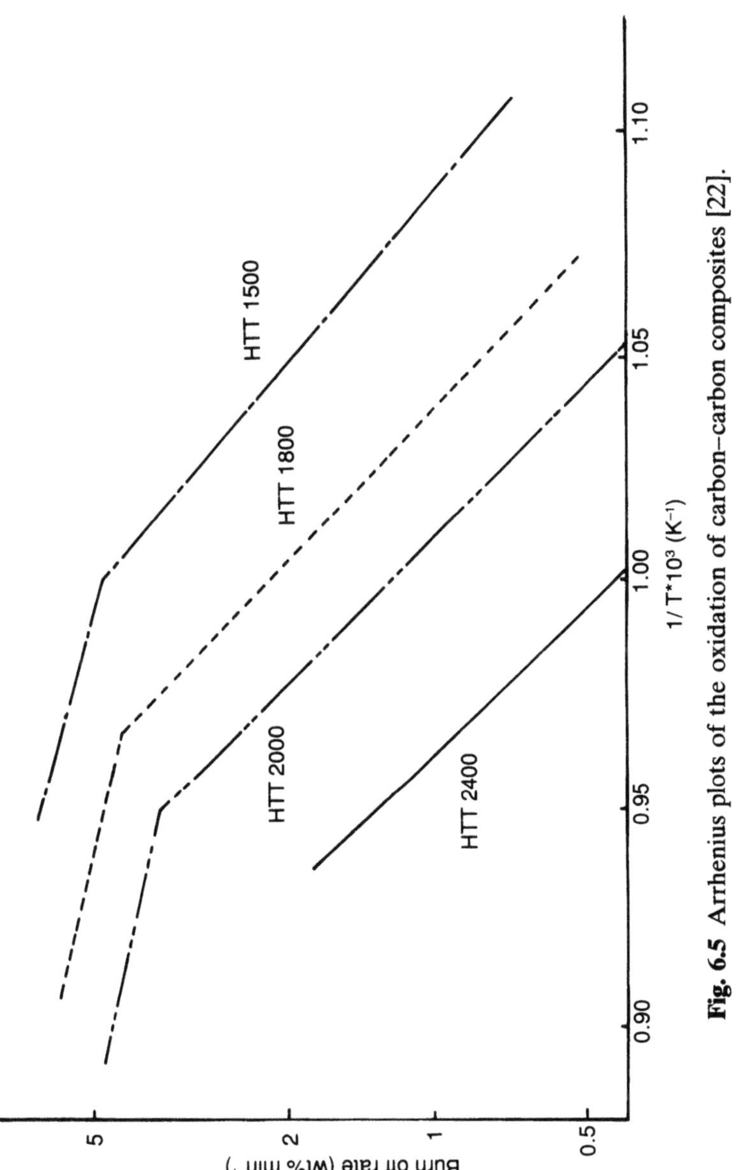

Fig. 6.5 Arrhenius plots of the oxidation of carbon–carbon composites [22].

If it is assumed that Δl is proportional to the square root of time ($t^{\frac{1}{2}}$)

$$\frac{\mathrm{d}w}{\mathrm{d}t} = \frac{R}{n} K_s C_r^m \phi \alpha' t. \tag{6.9}$$

Differentiation of equation (6.9) yields

$$\frac{\mathrm{d}^2 w}{\mathrm{d}t^2} = \frac{R}{n} K_s C_r^m \phi \alpha', \tag{6.10}$$

showing that the rate of oxidation accelerates with time. If equation (6.9) is integrated

$$W = \frac{R}{n} K_s C_r^m \phi \alpha'' t^2. \tag{6.11}$$

The weight loss due to oxidation is thus proportional to t^2. Curves of oxidation follow a parabola and may be described by a quadratic equation. This result has been illustrated practically by Fischbach and Uptegrove [26] (Fig. 6.6).

The rate of oxidation of carbon fibre/isotropic carbon matrix composites in the region 650–850 °C is quicker than that of both pyrolytic graphite and non-reinforced isotropic (glassy) carbon. As one would expect, oxidation commences at the regions of highest energy, which are manifested as the presence of lots of edge sites and porosity that is to say at the fibre/matrix interface. The reaction then proceeds to regions of laminar, optically anisotropic carbon matrix, optically isotropic 'glassy' matrix, fibre lateral surface, fibre ends and, finally, fibre cores [26].

Microscopic observations indicate that oxidation is governed by the structural defects or by stress accumulation within the matrix as a result of carbonization shrinkage. The general 'rule of thumb' is that the rate of oxidation is increased by increase in operating temperature and reduced by increased HTT of the composite. The latter observation is interpreted as being due to a reduction in the degree of retained impurities, relaxation of carbonization stresses [27] and reduction of reactive edge sites by annealing, despite the increased fraction of open pores. It should be noted that the relative reactivity of the various components of the system do not always follow this idealized pattern. At very high rates of burn-off, for example, the fibres may appear more prone to oxidation than the matrix [24], depending on the type of fibre and matrix.

The surface area S_v increases with the oxidation of carbon–carbon composites as a direct consequence of the preferential oxidation of the matrix. The rate of oxidation therefore accelerates with respect to time in the zone of the chemical rate-controlling step. Equation (6.5), on the other hand, shows that in the zone of the oxygen diffusion-controlling step, the surface area does not affect the rate of reaction. The parameters C_g, C_R,

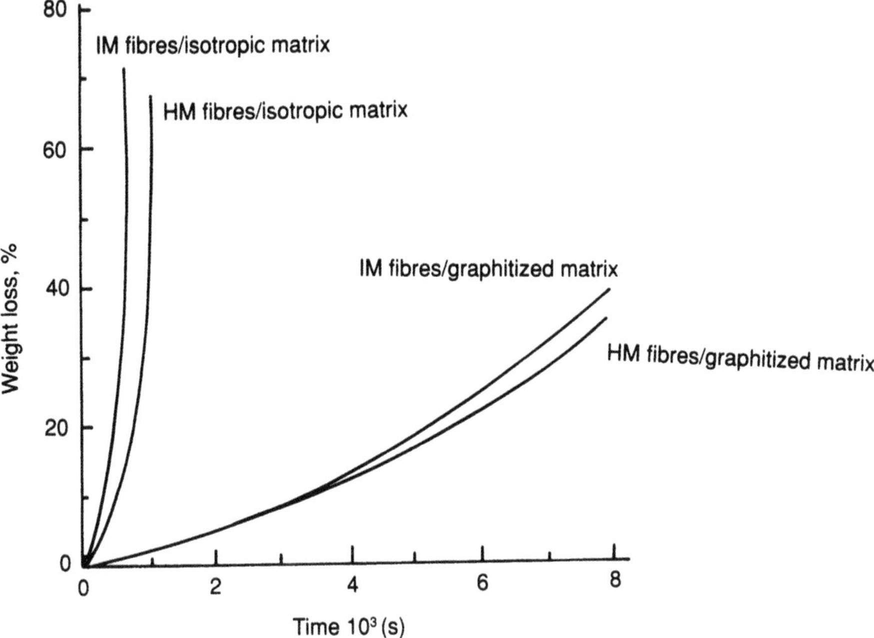

Fig. 6.6 Isothermal oxidation behaviour of carbon–carbon in dry air at 650 °C [26].

D_f and δ are not related to the surface area and may therefore be assumed to be constant. When equation (6.5) is integrated

$$W = \frac{(C_g - C_r)}{\delta} D_f t. \tag{6.12}$$

That is to say, the oxidation loss is proportional to time and may be represented by a linear curve. The oxidation rate does not accelerate with time as shown in Fig. 6.7 [25].

Although investigators generally believe the oxidation of carbon–carbon in air to be controlled by the reaction itself at low temperatures and gaseous transport at higher temperatures, the critical temperature separating the two controlling steps is not well defined or exactly determined. Two very serious problems arise; firstly, the reaction surface changes with time, making it difficult to express the rate of reaction as a simple function of time, temperature and macroscopic surface area of the composite. Secondly, open pores and microcracks are generated during the oxidation process. As a result it is difficult to express the rate of transport of the gaseous species as a simple function of the initial porosity and the labyrinth factor of the composites. Table 6.1, for example, shows the results of Chang and Rhee [27] from their investigation into the increase of the surface area of carbon derived from phenolic resin. The surface area is observed to increase markedly after a 5% loss in weight due to oxidation.

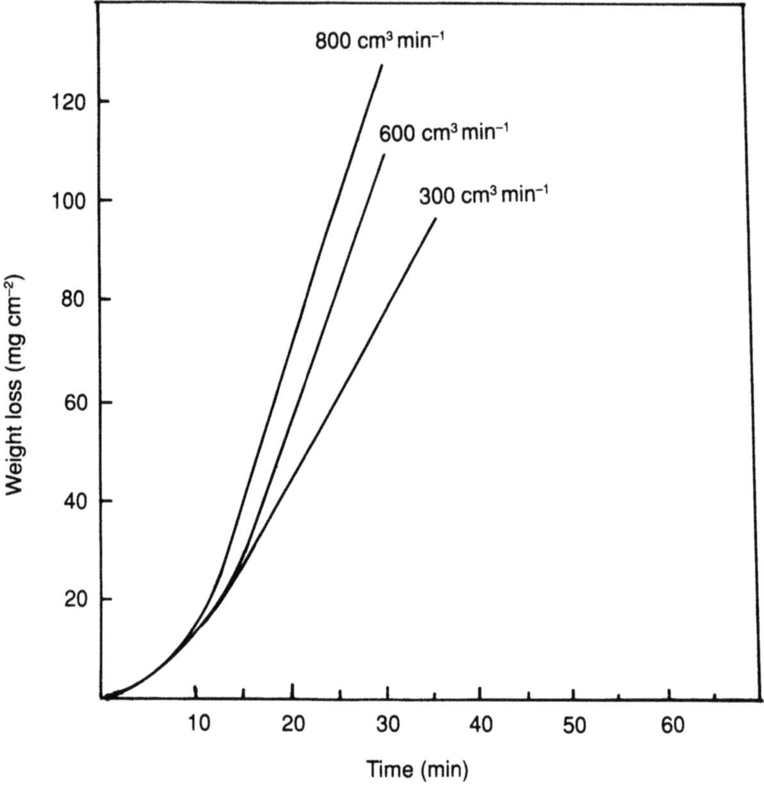

Fig. 6.7 Oxidation of carbon-carbon and the effect of the flow rate of air on the oxidation rates at 600 °C [25].

Table 6.1 Surface area of carbon–carbon (m^2 g^{-1}) after 5 and 10% oxidation

HTT (°C)	*Surface loss of carbon–carbon* (m^2g^{-1}) *after 5 and 10% oxidation* (%) 0	5%	10%
1000	12.6	599	854
1400	1.3	107	114
1800	2.1	24.3	20.5
2400	2.4	4.6	3.1

The rate of increase is clearly very much larger for carbon heat-treated at lower temperatures. Further work by Chang and Rusnak [28], on a carbon–carbon brake material, re-emphasized the general observation that the rate of oxidation increases with increased surface area. The surface area is influenced by the extent of graphitization, degree and shape of porosity, active sites and impurity levels. It is thus virtually impossible to derive a simple relationship between rate of oxidation and surface area.

When the reactant gas flow rate is increased, the thickness of the stagnant layer close to the solid phase is decreased. Chang and Rusnak observed the flow rate to have very little effect on oxidation rate at 650 °C but that the rate of reaction increased significantly with flow rate at 700 °C [28]. When the rate increases with increase of flow rate, diffusion control is generally assumed. They interpreted their results, therefore, as backing up the 'low-temperature chemical/high-temperature diffusion control hypothesis'. In reality, however, the oxidation behaviour is far more complex as illustrated by the results of Thrower and Marx [29]. The oxidation rate of graphite at ≈900 °C was found to be increased by increase in flow rate from 9 to 13 $cm^3 s^{-1}$ but decreased by further increase to 15 $cm^3 s^{-1}$. The result was postulated to be due to the change in surface roughness resulting from increased flow rate. What becomes apparent, is that the rate being controlled by diffusion below a threshold temperature and diffusion above is not a general conclusion but rather an empirical observation. Far more detailed work is required before a general rule describing the rate-controlling steps of the oxidation of carbon–carbon composites is established. In either rate-controlling zone, the rate of reaction is found to be proportional to the partial pressures of oxidizing gases.

Thrower and Marx also measured the effect of an applied compressive stress upon the rate of oxidation of graphite. Their results showed there to be no major effect upon oxidation [29]. A tensile stress, on the other hand, was found to accelerate oxidation as a result of the increase in open porosity and microcrack density [30,31]. Peng was furthermore able to show a decrease in tensile strength of polycrystalline graphite as a direct result of oxidation [32]. He attributed his findings to pore enlargement and component debonding.

6.3 FUNDAMENTAL CONCERNS IN THE OXIDATION PROTECTION OF CARBON–CARBON COMPOSITES

In developing a successful oxidation protection system for carbon-based materials there are a number of factors which must be considered, as depicted schematically in Fig. 6.8. The primary aim is to apply a coating system which isolates the composite from the oxidizing environment. This

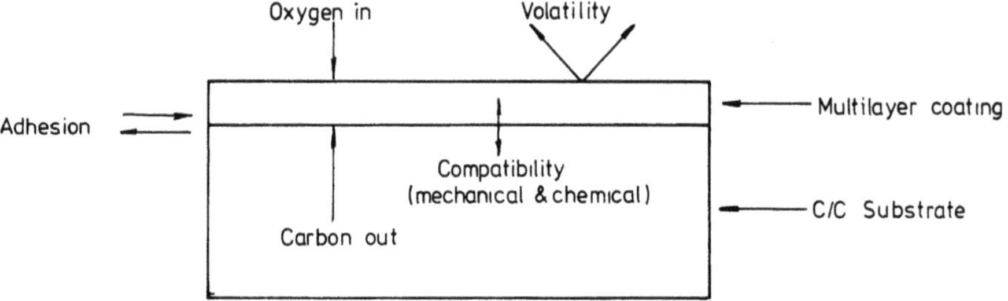

Fig. 6.8 Considerations when designing a carbon–carbon oxidation protection system.

coating system must have at least one major component which acts as an efficient barrier to oxygen. The primary oxygen barrier must ideally exhibit low oxygen permeability and ought to encapsulate the carbon with few or no defects through which the oxidizing species can ingress. Additionally, it must possess low volatility to prevent excessive ablation in high-velocity gas streams. A good level of adherence to the substrate must be achieved without excessive substrate penetration. The internal layers must also prevent outward diffusion of carbon to avoid carbothermic reduction of the oxides in the outermost layers. Finally, all of the various interfaces must exhibit chemical and mechanical compatibility.

In designing protection systems for high-performance, structural carbon–carbon for long periods of operation, mechanical compatibility, i.e. the avoidance of coating spallation, becomes the overriding issue. Figure 6.9 [33,34] compares the thermal expansion behaviour of a number of refractory ceramics with those of carbon–carbon (in-plane). The composite has a considerably lower CTE within fibre-reinforced lamina than any ceramic exhibiting a symmetric crystal structure. Any applied coatings therefore contain microcracks since the coating process is carried out at elevated temperatures. We may therefore schematically represent the relationship between protection requirements and system performance (Fig. 6.10) [35].

The oxidation threshold for carbon–carbon is around 370 °C, which may be improved to approximately 600 °C by the incorporation of refractory particulate inhibitors. The 'intrinsic protection range' of the coating is defined by the microcracking temperature and the limiting use temperature of the coating. In this range, the cracks are mechanically closed and sealed by the oxidation products. Flaws always exist in the primary oxygen barrier due to fabrication imperfections, thermal expansion mismatch with the substrate, or as a result of service stresses. To date, the most successful solution to the cracking problem has been the employment of a sealant glass to fill any cracks in the primary coating. For a successful full range of oxidation protection, the glass sealant must be capable of operating from ≈600 °C to the microcracking temperature of the primary oxygen barrier.

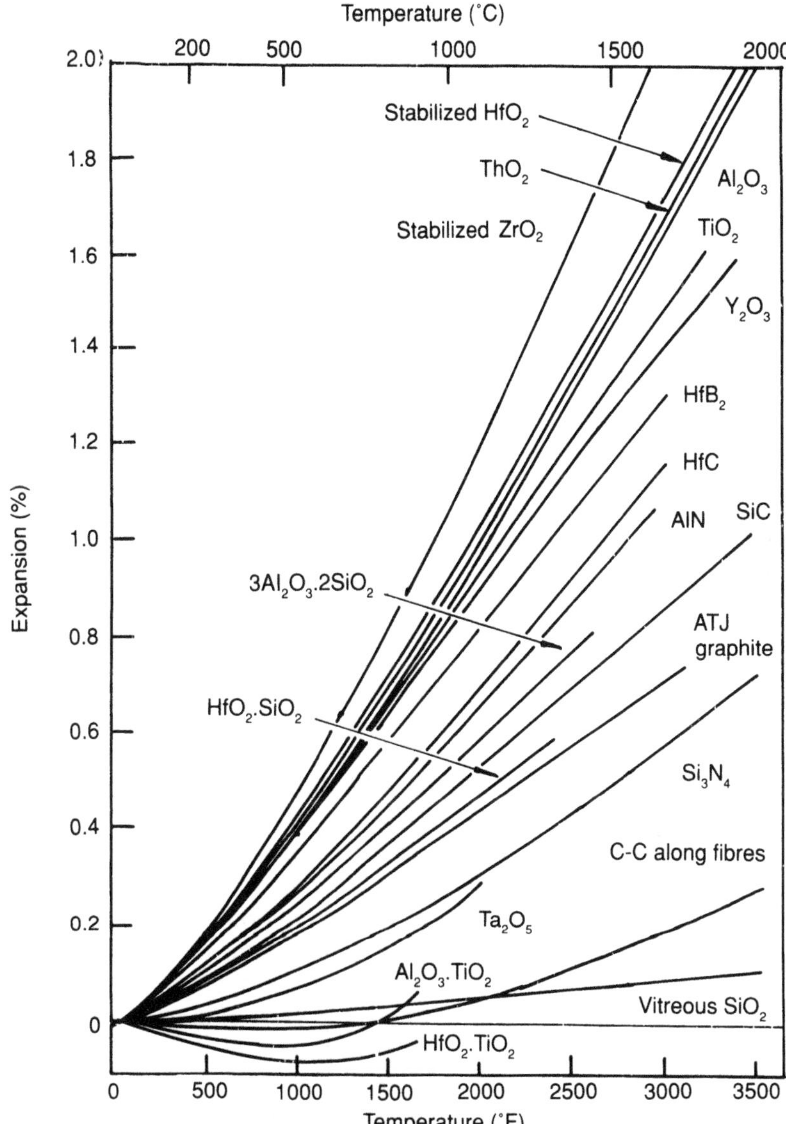

Fig. 6.9 Comparison of the thermal expansion characteristics of a number of refractory materials [33, 34].

In choosing the primary oxygen barrier, one would ideally prefer a material which forms an adherent *in situ* oxide so that oxygen is gettered at free surfaces. The oxidation kinetics of several candidate refractories are compared in Fig. 6.11. The kinetics of oxide growth are considerably slower for the silicon-based ceramics than those based on aluminium, hafnium or zirconium [36–42]. The silicon ceramics exhibit the most attractive thermal expansion compatibility with carbon–carbon and the lowest oxidation rates;

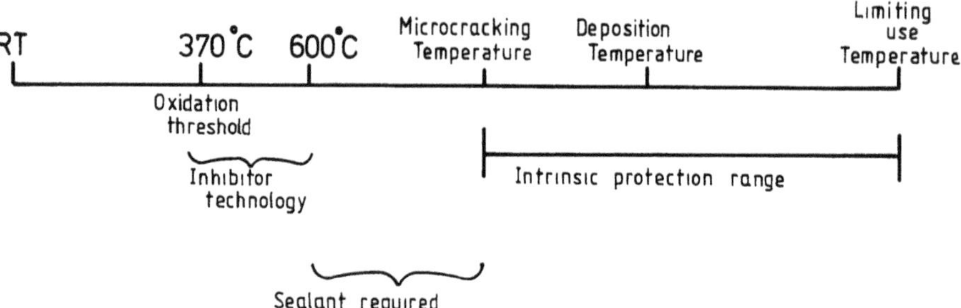

Fig. 6.10 General oxidation behaviour of carbon–carbon composites coated with oxidation protection systems.

thus the amorphous SiO_2 scales which grow during oxidation have very low oxygen diffusivity. Thin films of these materials can thus be utilized for protection without concern for total conversion of the ceramic coating or oxide spallation. They are limited at high temperatures, however (>1700 °C), by dissociation of SiO_2, and at low temperatures because the silica glass is too viscous to flow and effectively seal cracks. It is therefore appropriate to discuss oxidation protection in three different temperature regimes: below 1500 °C, 1500–1800 °C and above 1800 °C.

6.4 PROTECTION AT TEMPERATURES BELOW 1500 °C

Ceramic coatings such as silicon carbide and silicon nitride are excellent primary barriers to oxidation. They are both refractory and oxidation-resistant due to the formation of a silica skin upon oxidation. Silica exhibits a relatively low vapour pressure to temperatures as high as 1650 °C as well as low oxygen diffusivity [43–45]. If cracks develop in the primary coating self-sealing can occur, because the silica skin exhibits a relatively low viscosity at elevated temperatures allowing it to flow effectively into the cracks, preventing oxidation of the underlying substrate. At temperatures below 1150 °C, however, the viscosity of the silica glass is too high for it to afford any crack-sealing properties. Additionally, thermal cycling of the composite results in cracking and spallation of the oxidation protection system as a result of the mismatch in the thermal expansion of the substrate and coating.

The most successful approach to solving this problem, to date, involves the introduction of boron or a boron-containing additive into the carbon–carbon substrate. The addition of boron has been found to retard the oxidation kinetics of a number of carbon materials including carbon–carbon composites [46,47]. Oxidation of boron results in the formation of boric oxide, B_2O_3, which has a lower melting-point and viscosity than silica [48]. Additives such as zirconium boride, boron and boron carbide particles,

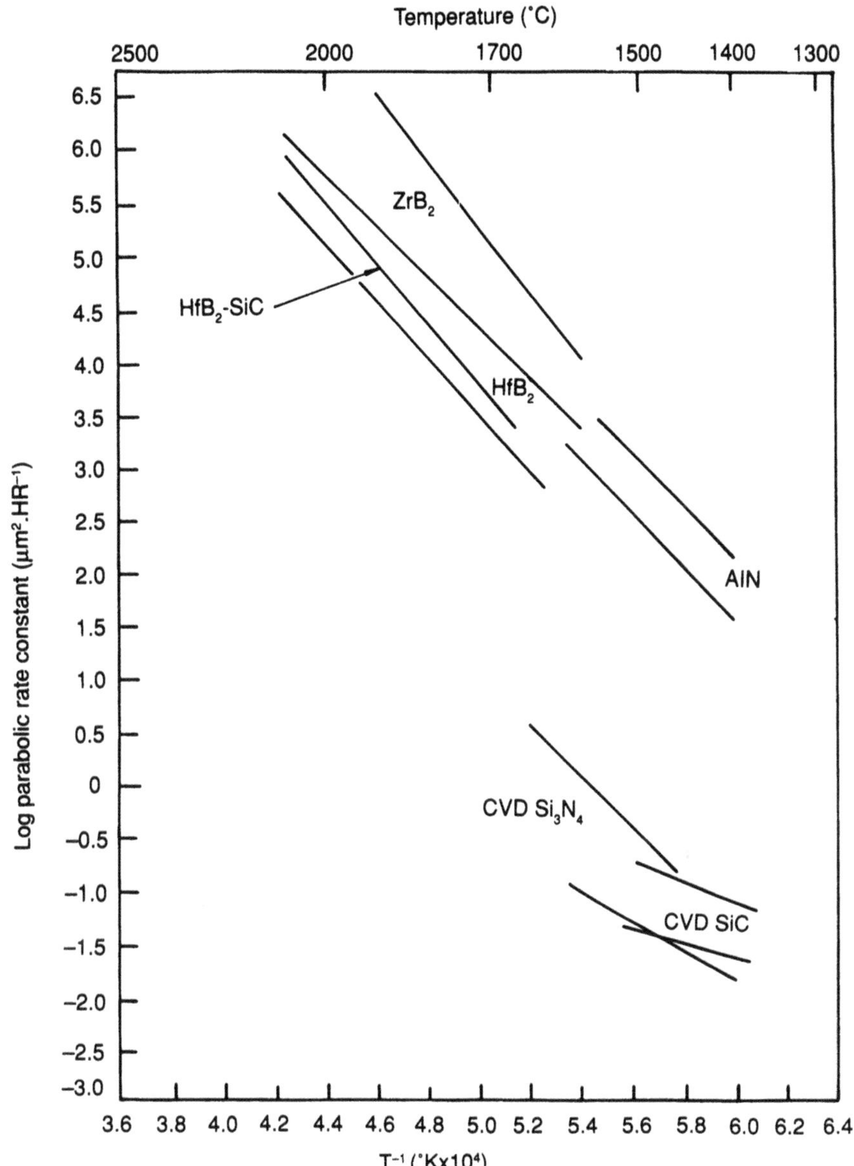

Fig. 6.11 Oxidation of refractory ceramics [36–42].

oxidize in air spreading a borate glass within the composite. A number of so-called 'inhibited' prepregs, for example, are available commercially for this purpose. The inhibitors only become effective, however, after appreciable fractions of the carbon have been gasified [49]. Furthermore, boric oxide exhibits higher vapour pressure and oxygen diffusivity than silica [18,50,51]. It is thus generally used as a secondary oxidation barrier beneath the primary coating which acts to minimize its depletion.

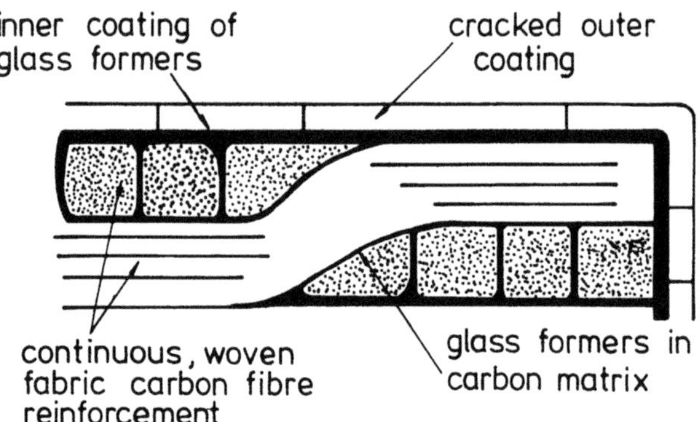

Fig. 6.12 Schematic diagram of carbon–carbon oxidation protection system.

A protection system, based upon the aforementioned principles, has been developed by GA Technologies in the USA. The material, denoted VLG-25/IMCC-3, has shown oxidation resistance for hundreds of hours of cycling between room temperature and 1370 °C and is shown schematically in Fig. 6.12 [10].

The composite materials system has also been successfully operated in stressed oxidation tests in the same temperature regime, and high flow rate burner and rig tests at temperatures of up to 1540 °C gave no indication of failure. It consists of a carbon–carbon composite substrate with borate glass former inhibitors, an inner coating of borate glass formers and an outer coating of CVD-deposited silicon carbide.

The problems associated with oxidation-protected carbon materials of this genre are moisture sensitivity of borate glasses, spallation of the carbide coating under thermal cycling and the long-term chemical compatibility between SiC and the borate glass. The exposure of borate glasses to ambient moisture in the air causes a gradual hydrolysis, resulting in swelling and crumbling of the glass. Hydrolysis of the borate glass sealant may cause coating spallation at room temperature due to glass swelling, and spallation during heating as a consequence of the rapid release of mositure. The high volatility of hydrated borates can add to the disruption on heating, resulting in glass depletion at relatively low temperatures in moist environments.

The solution generally employed is to modify the borate glass composition to improve moisture durability while retaining its attractive viscosity and carbon-wetting characteristics. As an example, it has been shown that additions of alkali significantly reduce the ambient moisture sensitivity of B_2O_3 [52,53]. Lithium appears to have a particularly strong effect [10]. In addition, lithia Li_2O is the most thermodynamically stable alkali oxide and is compatible with carbon in the same temperature range as boric oxide [54]. Studies show that lithia tends to reduce the viscosity of B_2O_3 at high temperatures, although this may be corrected by the addition of a third oxide which will also improve moisture resistance [48,52,55]. Li_2O additions also increase the surface tension of B_2O_3 so that the carbon wetting may be somewhat impaired as compared with a 'pure' borate glass [52,56].

Spallation of the silicon carbide primary oxidation barrier occurs as a result of its thermal expansion mismatch with the carbon–carbon substrate. The problem may be alleviated slightly by replacing SiC with a coating of silicon nitride which has a lower CTE (3.2 as opposed to $4.9 \times 10^{-6}\ °C^{-1}$). It is further possible to reduce the thermal expansion coefficient by vapour depositing a modified primary coating of either SiC or Si_3N_4 containing a dispersion of low expansion particulates of, for example, silica or boron nitride (BN). The first carbon–carbon composites to be deployed in service consisted of rayon-based fibres as the reinforcing phase. As discussed earlier, thermal mismatch was not such a great problem with the rayon fibres as those derived from PAN as they have around double the CTE. It is not inconceivable that one ought to be able to 'engineer' both PAN fibres and carbon matrix to increase the CTE of the substrate, but it is doubtful whether this could be done without adversely affecting mechanical properties. If one has a highly oriented carbon fibre – PAN or pitch – it will have a low, or negative CTE. Only poor orientation fibres have a high CTE and hence poor mechanical properties.

At high temperatures, a chemical incompatibility arises between the borate glass and the outer coating as a result of fluxing of the protective layer of SiO_2, formed on the primary protective coating, by the glass. Additions of silica and lithia, etc., used to improve moisture durability and increase the high-temperature viscosity of the borate glass, will also serve to retard the interaction with the outer layer by lowering the chemical driving force for interdiffusion.

The factors governing oxidation protection below 1500 °C are well understood and based on proven scientific principles backed up by experimental evidence. A number of systems have been employed in service with a good deal of success, the most notable being that on the space shuttle. A great deal of optimization and development work remains to be done, however, and one would certainly expect a number of improvements in the future.

6.5 PROTECTIVE COATINGS FOR THE 1500–1800 °C RANGE

The reduction of borate glasses by carbon limits oxidation protection systems of the type developed by General Atomic (VLG-25/IMCC-3) to temperatures below 1500 °C. Thermodynamic analysis (Fig. 6.13) [54], shows the total CO pressure for the reduction of B_2O_3 by carbon, to form B_4C and CO, to be approximately 0.1 MPa (1 atm). Since CO pressures at higher temperatures are greater than the total ambient pressure, 1500 °C may be considered as the upper limit for boric oxide in contact with carbon.

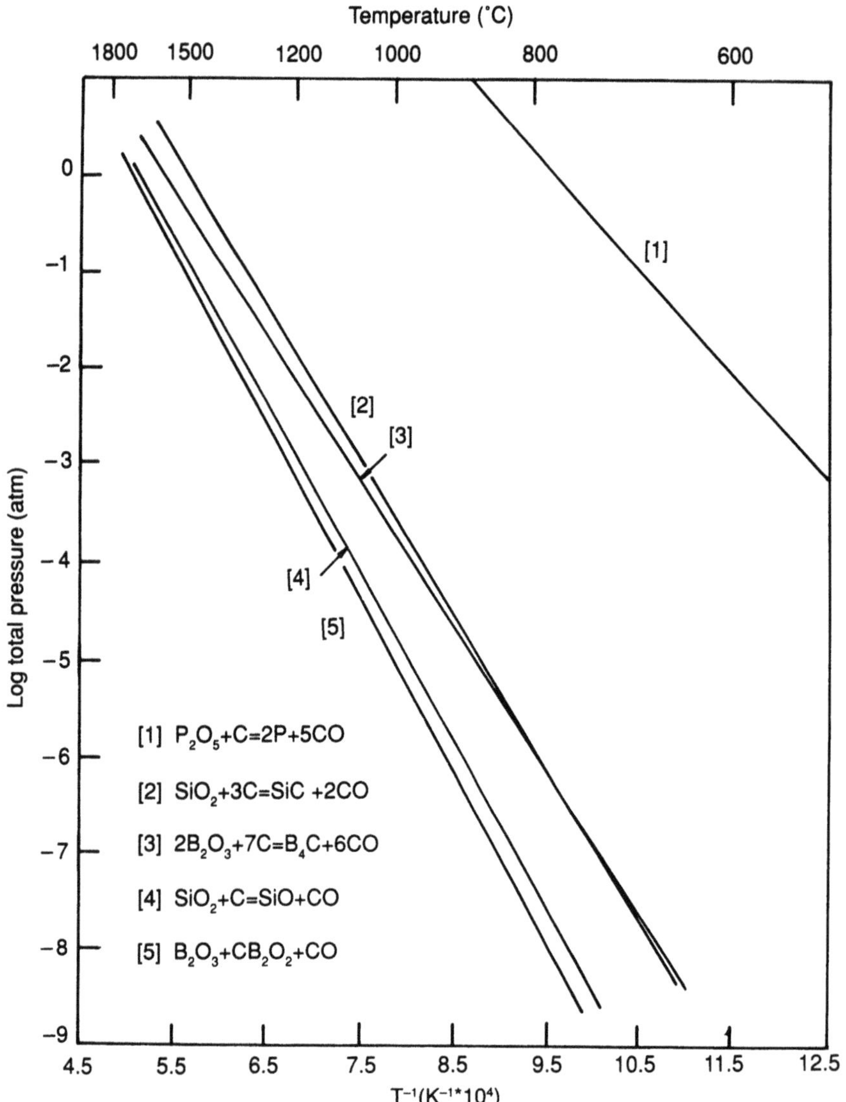

Fig. 6.13 Carbothermic reduction of glasses [54].

It is important to note that the data presented in Fig. 6.13 show this same temperature limit to exist for the reaction of silicate glasses with carbon.

In discussing oxidation protection at elevated temperatures it is first necessary to define the upper temperature limit of silicon-based ceramic coatings. Experimental work has shown the upper use temperature of CVD coatings of SIC and Si_3N_4 to be defined by reactions at the interface between the SiO_2 scale and the underlying ceramic [36,57]. A thermodynamic analysis of the stability temperature of SiC and Si_3N_4 is shown in Fig. 6.14

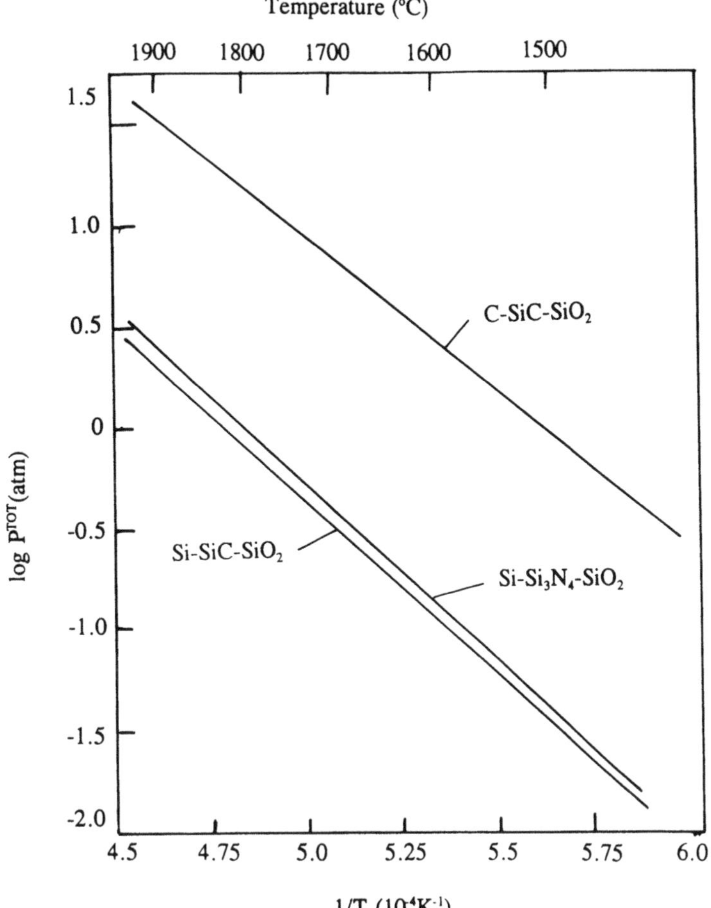

Fig. 6.14 Total gas pressures at three equilibrium conditions in Si–O–C and Si–O–N systems [35].

[35]. It can be seen that, if the silicon activity is maintained at unity, SiO_2 films will be stable on SiC or Si_3N_4 up to approximately 1800 °C. At higher temperatures the reaction product gas exceeds 1 atm, thus disrupting the protective scale. Strife and Sheehan [35] have clearly demonstrated, in the case of silicon nitride vapour-deposited films, that exposure to flowing air at 1760 °C produced a stable SiO_2 scale growth, whereas oxidation at 1840 °C caused gas generation and bubble formation in the silica. It has further been shown that the temperature limit for the stability of B_2O_3 on B_4C with unit boron activity is roughly 1760 °C [10]. One may conclude, therefore, that the temperature limit of the protective oxide coatings may be enhanced by several hundred degrees provided they are in contact with carbides rather than carbon.

Borate glass sealants are employed in oxidation protection systems to provide protection below the intrinsic protection range by the sealing of microcracks in the outer protection layer. This is an especially important consideration in the case of components which are subject to thermal cycling between ambient and their designed operating temperature. From the standpoint of chemical compatibility, it ought to be possible to use borate glass sealants in conjunction with boron carbide inner coatings up to temperatures of 1760 °C. The use of borate glasses at such high temperatures is unfortunately not possible due to volatilization, high oxygen permeability and poor compatibility with SiC and Si_3N_4 outer coatings.

The glass, despite being primarily contained beneath the outer coating, would be very prone to volatilization over extended periods at temperatures above about 1300 °C. It is difficult to imagine a useful protection period of over 10 h using borate glasses at or above this temperature. Furthermore, as one may easily deduce from Fig. 6.15 [44,50]. B_2O_3 has a relatively high permeability to oxygen while silica forms an excellent barrier to the diffusion of oxygen. It thus becomes apparent why silicon ceramics, which oxidize to produce protective layers of SiO_2, are oxidation resistant to temperatures above 1650 °C, whereas boron carbide and nitride are limited to much lower temperatures. Indeed, it is postulated that the rates of oxidation of SiC and Si_3N_4 remain very slow up to temperatures at which the stability of the oxide is the performance-limiting factor [36].

The high-temperature limitations of borates require that they be replaced for operation in the 1500–1800 °C region. The obvious choice is to use a refractory silicate glass in contact with a silicon carbide inner coating. The vapour pressure of SiO_2 is several orders of magnitude lower than that of B_2O_3 in this region [10]. Silica also offers lower oxygen permeability and improved chemical compatibility with the outer CVD coatings. The major problem associated with silicate glasses is their high viscosity at low temperatures. The viscosity of 'pure' SiO_2 glass is generally an order of magnitude or more higher than pure B_2O_3 and borate glasses, thus limiting the flow necessary to allow for effective crack sealing and spreading on internal surfaces.

A logical approach to the viscosity problem is to use highly fluxed silicate glasses. The obvious choices for additions are lithia and boric oxide, since their strong fluxing action is coupled with thermodynamic stability. Alkali additions are the stronger flux, reducing the viscosity of SiO_2 into the borate range at temperatures above 800 °C. All silicate glasses are found to exhibit viscosities which are strongly dependent on temperature, being very high below 800 °C. The viscosity behaviour at 800 °C achieved by fluxing SiO_2 with Li_2O therefore probably represents the best achievable from this glass system.

As is often the case in materials science, altering or blending a material to improve one property must be done to the detriment of another. The fluxing of silicate glasses with oxide additions such as B_2O_3 and Li_2O is no

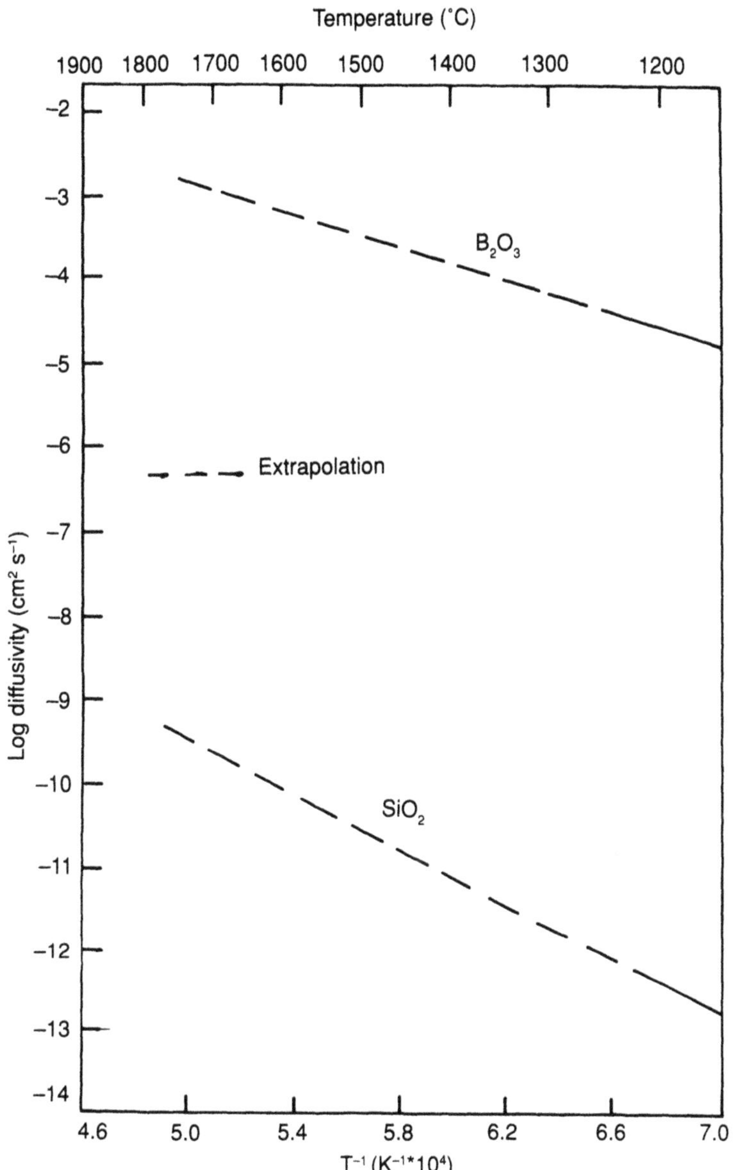

Fig. 6.15 Oxygen diffusivity in SiO_2 and B_2O_3 glasses [44,50].

exception to this rule. Both additions compromise the advantages pure SiO_2 presents over borate glasses in that they reduce moisture durability and increase oxygen permeability. Damage due to mositure uptake, chemical incompatibility with outer coatings and accelerated oxygen transport to the substrate might therefore result. Both B_2O_3 and Li_2O have significantly higher vapour pressures than SiO_2 (although that of Li_2O is an order of magnitude lower than that of B_2O_3) and thus may present volatility problems over long periods at high temperatures.

The silicon ceramics generally favoured as outer coatings in carbon–carbon oxidation protection systems are limited in temperature of operation to around 1800 °C. Glass sealants employed to repair cracks in the outer coating are reduced by the carbon–carbon substrate at temperatures of 1500 °C and above, but may be used to higher temperatures provided a carbide inner layer is employed. Borate glass sealants ought to be chemically compatible with such interfacial carbide layers up to around 1760 °C, but the relatively high vapour pressure of B_2O_3 tends to dictate a limit of 1650 °C or lower. Borates tend to be protective down to the temperatures generally associated with the onset of carbon oxidation. Glasses based on silica provide a temperature advantage over borates, the limit again being set by volatility and/or chemical compatibility. Additionally, silicate glasses are thought to possess improved moisture durability, oxygen permeability and chemical compatibility with outer coatings. The major drawback with silicate glasses is a high viscosity at low temperatures which reduces protection at those temperatures. Fluxing the silicate with oxide additions can reduce the low-temperature viscosity but at the expense of high-temperature performance.

6.6 OXIDATION PROTECTION AT TEMPERATURES IN EXCESS OF 1800 °C

Outer coatings of silicon ceramics cannot be used in long-term protection of carbon–carbon at temperatures above 1800 °C because reaction product gases disrupt the protective SiO_2 layer whence oxidation resistance is derived. Glass sealants cannot be used in contact with the composite substrate or interfacial carbide layers for the same reason. Protection of carbon to very high temperatures thus requires the understanding of the fundamental principles governing behaviour of materials at those temperatures. This will necessitate the development of coating systems very different and far more sophisticated than those employed at lower temperatures.

The most oxidation-resistant refractory materials at temperatures in excess of 1800 °C, identified to date, are hafnium boride (HfB_2) and mixtures of the same with silicon carbide [37, 38]. A great deal of research is presently under way, assessing these materials in short-term protection applications at temperatures in which silicon ceramics cannot operate. Figure 6.11 shows hafnium boride materials to oxidize in a parabolic manner but at a fairly rapid rate. Calculations have shown that 100 h at 1927 °C would produce approximately 2 mm of oxide scale on the surface of a CVD-deposited HfB_2–SiC material [41,42]. An outer coating of this material would therefore oxidize through and become an oxide coating within a very short period of time. Similar observations are made for refractory carbides such as HfC.

The high rates of oxidation of refractory carbides and borides preclude their use in long-term protection above 1800 °C. The only viable solution is to use refractory oxides as the primary oxidation barrier. The use of these oxides will, however, create problems of physical and chemical compatibility. When selecting an oxide coating it is necessary to consider its melting-point, vapour pressure and coefficient of thermal expansion. The vapour pressures of a number of appropriately high melting oxides are compared with those of selected carbides and nitrides in Fig. 6.16 [10]. The data show zirconia (ZrO_2), hafnia (HfO_3), yttria (Y_2O_3) and thoria (ThO_2) to possess the necessary thermal stability for long-term use above 2000 °C, while alumina (Al_2O_3) could be used in the 1800–2000 °C range.

Figure 6.9 shows how all but a few oxides have large CTE mismatches with carbon materials. The spallation of coatings during thermal cycling therefore presents a serious problem [33]. A number of hafnium, zirconium and thorium titanates and silicates exhibit both high-temperature stability and low CTE [33,34]. All of these materials have highly anisotropic crystal structures and microcrack at low temperatures. Pure silica glass has a very low CTE but devitrifies in the 1000–1650 °C temperature range. The resulting cristobalite phase transforms to a modified structure at low temperatures with an associated volume contraction of around 3% [58]. Such disruptive behaviour, when combined with viscosity and vapour pressure characteristics, indicates silica to be an inappropriate choice as outer coating since it would erode rapidly in high gas flow conditions.

The rapid oxidation of HfB_2 makes the choice of effective oxygen barrier a serious concern. The high refractory oxides all have high oxygen permeabilities (Fig. 6.17). Silica, on the other hand, is one of the least permeable oxides [45,59]. One might therefore propose a dual system, consisting of a refractory oxide outer coating to act as an erosion barrier, with an inner SiO_2 glass coating as an oxygen barrier and sealant for cracks in the outer coating. Iridium has been used for some time to protect graphite from oxidation [8, 9], being an efficient barrier to both oxygen and carbon diffusion at temperatures up to 2100 °C. The noble metal is, as previously discussed, impractical, except in low-volume specialized applications, due to high costs and limited availability.

The dual coating of refractory oxide over silica glass will, unfortunately, be chemically incompatible with carbon or carbides. The SiO_2 would require to be separated from the substrate by a layer of refractory oxide. The inner oxide layer, to ensure carbon compatibility, ought to be in contact with a refractory carbide interlayer between it and the carbon–carbon composite. Figure 6.18 shows the carbides of tantalum (TaC), titanium (TiC), hafnium (HfC) and zirconium (ZrC) to have appropriately low

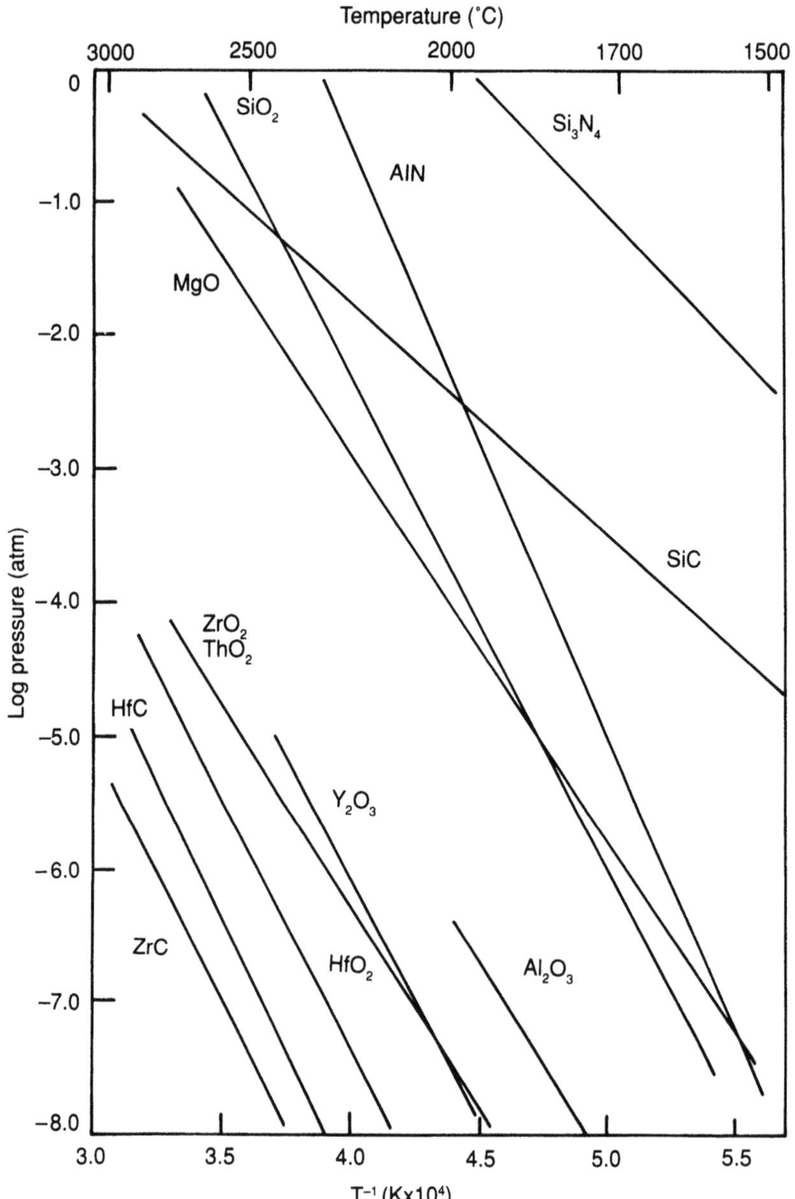

Fig. 6.16 Vapour pressures of ceramic materials [10].

carbon diffusivities [60]. The four-layer protection system would therefore consist of: refractory oxide/modified SiO_2 glass/inner refractory oxide/refractory carbide (Fig. 6.19). A multi-layer coating such as this, while possessing the required chemical stability, would present serious problems in deposition of consistent quality. In addition, the high CTE of the

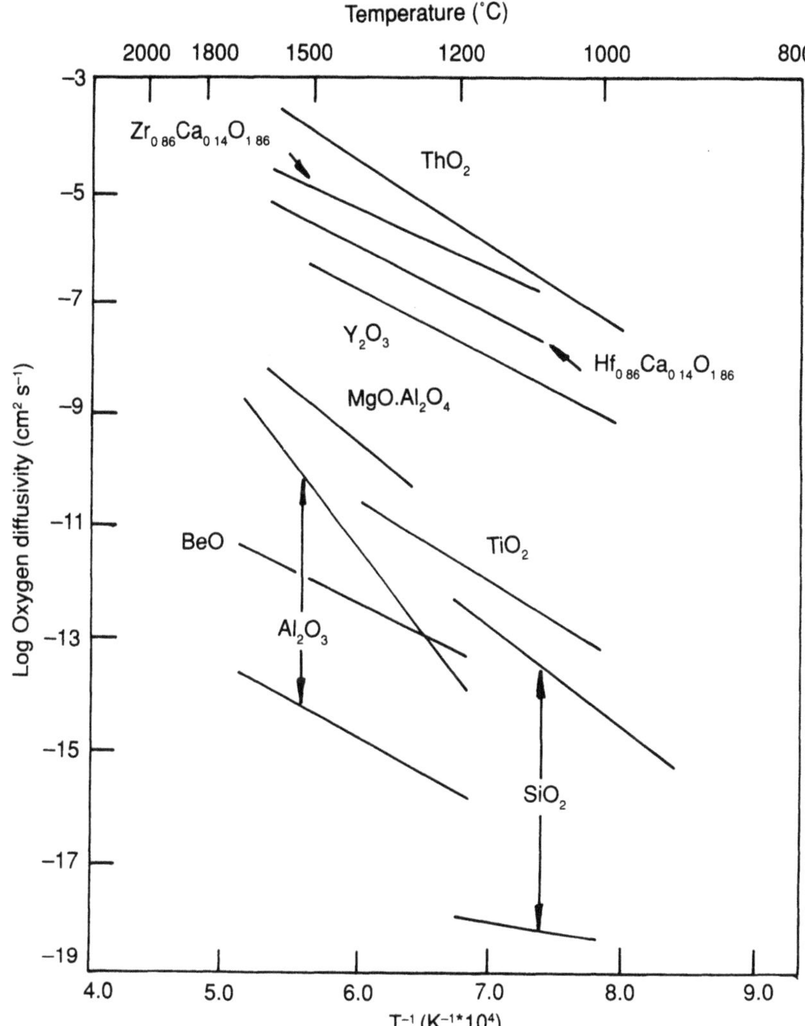

Fig. 6.17 Apparent oxygen diffusivities in silica glass and polycrystalline oxides [45,59].

protection system compared to the substrate will create severe problems of mechanical compatibility in thermal shock situations.

6.7 SUMMARY

The excellent high-temperature properties of carbon–carbon composites make them prime candidates for use in a whole host of refractory engineering structures. A great number of these applications, such as jet engines, require the component to operate in an oxidizing environment. Their

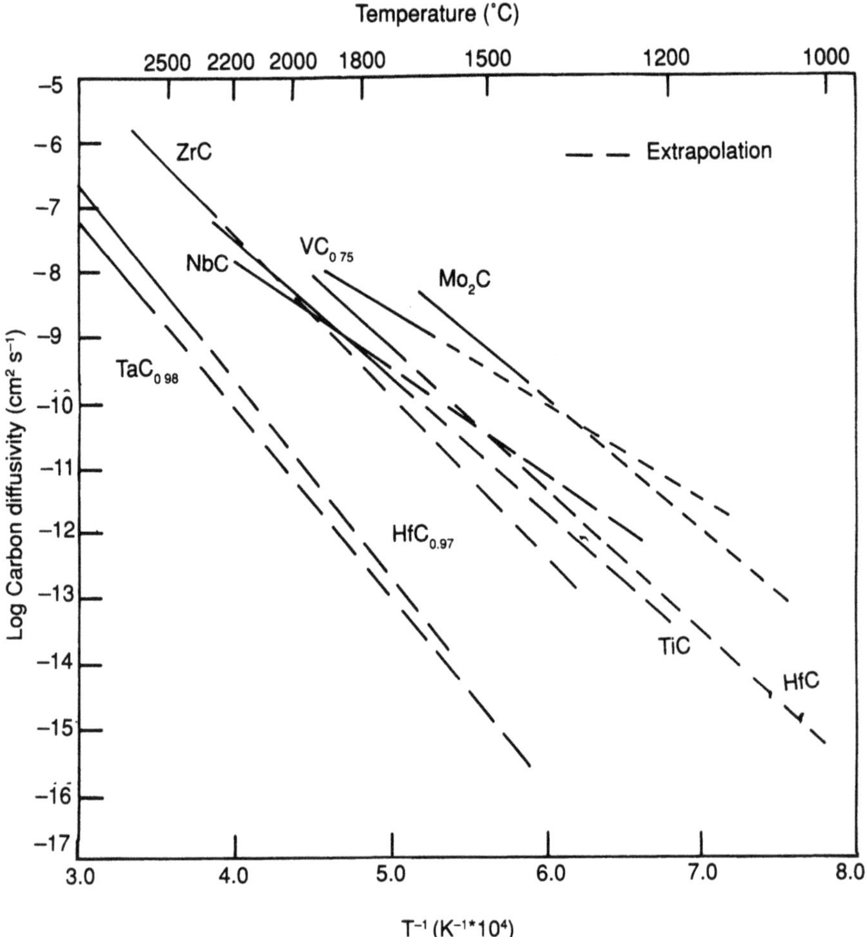

Fig. 6.18 Apparent carbon diffusivities in polycrystalline carbides [60].

usage is thus grealty limited unless they can be protected in some way from oxidation. Carbon–carbon will oxidize at temperatures as low as 370 °C in air. Experimental work has shown the oxidation reaction at low temperatures to be controlled by the reaction of oxygen with active sites on the carbon's surface. At higher temperatures, the rate-limiting step is the diffusion of oxygen through the boundary layer at the surface of the composite. The temperature at which the reaction changes to being governed by one regime or the other varies considerably from material to material within the 600–800 °C range. The oxidation mechanisms which occur, in what are extremely complex multiphase materials, are virtually impossible to model to any degree of accuracy with simple mathematical relationships. The majority of information tends, therefore, to be empirical, based on experimental observations. Oxidation attack tends to occur

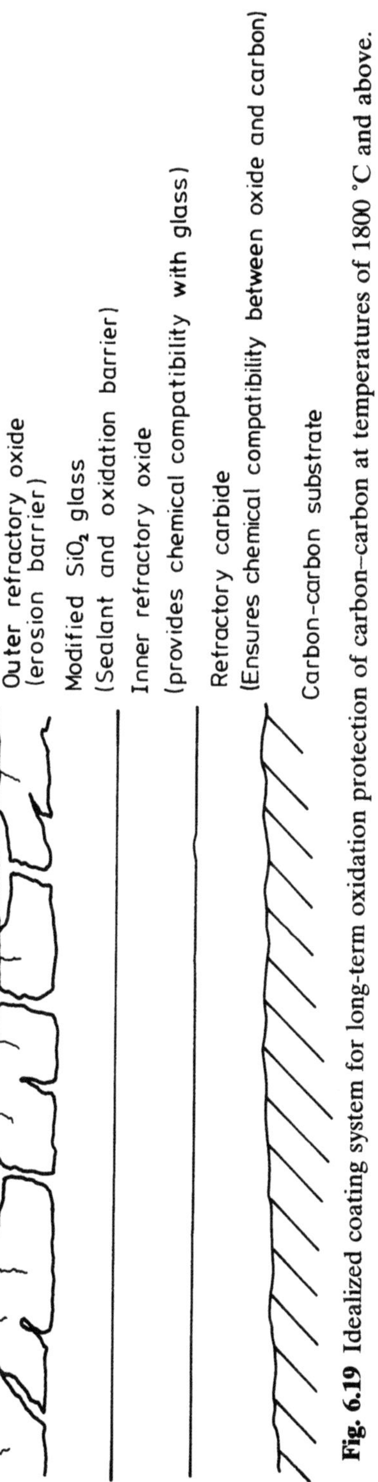

Fig. 6.19 Idealized coating system for long-term oxidation protection of carbon–carbon at temperatures of 1800 °C and above.

preferentially in high-energy regions such as the fibre/matrix interface, proceeding to laminar, isotropic and finally fibrous carbon. Increasing the HTT of the composite increases the temperature of the onset of oxidation and decreases the oxidation rate. This is believed to occur as a result of a reduction in reactive edge sites, defect concentration and residual carbonization stresses.

The only solution to the problem of carbon oxidation is to isolate the material from the high-temperature gas. This is achieved by encapsulating the composite in a refractory coating system. The coating must possess low permeability to oxygen, chemical compatibility with the substrate, good adhesion without excessive penetration and a comparable CTE to avoid spallation under thermal cycling. The onset of oxidation of carbon–carbon can be extended to around 600 °C by the use of inhibitors. These are ceramic materials placed within the composite substrate. Upon oxidation, oxide glasses are formed which flow within the structure forming an oxygen barrier, the most widely used being borates. Inhibited carbon–carbon is limited to 600 °C by the volatilization of the protective glass. Furthermore, some of the carbon is always lost because of the finite time period prior to the glass becoming effective, which depends upon its viscosity.

The driving force for the oxidation protection of carbon–carbon composites was the development of the leading edges and nose cone of the space shuttle. All of the successful systems which have evolved use a primary protective coating of a silicon ceramic. Coatings such as silicon carbide or nitride, when oxidized, form a protective surface scale of SiO_2 which acts as a barrier to the diffusion of oxygen. The coatings are applied by CVD, pack cementation or direct reaction with the substrate. All of the materials used as coatings exhibit higher coefficients of thermal expansion than the carbon–carbon substrate.

The problem of thermal mismatch between the composite and its protection system has been exacerbated in recent years with the move from low strength and modulus rayon-based fibres towards PAN fibres which generally possess a CTE of around half that of their predecessors. Oxidation-protective coatings are therefore generally flawed and cracked as a result of differential thermal expansion between themselves and the composite on cooling from the coating temperature. Borate glass forming materials are placed within the substrate and coating to enable cracks and other flaws within the coating to be sealed. This is done in order to prevent loss of carbon and reduce spallation of the outer coating during thermal cycling. Systems of this type have been successfully tested for hundreds of hours of operation. Much optimization remains to be done however as, for example, in the modification of the borate glass sealants to reduce their susceptibility to moisture and volatilization, and the lowering of the CTE of silicon ceramics with particle additions.

Oxidation protection at temperatures above 1500 °C demands more sophisticated materials systems to be effective. At this temperature both silicate and borate glasses, used as sealants, are reduced by the carbon substrate. To increase the temperature of operation it is necessary to introduce an inner protective layer of a carbide between the carbon and the glass. Borate glasses are prone to volatilization at around 1500 °C and require to be replaced. Silicate glasses offer lower oxygen diffusity and better chemical compatibility over borates at high temperatures, but a high viscosity at low temperatures reduces their effectiveness as sealants. Additions of alkali or B_2O_3 can lessen this problem but at the expense of moisture vulnerability and high-temperature performance.

Silicon protective coatings are limited in temperature of operation to ≈1800 °C. At or above this temperature the products of reaction disrupt the protective oxide scale, thus rendering long-term protection impossible. Much research is ongoing into addresssing the fundamental scientific and engineering problems posed at such high temperatures. Just how effective the work has been is difficult to assess owing to the extremely classified nature of the military research undertaken. The solution demands complex multi-component systems of four or more different coatings to permit protection against oxidation and erosion while maintaining chemical compatibility of the system not only with the substrate but also with itself.

One of the major obstacles to long-term protection and a truly 'reusable' composite materials system, is the spallation of coatings due to thermal mismatch with the substrate under cyclic conditions. The most feasible way to avoid this occurrence would be to replace the discrete coating/composite interface with a functionally graded system. In such an artefact, there would be a hybrid region between protection and composite, thus minimizing thermal expansion differences.

REFERENCES

1. Johnson, H. V. (1934) US Patent 1,948,382.
2. Klein, A. J. (1986) *Adv. Mat. Proc. inc. Metal Prog.*, **11**, 64.
3. Zeitsch, K. J. (1967) Oxidation resistant graphite base composites, in *Modern Ceramics* (eds J. E. Hove and W. C. Riley), Wiley, New York, p. 314.
4. Bortz, S. A. (1971) Testing of graphite composites in air at high temperatures, in *Ceramics in Severe Environments* (eds W. W. Kriegel and H. Palmour), Plenum, p. 49.
5. Gloedstein, E. M., Carter, E. W. and Klutz, S. (1966) *Carbon*, **4**, 273.
6. Chown, J., Dencon, R. F., Singer, N. and White, A. E. S. (1963) Refractory coatings on graphite, in *Special Ceramics* (ed. P. Popper), Academic Press, p. 81.
7. Fitzer, E. (1978) *Carbon*, **16**, 3.
8. Criscione, J. M., Mercuri, R. A., Schram, E. P., Smith A. W. and Volk, H. F. (1974) *Rpt ML-TDR-64–173 Part II*, US Army, Washington, DC.

9. Nat. Acad. Sci. and Eng. (1970) *Pub. no. 15BMO-309–01769–6*. Washington, DC.
10. Sheehan, J. E. (1987) *Proc. 4th Ann. Conf. 'Recent Research Into Carbon–Carbon Composites'*, Southern Illinois Univ., 5 May.
11. NASA (1981) *Tech. Briefs*, **6** (2), MSC-18898.
12. Shuford, D. M. (1984) US patent 4,471,023.
13. Shuford, D. M. (1984) US patent 4,465,777.
14. Ehrenreich, L. C. (1978) US patent 4,119,189.
15. Saito, M. K. and Kogo, Y. L. (1980) US patent 4,197,279.
16. Holzl, R. A. (1985) US patent 4,515,860.
17. Vasilos, T. (1986) US patent 4,559,256.
18. McKee, D. W. (1986) *Carbon*, **24**, 737.
19. Taylor, H. S. and Neville, H. A. (1921) *J. Am. Chem. Soc.*, 2055.
20. Jalan, B. P. and Ras, Y. K. (1978) *Carbon*, **16**, 175.
21. Jalan, B. P. and Ras, Y. K. (1972) *Met. Trans.*, **3**, 2465.
22. Yasuda, E., Kimura, S. and Shilsusa, Y. (1980) *Trans. JSCM*, **6**(1), 14.
23. Walker, P. L., Rusinko, F. and Austin, L. G. (1959) *Adv. in Catalysis*, **11**, 164.
24. Goto, K. S., Han, K. H. and St Pierre, G. R. (1986) *Trans. Iron Steel Inst. Jap.*, **26**, 597.
25. Han, K. H., Ono, H., Goto, K. S. and St Pierre, G. R. (1987) *J. Electrochem. Soc.*, **134**(4), 1003.
26. Fischbach, D. B. and Uptegrove, D. R. (1977) *Proc. 13th Biennial Conf. on Carbon.*
27. Chang, H. W. and Rhee, S. K. (1978) *Carbon*, **16**, 17.
28. Chang, H. W. and Rusnak, R. M. (1979) *Carbon*, **17**, 407.
29. Thrower, P. A. and Marx, D. R. (1977) *Proc. 13th Biennial Conf. on Carbon.*
30. Knefeld, R., Linkenheil, G., Glaude, P. and Karcher, W. (1972) *Proc. Carbon 72 Conf.*, Baden-Baden, FRG.
31. Knefeld, R., Linkenheil, G. and Karcher, W. (1973) *ORNL-CONF – 730601*, 88.
32. Peng, T. C. (1977) *Proc. 13th Biennial Conf. on Carbon.*
33. Lynch, J. F., Ruderer, C. C. and Duckworth, W. H. (1966) *US Air Force Mat. Lab. Tech. Rpt no. AFML-TR066–52.*
34. Lynch, J. F. and Morosin, B. (1972) *J. Am. Ceram. Soc.*, **55**(8), 409.
35. Strife, J. R. and Sheehan, J. E. (1988) *Am. Ceram. Soc. Bull.*, **67**(2), 369.
36. Schriroky, G. H., Price, R. J. and Sheehan, J. E. (1986) *GA Technologies Rpt no, GA-A18696.*
37. Bock, P., Glandus, J. R., Jarrige, J., Lecompte, J. P. and Mexmain, J. (1982) *Ceram. Int.*, **8**, 34.
38. Laurenko, V. A. and Alexeev, A. F. (1983) *Ceram. Int.*, **9**, 80.
39. Hirai, T., Niihara, K. and Goto, T. (1980) *J. Am. Ceram. Soc.*, **63**, 708, 419.
40. Costello, J. A. and Tressler, R. E. (1986) *J. Am. Ceram. Soc.*, **69**(9), 674.
41. Kaufman, L., Clougherty, E. V. and Berkowitz-Mattuck, P. J. (1967) *Trans. Met. Soc. AIME*, **239**, 458.
42. Clougherty, E. V., Pober, R. L. and Kaufman, L. (1968) *Trans. Met. Soc. AIME*, **242**, 1077.
43. Schick, H. L. *Chem. Rev.*, **1960**, 331.
44. Sucov, E. W. (1963) *J. Am. Ceram. Soc.*, **46**, 14.

45. Harrop, P. J. (1968) *J. Mat. Sci.*, **3**, 206.
46. Woodley, R. E. (1968) *Carbon*, **6**, 617.
47. McKee, D. W., Spiro, C. L. and Lainby, E. J. (1984) *Carbon*, **22**, 507.
48. Napolitano, A., Mucedo, P. B. and Hawkins, E. G. (1965) *J. Am. Ceram. Soc.*, **48**, 613.
49. McKee, D. W. (1988) *Carbon*, **26**, 659.
50. Grigorev, A. I. and Polishchuk, D. I. (1973) *Fiz. Aerodisp. Sist.*, **8**, 87.
51. Greene, F. T. and Margrave, J. L. (1971) *J. Phys. Chem.*, **70**, 2112.
52. Mazurin, O. V., Streltsina, M. V. and Shavaiko-Shavaikovskaya, T. P. (eds) (1985) *Handbook of Glass Data*, Elsevier, Amsterdam.
53. Adams, P. B. and Evans D. L. (1978) *Mat. Sci. Res.*, **12**, 525.
54. Stull, D. R. and Prophet, H. (eds) (1971) *JANAF Thermochemical Tables*, 2nd edn, Nat. Bur. Stds Pub. NBS 37.
55. Riebling, E. F. (1964) *J. Am. Ceram. Soc.*, **47**, 478.
56. Kingery, W. D. (1959) *J. Am. Ceram. Soc.*, **42**, 6.
57. Schiroky, G. H. Kaae, J. L. and Sheehan, J. E. (1985) *Am. Ceram. Soc. Bull.*, **64**(3), 447.
58. Fleming, J. D. (1964) *Fused Silica Manual*, Final Rpt AEC Proj. B-153, Georgia Inst. Tech.
59. Freer, R. (1980) *J. Mat. Sci.*, **15**, 803.
60. De Poorter, G. L. and Wallace, T. C. (1971) Diffusion in binary carbides, in *Advances in High Temperature Chemistry*, (ed. L. Eyring), Academic Press, New York, p. 107.

Laboratory scale production and evaluation of carbon–carbon

7

7.1 INTRODUCTION

In earlier Chapters the principles and processing involved in the various industrially exploited methods of carbon–carbon manufacture have been discussed. The aim of this Chapter is to detail the production, testing and evaluation of the materials on a laboratory scale for research purposes.

7.2 RAW MATERIALS

Carbon fibres are produced by a large number of companies from the three basic precursors, rayon, PAN and pitch. Research quantities may be obtained relatively easily, especially from the larger organizations. The American company Amoco, for example, produce fibres from all three types. Having taken over the capacity of Union Carbide, one of the pioneers in this field, the most impressive products marketed by Amoco are their family of ultra-high modulus fibres made from pitch. Hercules in the USA and Toray in Japan specialize in very high quality intermediate and high modulus PAN-based fibres. Both companies market equivalent products and may be considered the leaders in the field. The Mitsubishi Chemical Corporation of Japan is unique in being the only major producer of fibres from coal-tar pitch, all other organizations in the pitch-based fibre business using petroleum pitch.

A large number of companies weave carbon fibres into a variety of felts, fabrics and 3-D preforms. The business is split between specialist weavers and larger companies such as Fiberite who perform weaving operations as a prelude to subsequent processing such as prepregging. The world leaders in the production of three-dimensionally woven stock and preforms are Aerospatiale/Hercules who produce both orthogonal and polar woven products which they subsequently densify.

The overwhelming majority of commercially operated chemical vapour deposition (CVD) operations use methane as the feedstock. In the laboratory a number of other sources of carbon may be used, such as propane and benzene. Bottled hydrogen is generally used as the diluent. One of the most common resin precursors used for structural carbon–carbon is the phenolic resol SC 1008 originally developed by Monsanto which can be readily obtained from Borden Resins. Furan resins such as those produced by the Quaker Oats Chemical Corporation are often used in reimpregnation/densification processes. Prepregs and moulding compounds are produced by a number of companies, Fiberite and BP Chemicals/US Polymeric being the most notable.

A whole host of impregnating carbon precursor pitches which are available for use in the 'conventional' carbon and graphite industries are equally apposite in carbon–carbon fabrication. Of special interest are the series of fractionated pitches produced by Ashland in the USA, for which a considerable experimental data base exists in the open literature on carbon–carbon use. Finally, for those interested only in the evaluation of existing products, a number of 'finished' materials may be obtained from the various manufacturers. A more complete listing of the companies involved in the various stages of carbon–carbon production can be obtained from Chapter 10. A number of 'exotic' thermoplastic precursors such as PEEK and PEI may be obtained commercially, while high char yield resins and oxidation inhibitors are generally proprietary, and must be synthesized in the laboratory.

7.3 DENSIFICATION OF COMPOSITES BY CVD

Chemical vapour deposition has been used for over 20 years for depositing carbon matrices in carbon–carbon composites. The same technology is appropriate for depositing silicon carbide or other matrices in ceramic–ceramic composites or carbon–ceramic composites. Furthermore, CVD is a well-developed technique for applying the graded oxidation-resistant coatings required for carbon–carbon to be subjected to long-time exposure to oxidizing environments.

Figure 7.1 shows a schematic diagram of a furnace capable of producing carbon–carbon and ceramic composite materials and coatings by CVD.

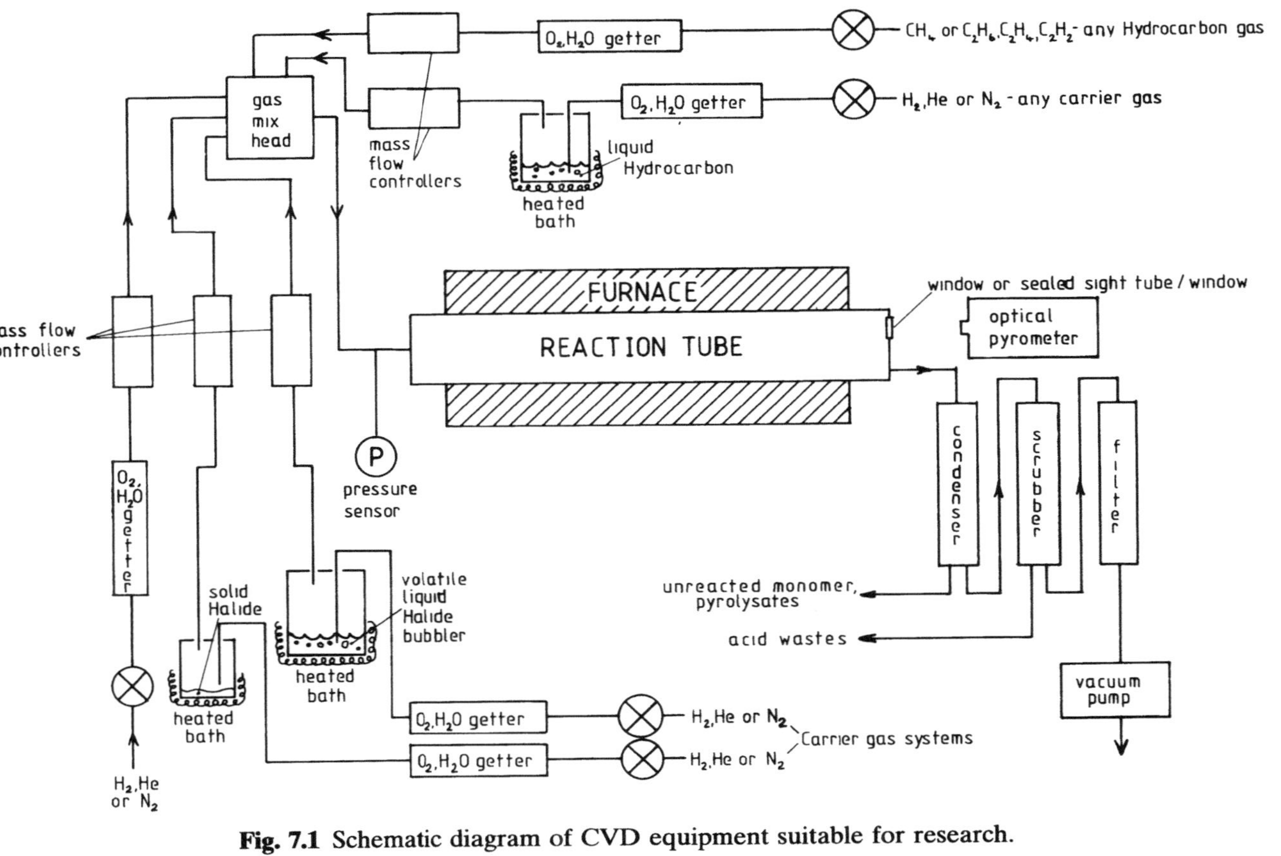

Fig. 7.1 Schematic diagram of CVD equipment suitable for research.

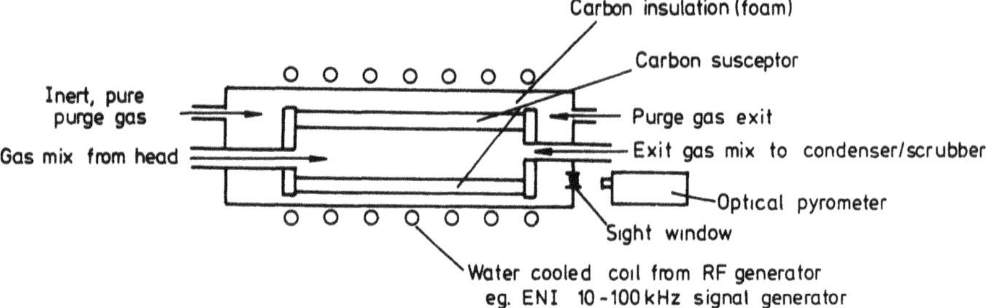

Fig. 7.2 Induction-heated furnace for use with CVD apparatus.

Starting materials can be a carbon felt preform or a pre-carbonized porous component made by resin/carbonization. The furnace contains five input lines to give flexibility for carbon deposition from gaseous/liquid hydrocarbon precursors and ceramic deposition from liquid/solid precursors such as ZrC–SiC alloys from $ZrCl_4/CH_3SiCl_3$, $/H_2$ and ZrB_2 from $ZrCl_4/BCl_3/H_2$. A conventional resistive heated tube furnace can be substituted by an induction-heated tube furnace of similar structure to the design in Fig. 7.2. A thermocouple such as W/WRe can be used to control the temperature in situations where only carbon is being deposited. Thermocouples cannot be used when acid halide gases are used for ceramic deposition, necessitating the use of a pyrometer/window system.

All of the parts, furnaces, mass flow controllers and valves, etc. are standard components and easy to obtain. It is possible therefore to construct such equipment from scratch although it is probably wiser to consult a specialist supplier. Modern instrumentation allows many variations in properties to be made which had been impossible in the past. All of the mass flow controllers/valves and furnace temperature settings should therefore be computer interfaced. A sophisticated computer program can then be used to control and operate the equipment using dual screen displays. Finally, an appropriate warning and alarm system is required to identify malfunctions, especially those which may result in permanent damage or safety concerns.

Most CVD technology has evolved empirically. Very little is known about the mechanisms of deposition and the relationships between reaction conditions, product microstructures and final properties. Furthermore, all of the proprietary information generated on low-pressure CVD of protective coatings and commercial products is zealously guarded by the various organizations involved. A great deal of 'trial and error' research will therefore be required of anyone entering this field, both at the R&D level and when scaling up to form a business.

7.4 FABRICATION OF THERMOSET RESIN LAMINATED PRECURSORS

7.4.1 Hand lay-up

The majority of composite structures, such as motor car body panels and racing kayaks, still use hand lay-up as the method of production. In the laboratory it is possible to use a variation on the process to make good quality 2-D precursor laminates. Using a rectangular matched metal mould, a woven fabric laminate plaque may be laid up using an excess of resin between two PTFE-coated aluminium sheets. Consolidation is best carried out under heat and pressure using a heated platen press. The resin content may be controlled using the applied pressure. Application of pressure via the platens results in excess resin being squeezed out of the mould. Experimentation allows the evaluation of pressure/resin content relationships, from which the desired laminate can be produced. The technique, for obvious reasons, is known as the 'leaky-mould' method.

7.4.2 Prepregging and filament winding

Virtually all of the structural carbon–carbon produced commercially by the thermoset resin route is of the 2-D lamellar type. In the laboratory, on the other hand, unidirectional (UD) material is often produced in order to minimize the number of variables in materials evaluation. A number of companies produce simple inexpensive machines for making UD carbon fibre prepregs. They all essentially consist of a drum winder with an impregnation system as shown in Fig. 7.3. The cylinder or drum is typically a 38 cm diameter steel tube with closed ends, driven by an electric motor with variable speed control, allowing approximately 1 m^2 batches of prepreg to be prepared. Thyristor heat controllers are used to vary the drum's temperature to allow the volatilization of solvents. The impregnation system relies on gravity or a positive pressure pump to deposit the resin on to the fibre just before the impregnation wheel (Fig. 7.3). The whole unit can be traversed along the length of the drum by a controlled-speed electric motor.

The 'wet winding' prepreg process is essentially a form of filament winding in which the tows are laid parallel to one another to produce a sheet material. Filament-winding machines facilitate the arrangement of impregnated tows at various orientations over a mandrel to form cylindrical or regular polygonol cross-sectional composite tubes. Structures so formed may be cured under ambient pressure at elevated temperatures or consolidated with a vacuum bag and autoclave cure.

Woven prepregs may be produced relatively easily by pultrusion through a solution bath (Fig. 7.4). Resin content is controlled by the process variables

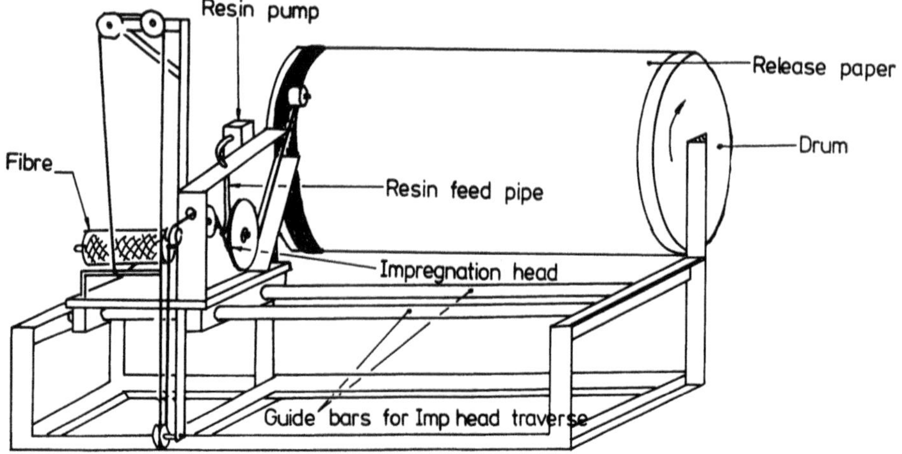

Fig. 7.3 Simple UD laboratory-scale prepregger.

Fig. 7.4 Solution prepregging of carbon fibre fabric (courtesy Ciba-Geigy).

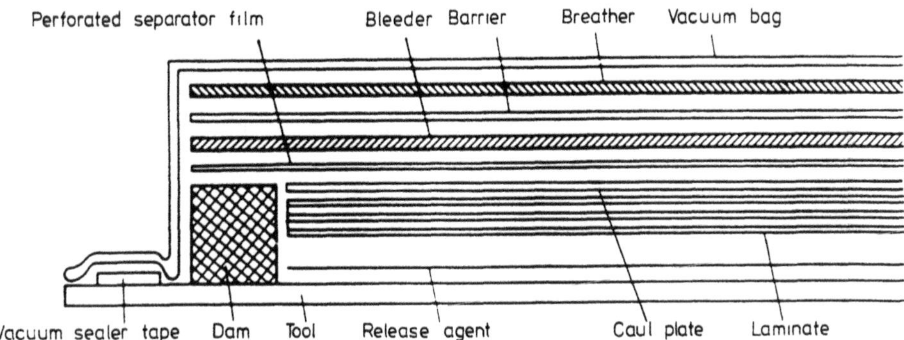

Fig. 7.5 Vacuum bag preparation.

of solution strength and line speed, etc. the most difficult parts of the operation being the provision of a consistent resin solids and volatile content, and subsequent removal and recovering of the generally flammable solvents. Finally, short-fibre brake-moulding compounds are made by a similar solution process save that two UD tows are pultruded through a circular die and then cut into approx 35 mm lengths. The composite is formed by compression moulding in a matched die and elevated temperature curing. All of the prepreging and winding processes lend themselves to the incorporation of additional phases such as particulate fillers and oxidation inhibitors.

7.4.3 Processing of thermoset prepregs

In the curing of thermosetting materials the prepreg resin is transformed from a semi-solid to a viscous liquid by the applied heat and the excess resin is squeezed out by the action of the applied pressure prior to the formation of a three-dimensional cross-linked molecular network known as gel. The chemical reactions which occur during curing are always exothermic. An applied pressure provides the force necessary to consolidate the plies and to compress vapour bubbles in order to obtain a product with minimum void content; alternatively, the voids diffuse out of the lay-up during resin flow. The vacuum bag lay-up sequence for a typical thermoset matrix composite (phenolic, epoxy, etc.) is shown in Fig. 7.5, which illustrates the fabrication of a flat plate laminate for materials evaluation. A shaped component would follow the same process save that a contoured male or female tool would replace the flat baseplate. The mould tool, or aluminium baseplate, is first thoroughly cleaned using acetone or some other degreasing agent. The surface is then 'released' twice with mould-release agent. A number of such products are available in the market-place; while their composition is generally proprietary, all operate as a suspension which leaves a thin coating of PTFE or silicone particles over the moulding surface. From experience of a number of different release agents, and lengthy discussions on the subject, it is suggested that the 'Frekote' range produced by the Dexter Corporation is one of the best available, especially the 'Frekote 44' product which may be considered as an 'industry standard'.

When making large numbers of laminates for mechanical property, etc. determination it is advisable to use a 'pre-released' baseplate. This is achieved by covering the metal with a self-adhesive film of PTFE-impregnated glass fibre fabric. Such a product can be obtained from Courtaulds or Airtech, Richmond in the USA. After each lamination the plate can be quickly cleaned and reused without the need for further release agent. The mould for a plate laminate is made in the form of a 'picture frame' using a self-adhesive rubber gasket material [1] specifically designed for the purpose. A rubber 'wall' is built up around an aluminium 'caul plate' cut to the same size as the prepreg sheets. The plate is removed and the laminate stack laid inside the mould. Two glass fibre cords are laid into the asssembly to facilitate removal of air once the vacuum is applied (see Fig. 7.5). The caul plate, a sheet of aluminium coated similarly to the baseplate, is then dropped into the mould above the laminate. In the case of curved mouldings the plate is replaced by a PTFE-coated glass fabric. The caul plate is secured to the rubber gasket using high-temperature 'tacky tape' to prevent movement during cure and damage to the vacuum bag. A continuous vent cloth is placed on top of the lay-up and extended over the vacuum line attachment. The purpose of the porous vent or 'breather' cloth is to provide a path for volatiles to escape in the applied vacuum and to achieve a uniform distribution of vacuum. A nylon bagging film is placed over the entire plate and sealed against the bagging adhesive, allowing enough material to permit conformation to all contours without being punctured. Finally, the vacuum line is attached and the bag checked for leaks.

Curing of the laminate takes place in an autoclave under vacuum. The autoclave is generally a large pressure vessel equipped with a microcomputer-interfaced temperature and pressure control system. The elevated temperature and pressure conditions are created by electrically heating a pressurized gas (usually air or nitrogen). Once locked in the autoclave the part is left under vacuum for around 30 min to recheck for leaks in the bag. Curing conditions are either provided by the manufacturer or may be determined experimentally. Research has shown that the best results (in terms of matrix properties and carbon yield) are obtained if the laminates are post-cured in an oven prior to pyrolysis in order to increase the degree of cross-linking [2,3] and removal of water in condensation polymerized resins. A large degree of matrix shrinkage occurs during pyrolysis of the composites which may result in delamination. In addition, delamination may occur due to pore pressures in post-cure. It is imperative,

Fig. 7.6 Low-pressure pyrolysis furnace for carbon–carbon production showing gas purification equipment.

therefore, that the precursor laminates be of the highest quality in order to minimize this effect. It is recommended that, after a panel is cured, it should be non-destructively inspected by the Ultrasonic C Scan [4] method. Should any material contain delaminations or an unacceptably high void content it should be rejected and a replacement produced.

7.4.4 Pyrolysis of carbon fibre reinforced thermosets

Carbon fibre reinforced thermoset laminates may be converted to carbon–carbon by controlled pyrolysis in an inert atmosphere. Figure 7.6 illustrates a furnace system capable of performing such a task on a laboratory scale. The furnace consists of a heated reaction tube through which a flow of inert gas (generally nitrogen or argon) may be passed (Fig. 7.7). A typical laboratory temperature profile would consist of heating at a linear rate of, say, 10 °C min^{-1}, holding for 2 h at the carbonization temperature (850–1200 °C) followed by a linear cooling at around the same rate. Larger components may require cycles of 5–10 days to 900 °C or more. Control is thus relatively easy to achieve using a simple three-term controller/programmer operating via a thermocouple. Within reason, it is preferable

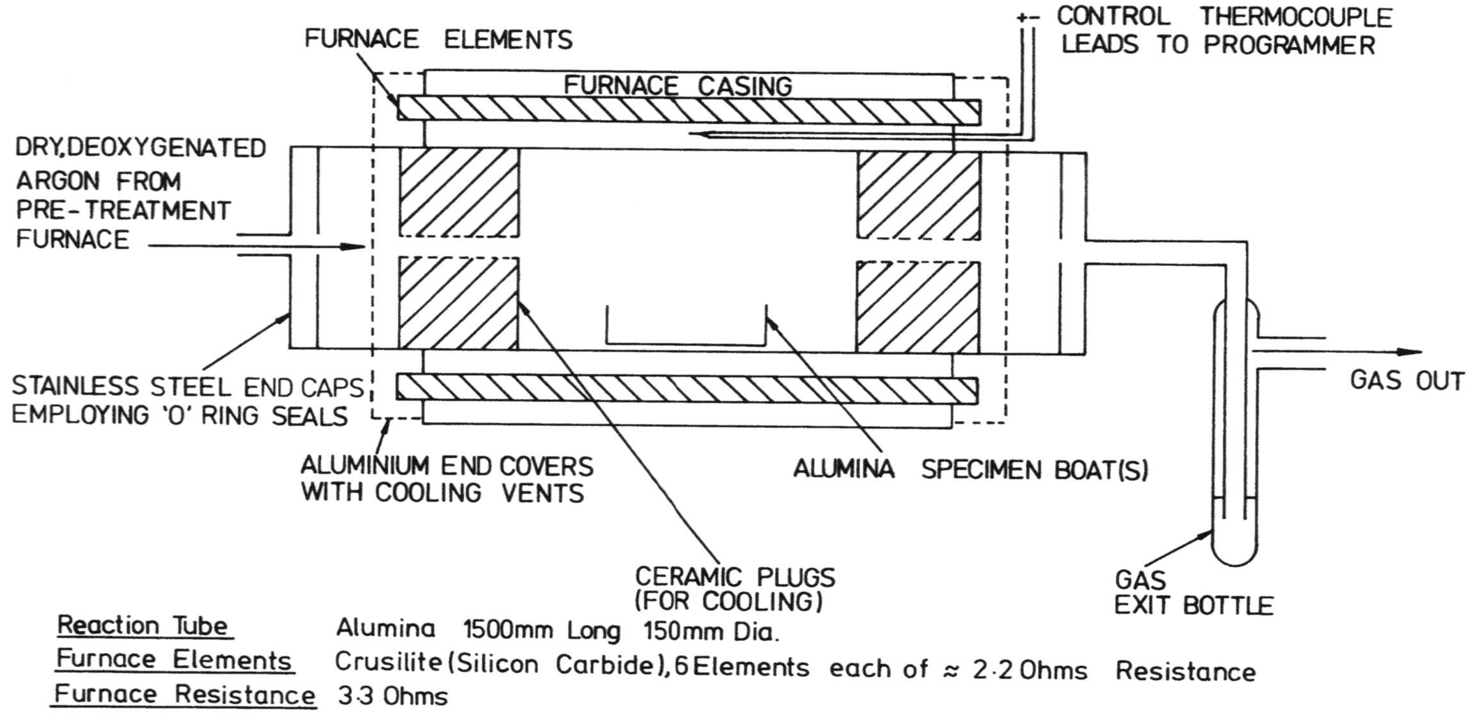

Fig. 7.7 Schematic diagram of ambient pressure carbonization furnace.

to use as slow a heating and cooling rate as possible to minimize the effects of shrinkage, cracking and thermal stresses which may build up during carbonization.

The presence of oxygen during processing is, as one would expect, extremely damaging to the carbon–carbon product. Despite bearing the tag of 'high purity', many of the gases supplied by manufacturers contain moisture and oxygen contamination in amounts which may be quite significant when the time period of the processing is considered. It is advisable, therefore, to ensure that gas is dry and free of oxygen before it enters the furnace. This is achieved by allowing it to pass through a series of drying columns containing a molecular sieve and into a pre-treatment furnace. This device is a simple tube furnace, held at around 900 °C using a single-term controller, containing titanium granules which act as an oxygen getter. Both gas purification devices can be clearly seen in Fig. 7.6. Other room-temperature oxygen getter systems can be used as well, such as supported catalysts.

7.4.5 Reimpregnation/densification cycles

The carbon yields obtained from the pyrolysis of thermosetting resins generally fall in the range 50–65% by weight. As a result, the carbon–carbon produced by this route contains large amounts of porosity, is of low density and hence possesses poor mechanical properties. To obtain a useful engineering material it is necessary to densify the cabonized material. Densification may be carried out by CVD of the porous laminate, or by vacuum reimpregnation with a resin or resin/pitch mixture followed by a further carbonization cycle. The process is repeated until the desired density is attained. Intermediate graphitization (Section 7.6) is often used to open up porosity and aid reimpregnation. A simple and inexpensive densification unit may be constructed in the laboratory using a vacuum desiccator, a one-way stopcock, a polyethylene funnel and a short length of tubing. The unit is assembled as shown in the sketch in Fig. 7.8. The desiccator may be made of glass or some form of polymer. Should a plastic unit be used, great care must be taken to ensure that it will resist the solvents which tend to splash around inside during the resin introduction step. Such a unit is only good for vacuum impregnations at room temperature with rather small parts such as mechanical test pieces, although scaling-up should not present too many problems.

7.5 PROCESSING THERMOPLASTIC PRECURSORS

The processing of thermoplastic composites involves heating the matrix to a temperature above its melting-point, application of a consolidation

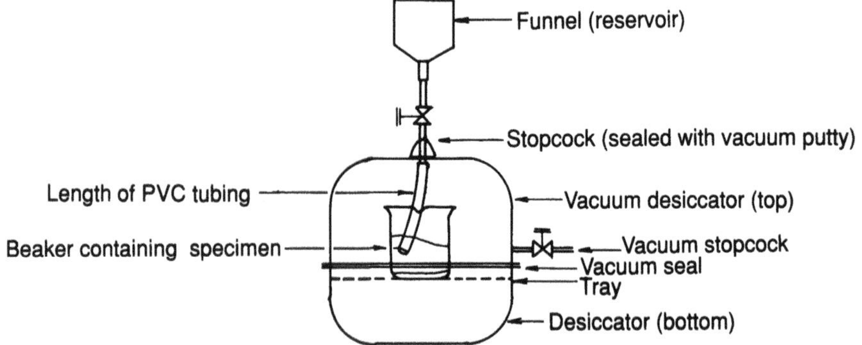

Fig. 7.8 Simple vacuum reimpregnation unit.

pressure and cooling. In the case of semi-crystalline polymers such as PEEK it is important to achieve the optimum degree of crystallinity and thus the cooling rate may be critical [5]. The cooling rate is not of concern when processing amorphous polymers such as PEI or pitches [6].

7.5.1 Impregnating with pitches

A woven or multi-directional carbon fibre structure may be vacuum impregnated with a molten pitch [7,8]. In some cases pressure is applied as part of the impregnation cycle to ensure penetration of all available porosity in the composite structure. The impregnated parts may be subjected directly to carbonization in an inert atmosphere to temperatures in the range 650–1100 °C [9]. The atmospheric pressure carbonization of pitch and other thermoplastics is extremely inefficient, as discussed in Chapter 5. The processing of such precursors is far better suited to the HIPIC process. In preparation for the HIPIC process, impregnation of the fibre preform structure is achieved by a hot pitch transfer technique [10]. The pitch is melted under vacuum in a heated reservoir. The fibre performs are placed in metal canisters and heated to the same temperature, again under vacuum, in an adjacent tank. Connecting pipes are employed to transfer the molten pitch to the metal cans containing the performs. The pitch transfer can be achieved with relative ease by backfilling the reservoir tank with nitrogen. Once the cans are full of liquid pitch the pressure is equalized by backfilling the preform tank with nitrogen in order to stop the transfer. The cans are closed off under vacuum by applying metal lids and then sealed with an electron-beam weld (Fig. 7.9).

Fig. 7.9 Electron-beam welded HIP can ready for processing.

7.5.2 Thermpolastic prepregs

Thermoplastic polymers are most easily prepregged in the molten state by a simple pultrusion process. A UD web of fibre tows, or a woven fabric is pulled through a 'melt pool' of the matrix material and then between 'nip' rollers which effect the impregnation. A scraper bar or similar device is used to remove any excess melt. The solidified material may then be collected by winding on to a spool. A schematic diagram of the process is shown in Fig. 7.10. Control of the process is achieved via the tension in the fibres, speed of pultrusion, matrix melt viscosity and the spacing between the nip rollers. A number of polymers may be prepregged in a simple melt condition. The materials useful in carbon–carbon production are generally high-temperature engineering resins. Polymers such as PEEK, Polyether sulphone (PES) and PEI, for example, have a melt consistency rather like chewing gum and must be 'thinned' with some form of solvent. The solvents needed and the dilution required are usually proprietary information.

Pitch has a very low melt viscosity and may be prepregged with ease. When solid, however, it is extremely friable and easily cracked off the fibres. The room-temperature toughess of pitch may be improved by adding graphite powder and milled carbon fibres to the melt. Such additions as well as making the prepreg handlable will also increase the carbon yield. A variation on the prepregging process involves the pultrusion of single or double tows through a circular die rather than nip rolls. The

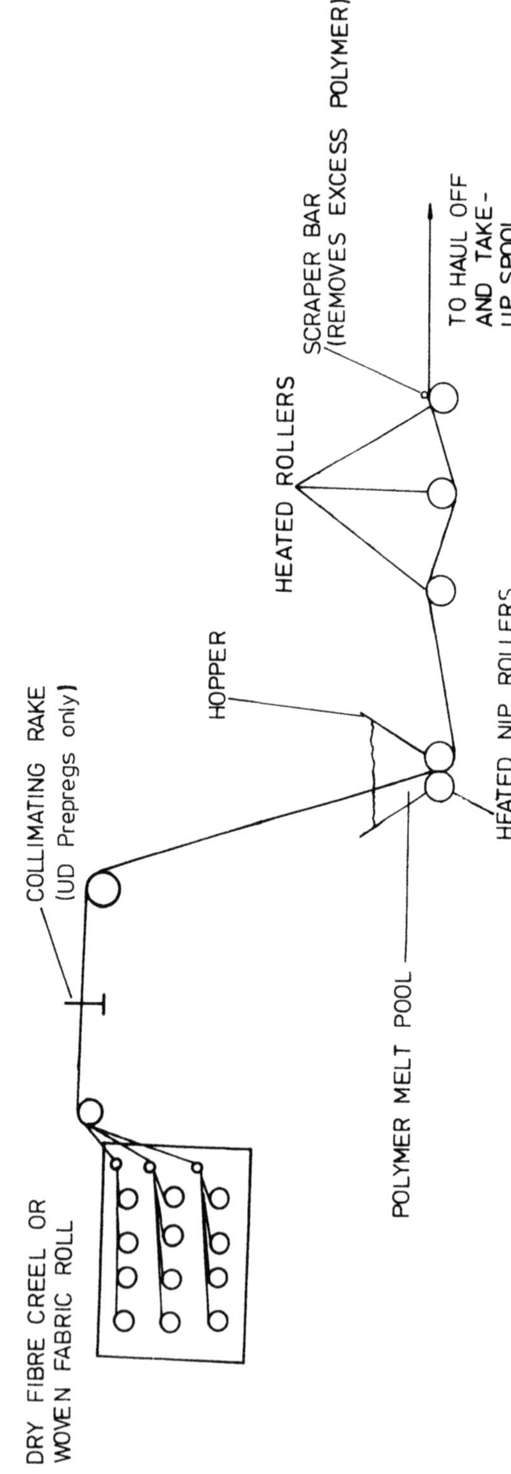

Fig. 7.10 Schematic diagram of thermoplastic matrix prepregger.

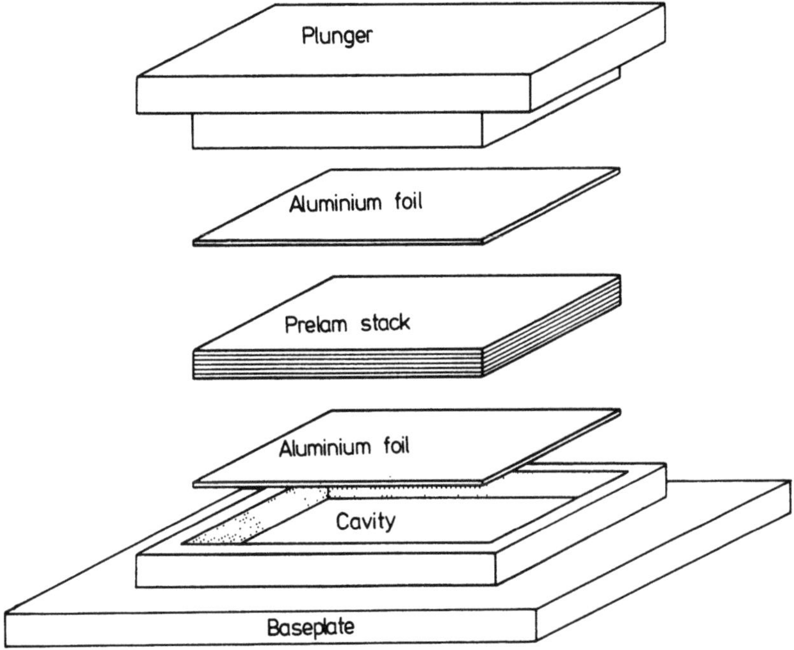

Fig. 7.11 Matched die moulding of thermoplastic.

product may then be chopped up to produce a thermoplastic equivalent of the thermoset 'matchstick' moulding compounds.

7.5.3 Lamination of thermoplastic composites

The production of laminates involves stacking layers of the prepreg which are then placed in a press, heating to above the melting temperature of the matrix under a low pressure, application of a consolidation pressure, and cooling rapidly. A prepreg stack is formed by cutting pieces of the prepreg in the desired orientation. Once the required number of plies have been produced they are laid up into a stack of layers and tacked together with a soldering iron. The preferred method for producing flat thermoplastic composite laminates is the matched-die moulding technique. The matched die is made of stainless or mild steel and consists of a female mould cavity and a male plunger (Fig. 7.11). The plunger fits into the cavity with the standard die-mould clearance. The moulding stack is then made up of the mould cavity, two pieces of aluminium foil in between which are placed the prepreg stack or moulding compound, and the plunger. The die and aluminium foil are liberally coated with release agent prior to processing. A slight variation on the above technique is the 'film stacking' process which can be used to manufacture woven reinforced laminates. The prepreg stack is replaced by a 'sandwich' of woven carbon fibre fabric interspaced with the matrix in the form of a film. The consolidation process is the same as before, save that longer time is required in order to allow for the melting of the film and impregnation of the fibres. Film stacking is a useful technique in making laminates to assess a fibre/resin combination not yet available in prepreg form, but is time-consuming and tends to produce poorer laminates due to incomplete impregnation and damage to the surface of the fibres during processing. Resin content of the composite is controlled by the relative amounts of fibre and film and the applied pressure. A number of technologies may be applied to produce curved geometry thermoplastic composites [11].

7.5.4 High-pressure carbonization

The majority of a HIPIC densification process is the same as that of an ambient pressure cycle save for the use of pressure during carbonization. A number of manufacturers such as ASEA and National Forge market laboratory-scale HIPs suitable for the processing of carbon–carbon (Fig. 7.12). A schematic of a complete HIPIC densification cycle using pitch is shown in Fig. 7.13. The major difference between an ordinary (if such a thing exists!) HIP unit and one suitable for carbon–carbon production arises from the severely corrosive nature of the outgassing which occurs during carbonization. This, coupled with a small but nevertheless noteworthy risk of hydrogen embrittlement of the pressure vessel, necessitates a special isolation chamber with a differential pressure control system (Chapter 5). The gas-pressure vessel is usually water-cooled with an internal furnace. The remainder of the system consists of gas storage, high-pressure gas transfer lines, compressor and controls (Fig. 7.14) [12]. Computer control via a PC interface is essential to maintain product quality [13].

The sealed metal can containing preform and excess pitch is placed in the pressure vessel. The temperature and pressure are raised at a programmed rate to effect the conversion to carbon–carbon. The thin metal can allows the isostatic gas pressure to be transferred to the workpiece in a cycle which may take typically 1–3 days. Once the cycle is complete the cans are removed and the sample machined to clean it up. The carbon–carbon artefact may be graphitized and/or reimpregnated and sealed in another can to undergo further cycles. Usually three or four HIPIC cycles are required to reach optimum density. Thermoplastic polymer matrix materials may be HIP carbonized without the need of metal cans as they have been shown to form their own hermetic seal [14]. Laminates based on PEEK, for example, may be processed to high-quality carbon–carbon in two or, in some cases only one, cycle.

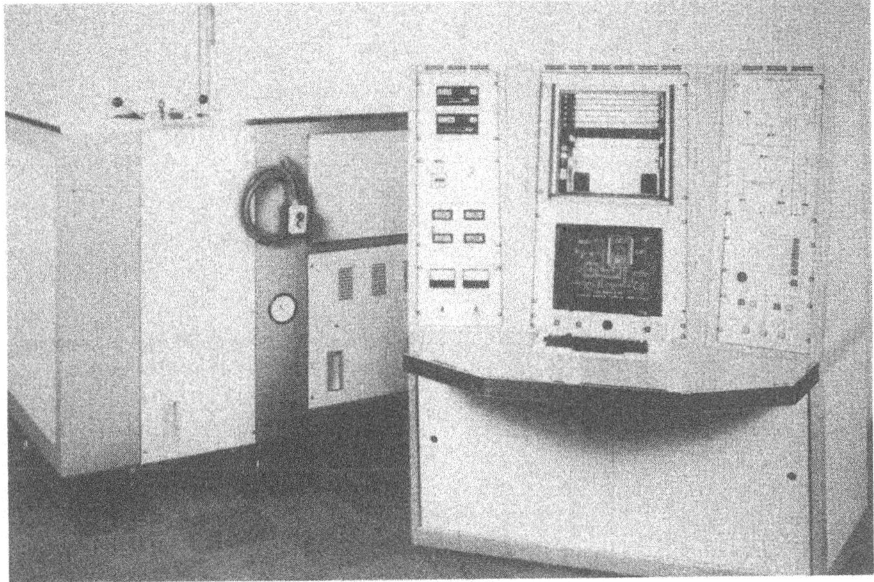

Fig. 7.12 Laboratory scale HIP unit suitable for making carbon–carbon composites.

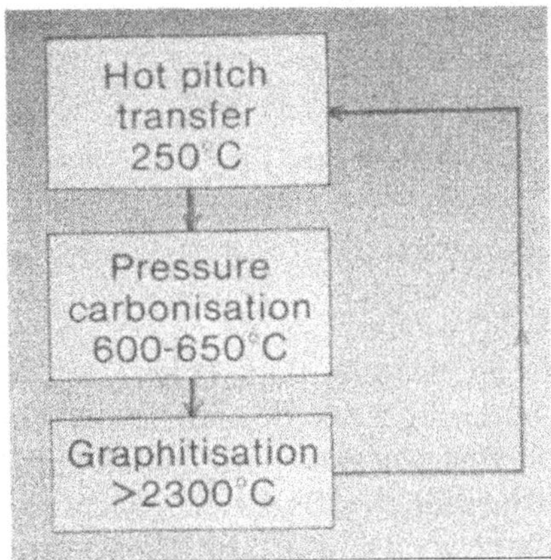

Fig. 7.13 Flow diagram of a complete HIPIC process.

7.6 GRAPHITIZATION OF CARBON–CARBON

The graphitization of materials which have passed through a liquid state on carbonization or the 'stress-induced' graphitization of fibre-reinforced isotropic carbon, is achieved by a controlled heat treatment to temperatures above 2300 °C. The furnaces to perform such a task consist of a heavy wall, seamless extruded aluminium or stainless steel chamber,

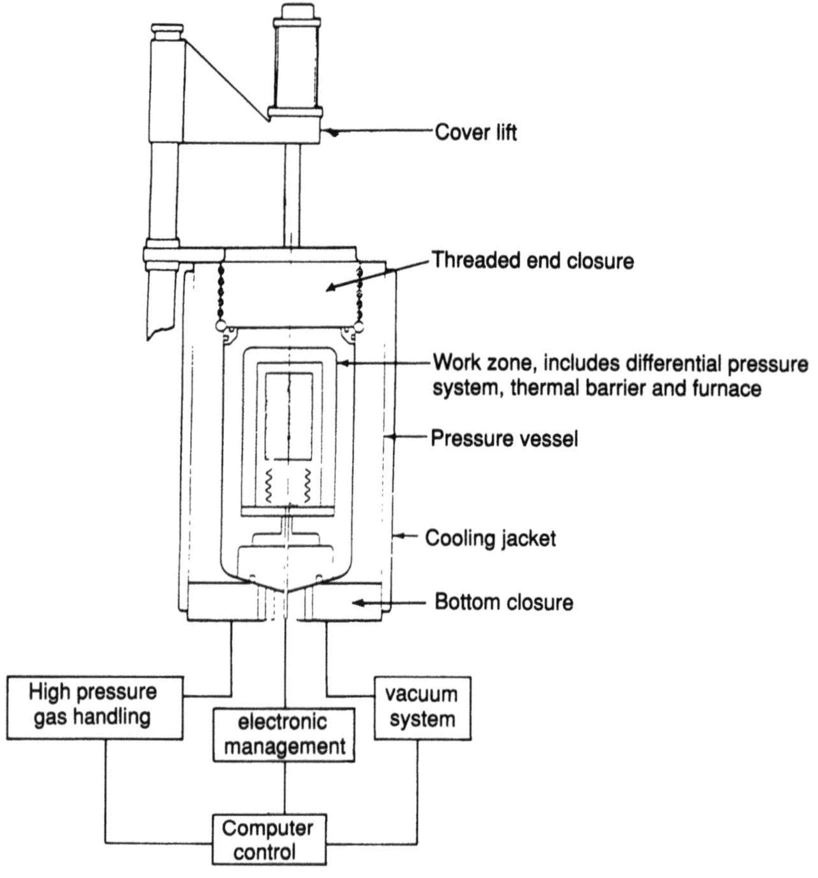

Fig. 7.14 Schematic diagram of laboratory-scale HIPIC unit.

water-cooled with an opening door (Fig. 7.15). The hot zone may be cylindrical, rectangular or cubic depending on the geometry of the perceived payload. Heating is achieved using graphite strip electrical elements or inductively heated graphite susceptors, which will allow continuous operation at temperatures of up to 2750 °C and short time usage at up to 3000 °C. A temperatures above 2800 °C the elements begin to sublime and their lifetime will therefore be finite at these temperatures. Operating environments are inert or dry reducing atmospheres with a maximum over pressure of 1 bar and full vacuum. The addition of a ceramic muffle tube permits working in oxidizing or wet atmospheres compatible with aluminium oxide (max. temperature ≈1850 °C) and zirconia (≈2200 °C). High-purity graphite felt is used as thermal insulation (Fig. 7.16). Temperature control is generally achieved using a three-term programmer/controller operating via a thermocouple at lower temperatures (<1400 °C) and a two-colour optical pyrometer over the high-temperature range. For maximum

Fig. 7.15 3000 °C capability graphitization furnace showing optical pyrometer, thermocouple retracting device and water-cooling coils over the main chamber.

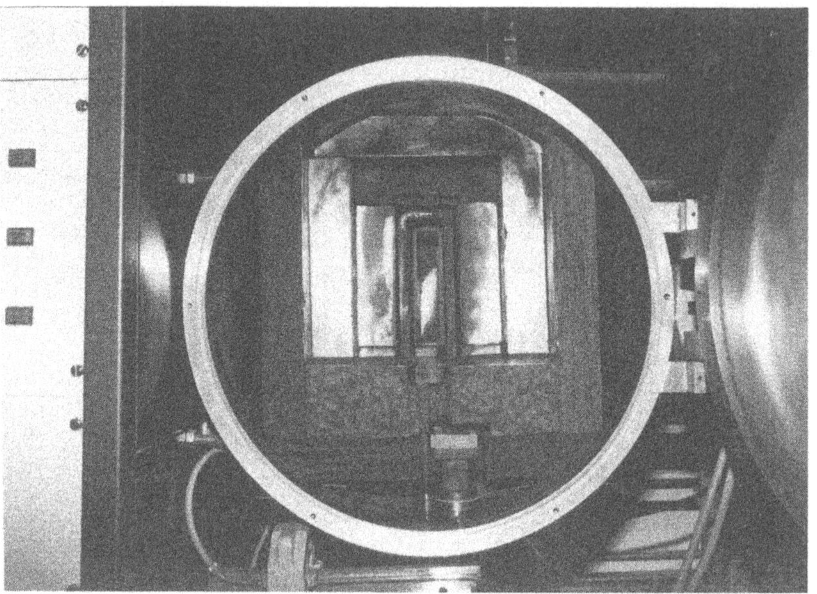

Fig. 7.16 End view of graphitization furnace with a rectangular hot zone.

accuracy it is advisable to place the thermocouple as close to the working zone as possible. To prevent damage to the thermocouple at high temperatures a retracting device is fitted which also triggers a 'bumpless transfer' to pyrometer control. A pneumatically operated device of this type can be seen underneath the furnace in Fig. 7.15. Prior to operation, air is removed from the furnace using a rotary pump capable of producing a vacuum of 10^{-2} bar followed by backfilling with an inert gas (argon or helium). This process must be repeated three or four times to ensure all traces of air are removed, thus preventing oxidation damage to furnace and workpiece during operation. Alternatively, the air can be displaced by flushing for extended times with nitrogen. The temperatures of operation involved make a furnace of this type potentially an extremely dangerous piece of apparatus. It is advisable therefore to install a safety interlock system coupled with water supply, inert gas supply, furnace body and contents over-temperature and over-pressure trips as 'fail-safe' devices.

7.7 MECHANICAL TESTING OF CARBON–CARBON

7.7.1 General considerations

All of the applications of carbon–carbon composites are either structural or require at least a degree of strength and stiffness to be successful. It is important therefore to determine the mechanical properties of the material. There are no standard procedures for the testing of carbon–carbon.

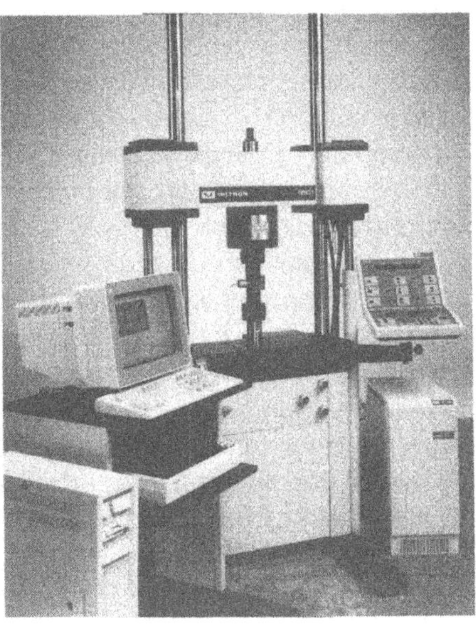

(b)

(a)

Fig. 7.17 Servomechanical (a) and Servohydraulic (b) universal test machines (courtesy Instron Ltd).

Having said that, a number of the tests detailed for the evaluation of fibre-reinforced polymer composites may be used [15,16]. Testing is carried out using a universal testing frame. Two types of frame are available, servohydraulic and servomechanical. Servohydraulic machines apply the load by means of a hydraulic actuator controlled by a servovalve and are best suited to dynamic testing of materials and components. A servomechanical tester applies the load via cross-head movements driven by one or two large screws. The servomechanical device is generally preferred in the 'static' testing of materials as a result of its inherently high degree of accuracy. This is especially true when testing extremely low strain-to-failure materials such as carbon–carbon. A number of manufacturers supply equipment of one or both types (Fig. 7.17) which are required to conform to the calibration standards BS 1610 [17] and/or ASTM E4 [15] and be equipped to record load, and deflection rate via cross-head or actuator movement. Accurate load alignment and position control are essential for satisfactory testing. Continuous or 'ramp' loading is preferred, with continuous load and strain recording.

The measurement of strain may be carried out using electrical resistance strain gauges or an extensometer. Foil strain gauges with 350 Ω resistance and a gauge length of 3–6 mm are preferred. Typically, two gauges, one each in the longitudinal and transverse direction, are bonded to the specimen using a cyanoacrylate adhesive [18]. Many researchers believe strain gauges to be far more accurate than extensometers. The technology of the latter has evolved considerably of late, such that they are now extremely precise especially when interfaced with the test equipment via a computer. Extensometers are far more easily attached to the specimen than strain gauges and can be used at very high temperatures. They are required to conform to ASTM E83 [15]. Universal testing frames are often known by the generic term of an 'Instron' machine after one of the best known companies in the field, in the same way that in the UK vacuum cleaners tend to be called a 'Hoover' irrespective of manufacturer. The company's model 4505 servomechanical and 8502 servohydraulic machines have a load capacity of 100 kN and can be used in load, displacement and strain control, operating via a well-researched software package through a personal computer, and are ideally suited to the testing of composites in general and carbon–carbon in particular.

7.7.2 Sample preparation

Considering their 'ceramic' nature, carbon–carbon composites are extremely tough. It is well known that a 2-D carbon–carbon material can be 'nailed' without fracturing. Photos of carbon–carbon composites with nails driven through them have, unfortunately, become something of a cliché in the literature so the author shall refrain from using one! Despite their apparent toughness carbon–carbon and indeed polymer matrix composites are extremely susceptible to damage during machining. Great care must be taken to avoid fraying and disorientation while preparing mechanical test pieces. Specimens are best cut using water-cooled, high-speed diamond-impregnated grinding wheels (Fig. 7.18). It is advantageous to secure the laminate to a backing plate of acrylic sheet using double-sided sticky tape. The acrylic plate serves as a carrier for the composite laminate, avoiding the fraying out of fibres, and allows the diamond saw to cut through the carbon–carbon without damaging the blade.

Alignment of the composite laminate is an important issue since the mechanical properties of these anisotropic materials are strongly dependent on the orientation of the fibres. Should the particular test piece require the fitting of end-tabs (Section 7.7.4) they should be bonded prior to machining.

7.7.3 Statistical analysis

Composite materials are notorious for their variability in mechanical properties. The variability is even more prevalent in carbon–carbon composites due to their inherent flaws and inhomogeneities. It is important, therefore,

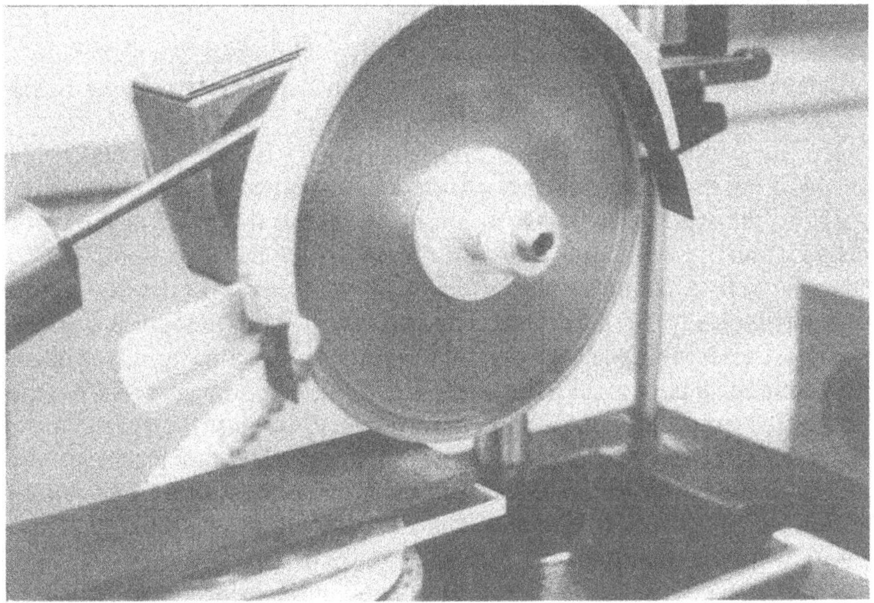

Fig. 7.18 Machining a carbon–carbon composite using a water-lubricated diamond saw.

to present the spread of results which is in itself a property of the material. Mechanical properties are given as the arithmetic mean ($\bar{x}$) of a sample of *n* specimens, where *n* must be greater than 10 and preferably around 30; then

$$\bar{x} = \sum^{n} \frac{x_n}{n}. \tag{7.1}$$

The standard deviation σ_{n-1} is then calculated using the following formula:

$$\sigma_{n-1} = \left(\frac{\sum^{n} (x_n - \bar{x})^2}{n-1} \right)^{1/2}. \tag{7.2}$$

The spread of results may then be expressed in terms of the 'coefficient of variation' which is defined as a percentage:

$$C_v = \frac{\sigma_{n-1}}{\bar{x}} \times 100\%. \tag{7.3}$$

7.7.4 Static testing

Static testing is used to evaluate the 'ultimate' or failure properties of materials in various configurations.

Tensile testing

The principal engineering properties of a carbon–carbon laminate are determined using a tensile test. The test may be carried out using either 0° UD or woven materials, providing the sample is axially orthotropic to obviate bending. Chamfered 'end-tabs' (Fig. 7.19) are attached to each end of the tensile specimen to ensure a uniform loading and prevent damage to the surface. A 30° taper is used to relieve the stress at the end-tab and ensure that failure occurs in the centre of the test piece. Tabs are made from a commercially available woven glass-cloth epoxy laminate sheet, typically 3.2 mm thick. The surfaces to be bonded are degreased using trichloroethane and are then lightly abraded using a grit-blasting technique. A pure aluminium oxide grit, of particle size range 80–120 mesh and free from contaminants, is employed at a low blast pressure of around 0.6 MPa. Loose particles are removed from the abraded surfaces by a short immersion in an ultrasonic cleaning bath followed by another degreasing operation. The end-tabs are attached using a suitable adhesive following the manufacturer's instructions [19].

The specimens must be carefully aligned in the jaws of the test machine to avoid induced bending of the specimen (this is best achieved using hydrautically operated grips) (Fig. 7.20). Width and thickness (w and t respectively) are be to measured at five positions on the specimen to an accuracy of 0.01 mm. Axial strain is measured using a 50 mm gauge length extensomether and width strain using a lateral extensometer at the centre of the gauge length. Alternatively a biaxial extensometer may be used. Tensile load, P, should be increased uniformly at a constant displacement of 1 mm min^{-1} or, more preferably, should the apparatus allow, a constant strain rate of 0.5% min^{-1}. Longitudinal tensile strength σ_{11} is given by

$$\sigma_{11} = \frac{P}{wt}. \tag{7.4}$$

The stress versus strain curve is sometimes non-linear with modulus changing slowly with increasing strain. The axial tensile modulus of the material is thus calculated as a spot value at a specified level of strain. The longitudinal tensile modulus is thus defined as

$$E_{11} = \text{secant modulus at } S\% \text{ longitudinal strain} \tag{7.5}$$

(cf. Fig. 7.21). Similarly, the major Poisson's ratio may be defined as

$$\nu_{12} = \frac{\text{transverse strain}}{\text{longitudinal strain}} \text{ at } S\% \text{ longitudinal strain}. \tag{7.6}$$

The data recorded should include longitudinal tensile strength (failure stress) modulus, the stress/strain curve to failure, strain to failure, major

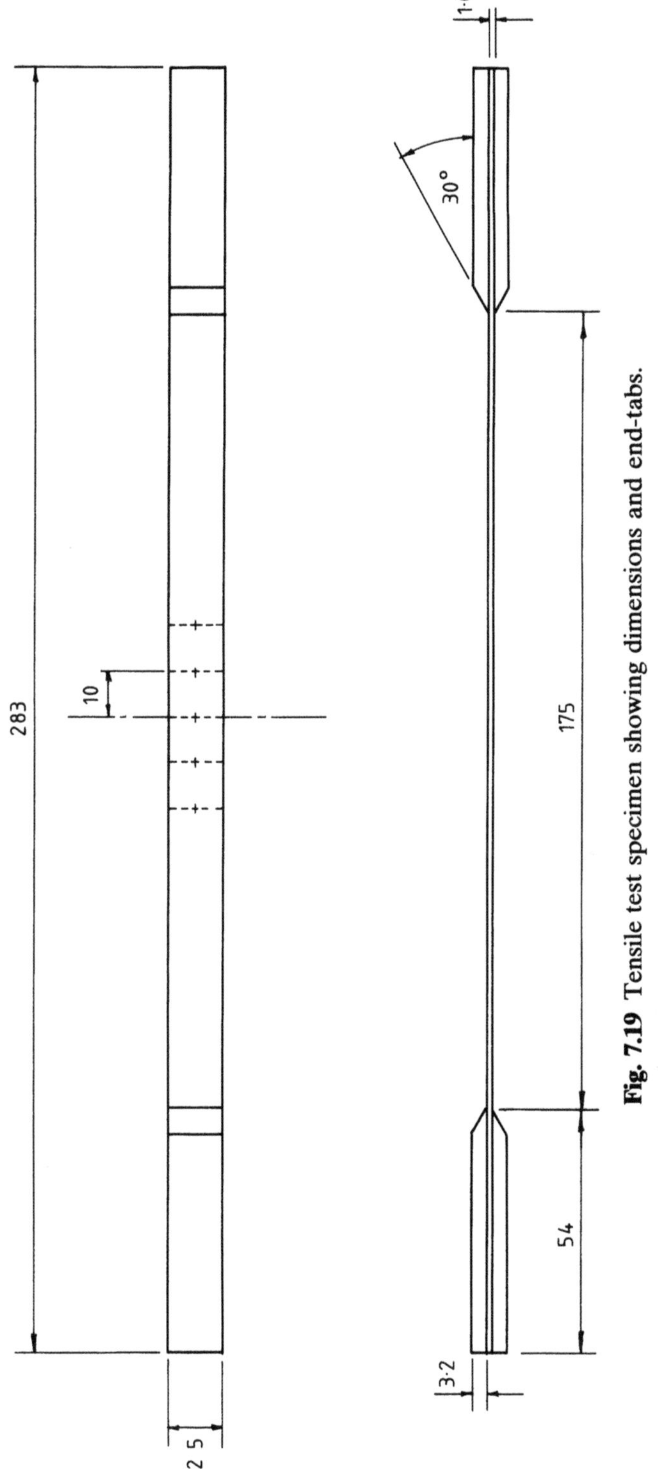

Fig. 7.19 Tensile test specimen showing dimensions and end-tabs.

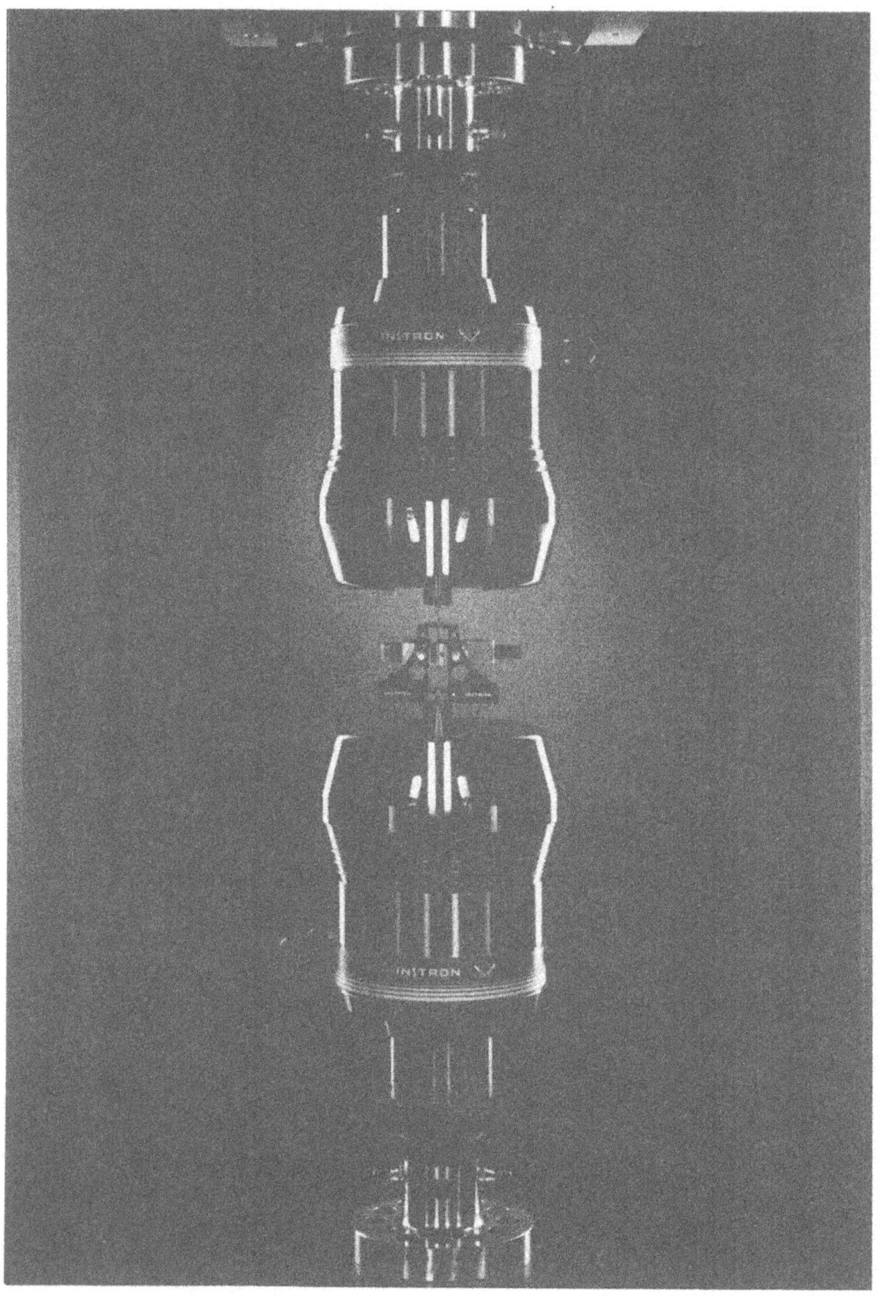

Fig. 7.20 Properly tabbed and aligned tensile specimen just prior to testing, complete with axial, averaging extensometer (courtesy Instron Ltd).

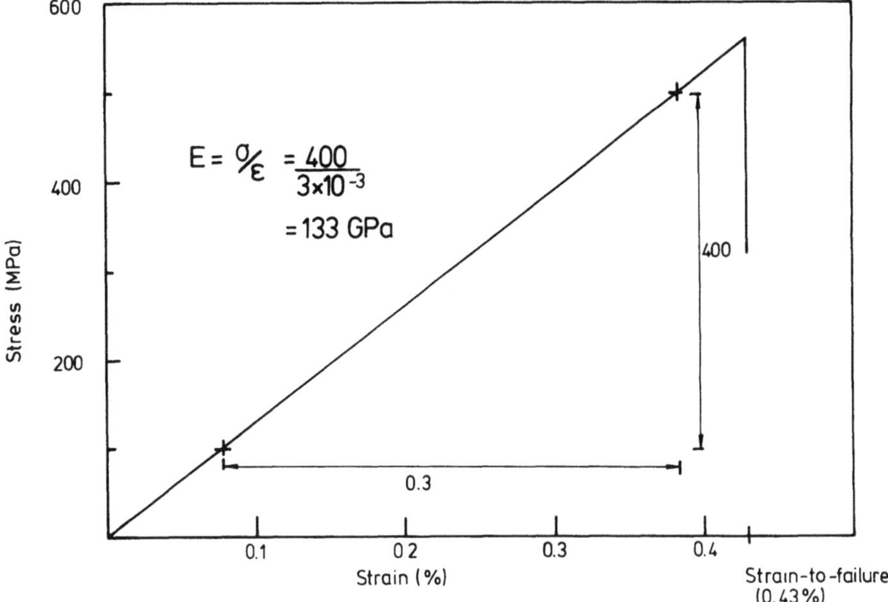

Fig. 7.21 Typical carbon–carbon composite stress–strain curve showing modulus calculation.

Poisson's ratio, and nominal and actual specimen depth and thickness. To be valid as design data, failure must occur in the central region of the specimen.

Flexure testing

The flexural properties of carbon–carbon laminates are evaluated using a three- or four-point test fixture. For a high modulus material such as this the three-point (Fig. 7.22) is generally preferred as the most direct and simplest technique to use. The procedure can be found in ASTM D-790 [15]. Flat specimens, machined with the same care and precision as previously described for tensile testing, are selected for measurements. The test may be carried out on either UD or woven material. The material direction under investigation must be orientated along the length dimension of the specimen. The test pieces require a span/depth (l/d) ratio high enough to minimize the influence of interlaminar shear deformation and to achieve failure in bending rather than shear [20]. The minimum l/d ratios for each specimen geometry are shown in Table 7.1. Woven fabric reinforced materials should be tested in both warp and fill directions. The 90° aligned specimens may be used as a qualitative/comparative estimate of the fibre/matrix bond strength.

The three-point flexure fixture should be properly mounted and aligned

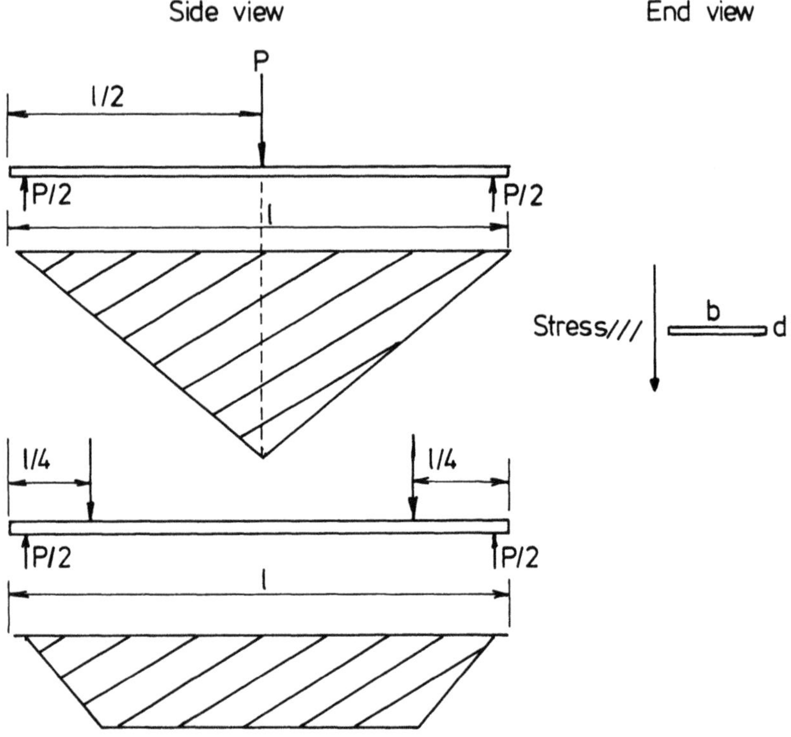

Fig. 7.22 Three- and four-point flexure test geometries showing stress distributions at the bottom surface.

Table 7.1 Span/depth requirements in flexure testing

Composite reinforcement	*Alignment of fibres to beam axis*	*Minimum l/d*
Unidirectional	0°	40/1
Unidirectional	90°	25/1
Woven	0°/90°	25/1

within a calibrated test frame (Fig. 7.23). Testing is carried out at a displacement of 1 mm min^{-1}. Strain readings are recorded continuously, being calculated from the displacement. The ultimate flexural strength σ_f, i.e. the stress in the outer fibre at failure, and the flexure modulus (Ef) are calculated as follows:

$$\sigma_f = \frac{3Pl}{2bd^2} \tag{7.7}$$

$$E_f = \frac{l^3 m}{4bd^3} \tag{7.8}$$

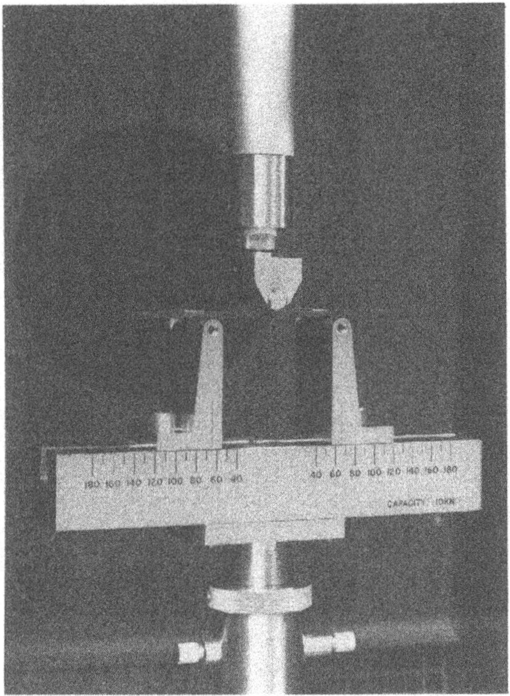

Fig. 7.23 Apparatus and sample set-up for three-point flexure test (courtesy Instron Ltd).

where the symbols represent the specimen dimensions in Fig. 7.22 and m is the slope of the tangent to the initial straight line portion of the load–deflection curve.

Shear deflections can seriously reduce the apparent modulus of highly anisotropic composites when tested at low span to depth ratio. It is recommended, therefore, that 0° UD specimens be tested at l/d of 50 : 1. Flexure testing imposes tension, compression and a degree of shear loading on the sample simultaneously. While the flexure specimens cannot therefore provide data suitable for design purposes, such tests are valuable in performing comparisons between materials or the effects of different processing and environmental conditions, etc. The four-point flexure test (Fig. 7.22) is sometimes used when testing carbon–carbon in order to minimize the local crushing between the central loading nose of the three-point loading procedure and to provide a uniform maximum stress field between the upper loading points. Loading at one-quarter of the span distance from each support is recommended [15].

In four-point flexure:

$$\sigma_f = \frac{3Pl}{4bd^2} \tag{7.9}$$

Fig. 7.24 Interlaminar shear test configuration (courtesy Instron Ltd).

$$E_f = 0.21\,\frac{l^3 m}{bd^3}. \tag{7.10}$$

Interlaminar shear strength

The interlaminar mode of fracture has aroused a great deal of attention in laminated composites since the early 1970s [21]. Since the introduction of fibre-reinforced composite materials subjected to severe service loading, it has become increasingly apparent that the interlaminar fracture mode is potentially the major life-limiting failure process [22].

One test frequently referred to in the literature (cf. ASTM D2344 [15]), which may be used to evaluate the interlaminar properties of composites, is the short beam interlaminar shear strength. Using a shortened span on a three-point flexural test fixture, with a span to depth ratio of 4 : 1, testing is performed as shown in Fig. 7.24. The short beam shear strength is calculated as follows:

$$\mathrm{ILS} = \frac{3P}{4bd}, \tag{7.11}$$

where ILS is the shear strength, P the breaking load, b the width of specimen and d the thickness. For the result to be meaningful, the failure

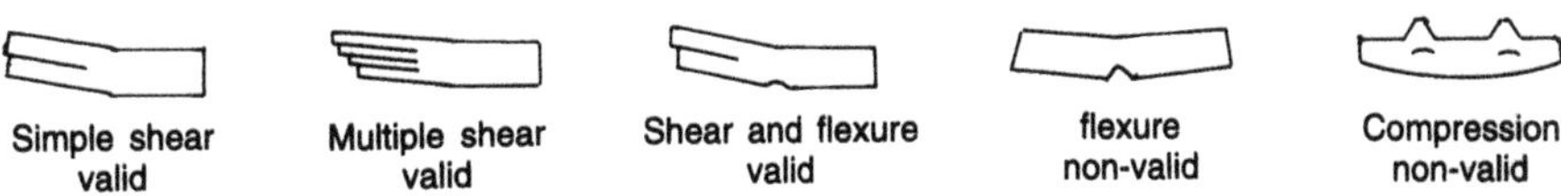

Fig. 7.25 Validity of the various failure modes possible in the short beam shear specimens.

mode must be either single or multiple shear, or plastic deformation with evidence of shear failure. The mode of failure must be recorded. Should failure occur by a non-shear mechanism, the result should be quoted as the 'short beam' rather than interlaminar shear strength. The various possible failure modes and their validity are shown in Fig. 7.25.

The interlaminar shear strength determined by this test method is useful for quality control and specification purposes. The data generated cannot be used as design criteria, but may be used in comparative testing. Carbon–carbon, especially that made by the resin pyrolysis route, in the 1- or 2-D form is extremely prone to interlaminar failure. The ILS test is therefore a useful method of characterizing the material. It should be remembered though, that flaws and defects in test specimens would generally disqualify the specimen from the test [23]. Since carbon–carbon composites usually contain an inordinate number of various types of flaws and inhomogeneities, the validity of such testing will always tend to be very questionable.

Iosipescu test

The major problem in shear testing, particularly in the determination of strength, is the influence of non-shear stresses on the properties measured. Such stresses are introduced by the specimen geometry and the test fixture. Iosipescu [24] proposed a test technique employing a double edge-notched test specimen. A region of pure shear stress is induced at the mid-point of the specimen by the application of two counteracting moments procured by two force couples (Fig. 7.26). A state of constant shear force is induced through the middle section of the test specimen, with the induced moments cancelling exactly at the mid-length, thereby producing a pure shear load at that point.

The average shear stress is given by

$$\tau = P/A, \tag{7.12}$$

where A is the cross-sectional area between the roots of the notches. The thickness of the specimen is not considered to influence the results of the test but practically it is found that specimens less than 4 mm thick are prone to instability due to twisting.

For UD specimens the fibres may be orientated either along the beam or perpendicular to the beam axis. The latter case provides the preferred configuration for an unambiguous failure as total separation occurs on

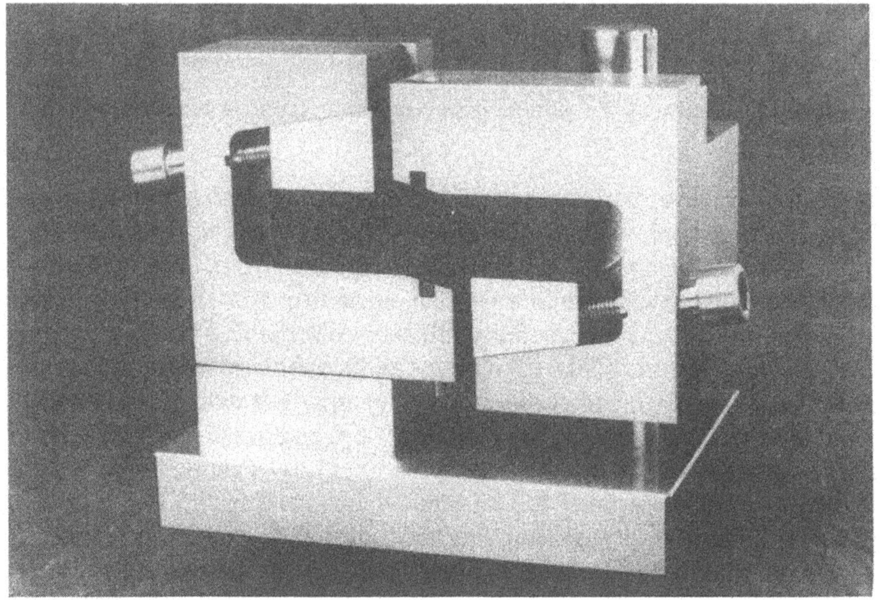

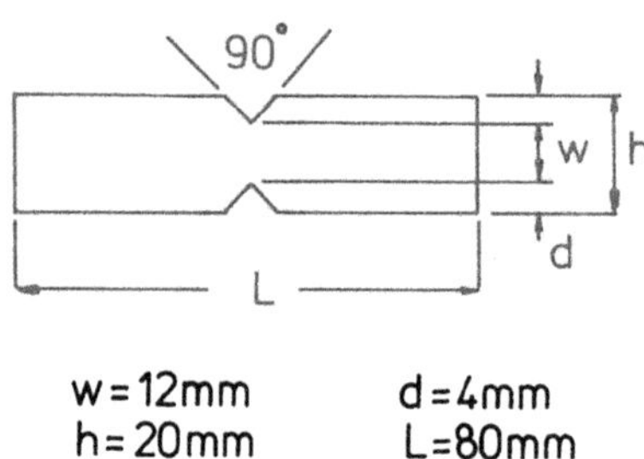

Fig. 7.26 Iosipescu fixture and test specimen geometry.

shear cracking. Despite great care being taken in diamond machining it is almost inevitable that sharp starter cracks will be introduced into the specimen at the roots of the v-notches. These microcracks will raise the local stress and introduce non-shear components. As a consequence, UD fibre orientation is observed to generate low strength results, although shear modulus data, calculated from the slope of the straight line portion of the stress–strain curve, should be identical to tests on beams with axial fibres.

Failure in specimens with axial fibres is harder to identify. Since small cracks may initiate at the roots of the v-notches running along the fibre direction. A small but definite load drop may be identified on the load deflection plot. A similar but smaller phenomenon is observed in composites with woven reinforcement.

Additional notes on shear testing

In all shear testing it is the matrix that is dominant in controlling the behaviour of the material. There are two principal classes: in-plane shear and interlaminar shear. A state of pure shear can rarely be established and the relatively low strength of the matrix and the interface renders the composite vulnerable to any extraneous local normal stress. The failure mode in a shear test therefore may not be a genuine shear failure. A further complication is the existence of areas and planes of weakness along which a specimen may fail preferentially, irrespective of the principal axes of the stress field. In most test situations those complications result in the measured property being notional rather than genuine.

When studying carbon–carbon composites, many of the techniques developed for polymer matrix materials are inappropriate due to the extreme brittleness of the matrix and weakness of the fibre/matrix interface. Unsatisfactory as the short beam ILS test may be, it is the best method currently available for evaluating the interlaminar shear strength of carbon–carbon. The Iosipescu test is considered to yield accurate modulus data and relatively good strength data. The procedure is likely to emerge as an ASTM standard for in-plane shear measurements in the near future, and is relatively easy to execute for carbon–carbon specimens. With this in mind an empirical set of acceptable and unacceptable failure modes has been produced (Fig. 7.27). Recent research has identified shear stress concentrations at the root of the v-notches, with the magnitude of the stress concentration being a function of the anisotropy ratio in the material. This will most likely lead to the introduction of correction factors for strength measured using this method [25].

7.7.5 Dynamic testing

Fatigue is a dangerous form of fracture which occurs in materials when they are subjected to cyclic or otherwise fluctuating loads [26]. It occurs by the development and progressive growth of a crack and is characterized by two equally unfortunate features. Firstly, as reported by Gotham [27], fatigue is a progressive weakening of a test piece or component with increasing time under load, such that loads which would be supported satisfactorily at short times produce failure at long times. Secondly, during the lengthening period of time that is required for fracture to propagate to the point of final failure, there may be no obvious external indication that internal fracture is occurring. Fatigue fracture is most familiar with respect to metals but no material is immune to this form of failure.

Fluctuating stress is generally far more harmful to a specimen than a steady stress of the same amplitude. Failure in fibre-reinforced composite materials proceeds in a progressive manner, similar to metals, though more complicated because of the extreme dissimilarity of the properties of the

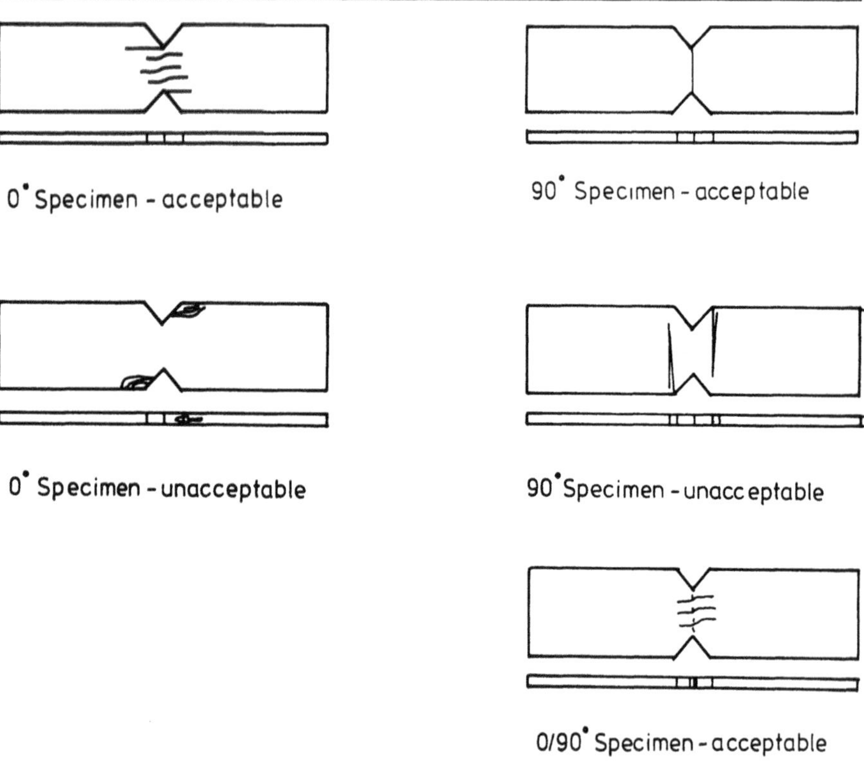

Fig. 7.27 Acceptable and unacceptable failure modes for Iosipescu four-point bend test.

constituents. There are at least three possible modes of failure. Matrix fracture, separation at the fibre/matrix interface and homogeneous fracture of both the matrix and fibres may all occur. In the case of brittle fibre/brittle matrix composites such as carbon–carbon, failure under dynamic loading is manifest in powdering ('dusting out') of the matrix. The appearance of the fracture surface depends on the nature of the stress cycle and the type of reinforcement. In laminates reinforced by fabrics or felts, the fracture surfaces produced by cyclic tensile stresses usually have a fibrous appearance and lie approximately normal to the stress axis. Should the stress cycle include compression, the fracture surface is usually fairly smooth and inclined at an angle to the stress axis. Fractures in UD materials are often delaminations, with the filament failure widely separated, producing a brush-like effect [28].

Fatigue data on non-metallic materials are extremely sparse and ill-understood. A considerable amount of work is being done to fill this gap but it is unlikely that designers will be provided with adequate information for many years to come. Only two internationally recognized test methods presently exist for plastic materials, whereas the fatigue testing of metals is well covered in the standards literature. Not surprisingly, there are no tests whatsoever to cover carbon–carbon.

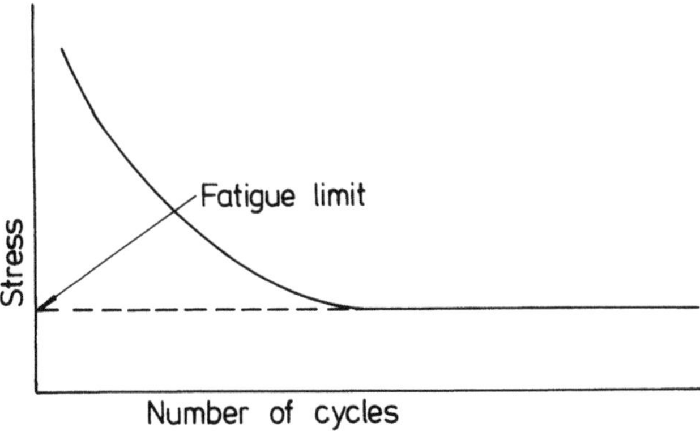

Fig. 7.28 Typical *S–N* fatigue curve.

A large number of cyclic deformation patterns are possible. The data are almost always displayed as the number of cycles to failure at a given maximum stress level. Such plots are known as *S–N* curves (Fig. 7.28). The fatigue life and fatigue limit may be derived from plots such as that in Fig. 7.28. The fatigue life, or endurance, is the number of cycles to failure at a specified level of stress. The greater the applied stress, the fewer cycles are needed to break the test piece. The fatigue limit is the stress condition below which the material may endure an infinite number of stress cycles without failure. The fatigue limit is not, unfortunately, as clear-cut as depicted in Fig. 7.28 and estimates based on statistical analysis are all that can be obtained for real materials.

The applied load may be tensile, compressive, flexural, shear or any combination of these. A sinusoidally varying stress is generally used as it is the easiest to generate. Modern testing equipment such as the Instron 8500 series of servohydraulic test machines operate an accurate system of 'closed loop' control allowing the generation of triangular and square-wave functions with ease. Testing may be carried out at constant cyclic stress amplitude or constant amplitude of cyclic strain. The latter method offers the advantage of ease in setting up the experiment but may create problems in that stiff materials are subjected to higher stresses than materials of low modulus. The constant strain test is therefore a more severe test for composite materials. Furthermore, once cracks develop the imposed stresses lessen such that the number of cycles to failure may become very large indeed. Once cracks start to develop in a constant stress test, on the other hand, failure usually follows very quickly.

Fatigure testing of carbon–carbon is best evaluated in tension or flexure. Flexure tests are the easiest to set up but will, of course, suffer from the usual problems associated with this test of mixed stress modes and tensile surface defect sensitivity. It is advisable that tensile tests are carried out such that all of the applied stress is tensile since the application of a compressive component would require unrepresentative short, squat

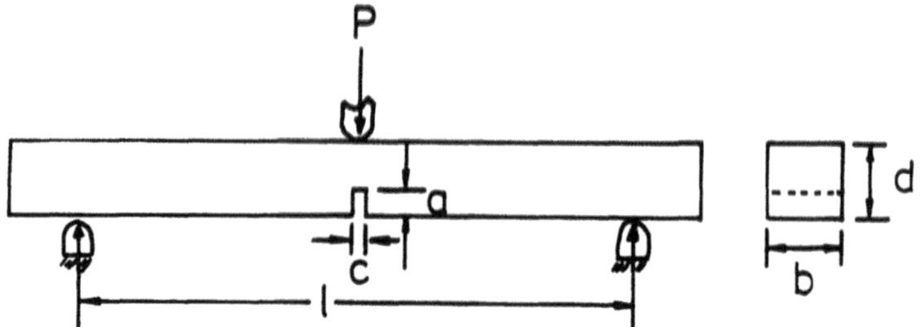

Fig. 7.29 Single edge notched beam fracture toughness test geometry.

specimens to eliminate buckling. As will be seen in Chapter 8, very little work has been carried out on the fatigue of carbon–carbon. For those who embark on this path it should be noted that the fatigue performance in contrast to metals is likely to be very dependent on the frequency of the applied stress as a result of the extreme brittleness of the matrix and weakness of the fibre/matrix interface.

7.7.6 Fracture toughness testing

Carbon–carbon composites have internal voids and flaws as a consequence of their fabrication processes. Fracture toughness is therefore a property which ought to be considered in the evaluation of a material for a specific application [29]. The toughness is best described as the ability of a material to resist crack propagation from a pre-existing flaw. Two fracture toughness parameters have been developed from linear elastic fracture mechanics, both of which deal with the onset of instability or crack initiation near the tip of a pre-existing flaw in ideally elastic and elastic–plastic materials [30, 31]. The toughness can be represented by the stress parameter known as the notch toughness K_c, or by the characteristic energy parameter known as the critical strain energy release rate, G_c.

The stress intensity factor describes the state of stress in the vicinity of the tip of a crack as a function of the specimen geometry, the crack geometry and the applied load. An analytical expression for K_{Ic} (K_c in mode I fracture) may be derived from a rectangular beam specimen tested in three-point flexure. The test (Fig. 7.29) 'borrowed' from monolithic ceramics testing employs a pre-machined crack, approximately half the depth of the specimen [32].

The crack is formed using a slow speed diamond wire saw [33]. K_{Ic} is then given by [34]

$$K_{Ic} = \left(\frac{Pl}{bd^{3/2}}\right)y, \tag{7.13}$$

where P is the load at failure, l the span, b the specimen breadth, d the specimen depth, a the crack length and y is an experimentally determined geometrical factor depending on specimen geometry and (a/d) ratio:

$$y = 2.9(a/d)^{1/2} - 4.6(a/d)^{3/2} + 21.8(a/d)^{5/2} - 37.6(a/d)^{7/2} + 38.7(a/d)^{9/2}$$

and

$$0 \le (a/d) \le 0.6. \tag{7.14}$$

When a crack is machined into a brittle material such as carbon–carbon there will inevitably be a degree of crack growth ahead of the cutting tool. It is therefore advisable, in the interests of accuracy, to measure the crack length a after the experiment is complete using a microscope.

The critical strain energy release rate is defined as

$$G_c = -\left(\frac{\partial U}{\partial a}\right)\delta_c, \tag{7.15}$$

where U is the strain energy, A the area of the new fracture surface and δ_c the midspan deflection at instability. In the single edged-notched beam (SENB) test geometry, equation (7.15) becomes

$$G_c = -\frac{1}{2bd}\left(\frac{P}{K}\right)^2 \frac{\partial k}{\partial(a/d)}, \tag{7.16}$$

where K is the bending stiffness of the specimen and $\partial K/\partial\,(a/d)$ is the slope of the K versus (a/d) curve evaluated for the crack length a at the onset of propagation. Since P can be measured experimentally and K versus (a/d) obtained by calibration [35], G may be determined directly using equation (7.16).

While the parameters K_c and G_c describe the resistance of a material to a pre-existing flaw, the work of fracture (γ_f) is a measure of the energy absorbed as a crack initiates and propagates through the material. In the case of a notched flexure test, the work of fracture is defined [36] as

$$\gamma_f = \frac{\bar{u}}{2b(d-a)}, \tag{7.17}$$

where $\bar{u}$ is the total area under the load–deflection curve and $2b(d-a)$ is the approximate area of the new surfaces created during fracture. The Linear Elastic Fracture Mechanics (LEFM) parameters K_c and G_c are found to provide an appropriate description of the ability of a material to resist crack propagation of an existing flaw under load. They are found to be independent of the (a/d) ratio, and may be considered 'material constants'. This is only true, however, in the case of 2-D composites tested in the interlaminar mode between the plies. Tests within the fibre plane (Fig. 7.30) and in 3-D composites result in delaminations which invalidate the tests. Having said that, it is the interlaminar failure in 2-D composites which causes the most concern with respect to unstable crack propagation in the matrix. Toughness testing thus offers a useful method of characterizing that behaviour. In the SENB geometry, K_c evaluation is generally preferred because it does not require the experimental determination of compliance curves.

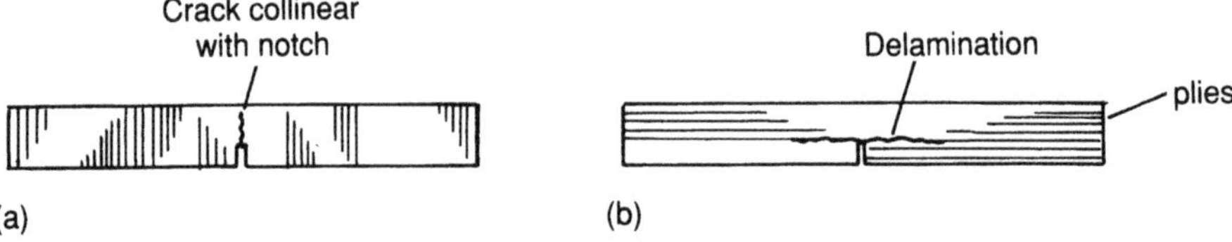

Fig. 7.30 Schematic representation of the failure modes observed in fracture toughness experiments: (a) fibres parallel to notch; (b) fibres perpendicular to notch.

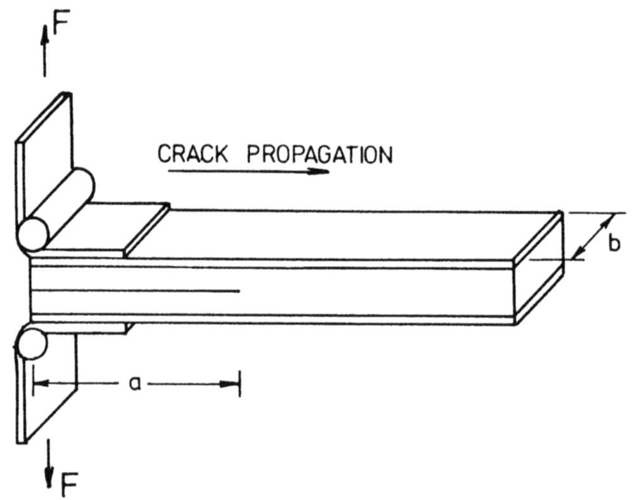

Fig. 7.31 The DCB test geometry.

The work of fracture, γ_f, is not independent of (a/d) and cannot therefore be considered a material constant. Calculated values of γ_f are significant not in a quantitative sense, but as a method of comparison between materials. The fracture toughness parameter G_c has recently been derived for carbon–carbon from an experimental relationship between the compliance and crack length, assuming G_c to be constant during crack propagation [33]. The relationship between fracture toughness and energy release rate during delamination has been investigated in carbon fibre reinforced polymer composite materials [37,38]. Double cantilever beam (DCB) specimens are used in a mode I test (Fig. 7.31). Data reduction is carried out by a compliance method based on elementary beam theory. A DCB test in which an interlaminar crack is propagated generally results in a force deflection trace of the type depicted in Fig. 7.32. With reference to Fig. 7.32, the critical strain energy release rate, calculated using the area method, is given by

$$G_{\mathrm{Ic}} = \frac{\Delta U_i}{2b(a_{i+1} - a_i)} = \frac{\frac{1}{2}(F_i\delta_i - F_{i+1}\,\delta_{i+1}) + \int F(\delta)\,\mathrm{d}\delta}{2b(a_{i+1} - a_i)} \tag{7.18}$$

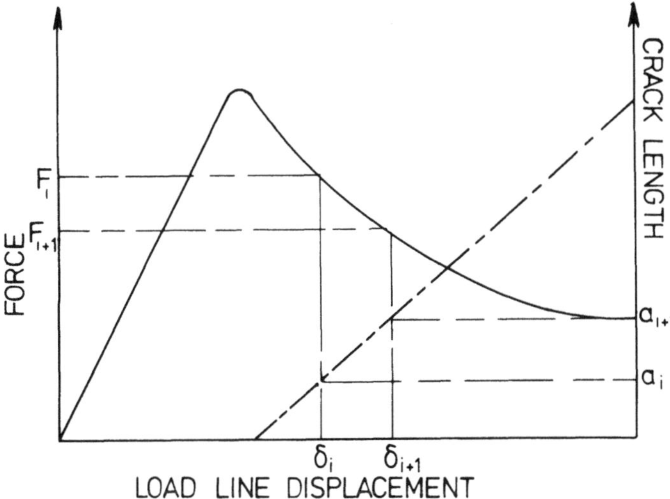

Fig. 7.32 Typical force–displacement–crack length curve recorded during the DCB test.

where U_i is the incremental energy, a the crack length, F the applied force and δ the load line displacement.

In carrying out DCB tests on carbon–carbon it was found that simply attaching hinges to the surface of the 1- and 2-D laminates often resulted in flexural failure of the two half-beams rather than interlaminar crack propagation. The problem can be rectified by bonding a 16-ply UD carbon fibre reinforced polymer composite to both sides of the carbon–carbon before attaching the loading hinges. This serves to restrict the flexural deformation in the beams and cause cracks to propagate along the beams in the required manner. Interlaminar starter cracks are best produced using a diamond-wire saw. Crack length measurement is achieved via resistive crack gauges affixed to one side of the test specimen. The signals from the testing machine and crack length measurement apparatus are interfaced with a microcomputer programmed to carry out calculations of G_{Ic} using equation (7.18).

Values of G_{Ic} are calculated from the gradients of the best-fit lines to the plots of the incremental energy (ΔU_i) versus incremental crack area ($2b(a_{i+1} - a_i)$). An example of such a plot is illustrated in Fig. 7.33. The full line is the least squares fit and the 95% confidence limits are shown dashed.

7.7.7 High-temperature mechanical testing

One of the major advantages of carbon–carbon composites is their excellent high-temperature mechanical properties. It is important therefore to be able to obtain high-temperature design data on the materials such that they may be exploited in refractory engineering applications. The flexural or bend test has been used extensively in evaluating the mechanical

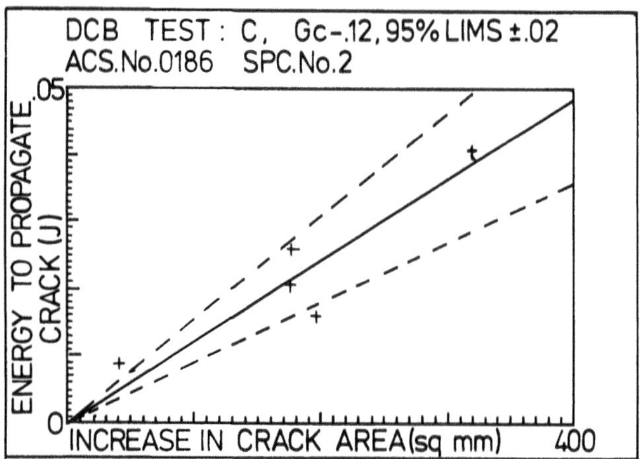

Fig. 7.33 Typical incremental crack energy versus incremental area plot obtained from DCB testing of carbon–carbon.

properties of carbon–carbon and other fibre-reinforced ceramics [39] at high temperatures because of its simple specimen geometry and the relatively inexpensive nature of the test. Testing is fairly straightforward and generally involves some form of three-point test fixture made in ceramic [40]. It has already been discussed how, despite its usefulness for the screening of composite materials, the flexure test cannot be employed to generate the strength and stiffness data required in design calculations.

Fibre-reinforced composites tested under flexural loading may fail in compression, tension, shear or a combination of fracture modes. At high temperatures creep and oxidation effects can further complicate the failure mode. Fracture of the specimen other than in tension results in significant errors in the flexural strength values calculated on the basis of elastic beam theory. Furthermore, flexural testing is more sensitive to surface flaws relative to volume flaws and may not yield the true strength statistics required for the design of components. The majority of the problems associated with the flexural test may be alleviated by uniaxial tensile testing of carbon–carbon. Under uniaxial tensile load the material experiences a uniform tensile stress in the tested volume and fractures when its ultimate tensile strength is reached. High-temperature tensile testing requires unique gripping, heating and specimen design. The stringent requirements of the test procedure result in uniaxial tensile testing being extremely difficult, time-consuming and expensive. Several gripping systems have been developed for testing monolithic ceramics in uniaxial tension [41–43] most of which employ tubular or cylindrical specimens and are unsuitable for flat, laminated carbon–carbon materials. A number of researchers have experimented with pin-grips and 'necked-down' samples, as used in metals testing, to promote failure within the gauge section [44]. Such techniques generally prove unsuccessful when testing carbon–carbon

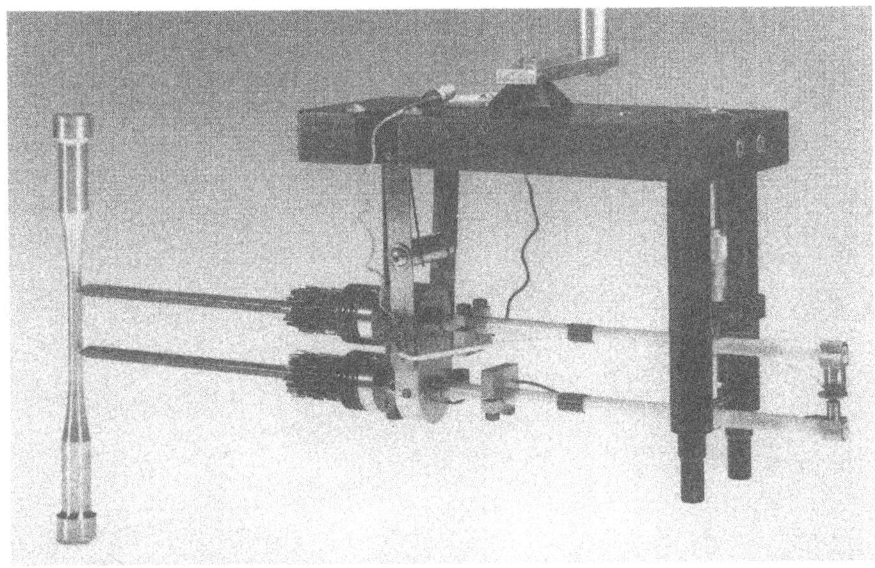

Fig. 7.34 High-temperature extensometer (courtesy Instron Ltd).

as they tend induce premature failure due to shear near the gripped ends of the sample [45,46].

High-temperature testing of carbon–carbon is most successfully carried out using straight section specimens identical to those used in room-temperature testing, but with deformable aluminium tabs bonded to the ends of the specimen, rather than tabs made from a glass reinforced composite. To date, testing has been carried out successfully at temperatures up to and including 1500 °C either in an inert atmosphere or vacuum [47].

A high-temperature tensile test system consists of a universal test machine, a furnace (either inductively heating the specimen, or electrically heated using, preferably, $MoSi_2$ elements), a high-temperature capacitive extensometer (Fig. 7.34) and temperature measuring equipment. A photograph of such a system is shown in Fig. 7.35. When designing the equipment, great care must be taken in the gripping mechanism. Very few materials can tolerate temperatures as high as 1500 °C and those that can are expensive and difficult to machine. They are also prone to failure from thermal cycling to the test temperature, so the lifetime of the equipment is likely to be very low. Chemical interaction may also occur between the grips and material under test, and finally, long extension rods reaching into a furnace make alignment difficult and are impractical for loading in the compression mode where lateral stiffness is important. To combat these difficulties, water-cooled grips that remain outside the furnace can be made of standard metallic materials and are easier to align and preload for tension/compression fatigue testing. The low temperatures also ensure no chemical interaction between the specimen and grip ends.

Fig. 7.35 Complete test set-up for high-temperature testing of flat laminated carbon–carbon (courtesy Instron Ltd).

The two main disadvantages of the 'cold' gripping technique are that the specimen must be much longer than the furnace whose design becomes critical in achieving an acceptable temperature gradient across the specimen especially when considering the high thermal conductivity of carbon–carbon. In contrast to the majority of other materials, carbon–carbon tends to increase in strength at high temperature. Great care must therefore be taken to avoid failure in the transition region between the hot and cold zone. Clearly all of the problems associated with the high-temperature testing of carbon–carbon are increased with the tendency towards higher testing temperatures.

7.8 MICROSCOPY

7.8.1 Optical microscopy

Carbon–carbon samples for optical microscopy are prepared by first cutting them to size using water-lubricated diamond-impregnated tools. The specimen is then mounted in a cold curing plastic resin, of which there are a great number commercially available specifically for this purpose. The specimens are polished to optical flatness using lapping wheels and a range of alumina and diamond pastes of sizes from 240 to 0.05 μm. The prepared specimens possess a surface of high reflectivity with little relief which can be examined using incident light illumination in the optical microscope. Cross-polarized light may be used to highlight areas of anisotropy as described in Chapter 1. A useful technique is to combine the microscope with a CCTV camera input to an image analyser which may be used to provide quantitative and statistical information on the various features observed.

7.8.2 Scanning electron microscopy

Scanning election microscopy (SEM) is best suited to revealing the topography of specimen. As a result it is an invaluable tool in fractography. Fracture faces of carbon–carbon samples used in mechanical testing are sectioned and fixed to aluminium stubs for viewing in the microscope. Despite being an electrical conductor, carbon–carbon materials are often found to produce electron beam charging in the SEM. It is therefore advisable to sputter-coat the specimens with a thin (≈200 Å) layer of gold prior to examination in the SEM. It is very difficult to exploit the increased resolution of the SEM with respect to optical microscopy since the specimens prepared for the latter, by definition, lack any appreciable topography. This problem may be overcome by oxidatively etching the samples prepared for optical microscopy in a manner analogous to the etching of samples in metallography.

A typical etching technique is to remove the specimen from its resin mounting and oxidize it in a chromic acid solution. Etching for 10 min at around 130 °C in nitric acid or, a 10% solution of $K_2Cr_2O_7$ in orthophosphoric acid has been found to be particularly useful [48]. Oxidation treatment of this type will preferentially attack the most thermodynamically unstable regions. A combination of optical and etched SEM microscopy is thus a very useful technique in identifying the various constituents of these complex multiphase materials [49].

7.8.3 Transmission electron microscopy

Transmission electron microscopy (TEM) has been used to characterize carbon–carbon in several modes [50]. Electron diffraction techniques (selected area diffraction, microdiffraction) have allowed the structural characterization of the materials in terms of crystallite size parameters, stack height, L_c, and width, L_a, and the interlayer spacing d_{002} [51,52]. Electron diffraction is useful in cases in which the longitudinal or transverse structure possesses structural heterogeneity. It is possible to obtain a selected area diffraction pattern from areas as small as 0.8 μm^2. The pattern may be analysed using a scanning densitometer. The diffraction peaks can be analysed by methods similar to those employed in X-ray diffraction [53]. The application of scanning transmission electron microscopy (STEM) permits microdiffraction from an area as small as 4 nm^2, providing an excellent technique for the study of microcrystallinity.

Dark field micrographs using the 002 reflection of the diffraction pattern have been used to estimate the crystallite size [54]. The most useful application of TEM is the study of the extremely complex fibre/matrix interface interactions in carbon–carbon [55].

Preparation of a thin sample of carbon–carbon composite for TEM analysis is an extremely complicated process. Many of the techniques employed such as microtomy are very aggressive and may result in damage to the delicate structures. The subsequent microscopy is thus very likely to be a study of the damage done in specimen preparation rather than the actual structure of the material! Specimens approximately 0.5 mm thick are cut from the bulk composite using a diamond saw. Discs of 3 mm dia. are then cut from the slice using a coring machine. The discs are thinned down by ion bombardment at an initial incidental angle of 45°. A four- or five-stage process is used to thin the specimen down to around 10 μm at the centre, with the milling angle lowered to ≈12° for the final stage. Thin sections of the fibres, matrices and the interface are analysed by electron diffraction, bright and dark field images.

7.9 DENSITY AND POROSITY MEASUREMENTS

The density and, indirectly, porosity of a carbon–carbon composite are generally used as a qualitative measure of the mechanical properties, efficiency of production and quality control. The complex structure of the material with its various phases, open and closed porosity, necessitates the use of a number of techniques to produce the overall ‘picture’.

The volume of the bulk specimen may be found using a water displacement technique. The bulk density of the composite is then calculated from

$$\rho = \frac{m}{v}, \tag{7.19}$$

where ρ is the density, m the mass and v the volume. If the carbon–carbon material has appreciable open porosity, however, water will enter it and the ‘bulk’ density will rise above its proper value.

Mercury porosimetry [56], whereby pressure is applied to force mercury into open porosity at the specimen's surface, is used to measure the size of the openings to pores and the open pore volume. The limitation of this technique is that it assumes a cylindrical pore geometry.

The true density of the material (i.e. accounting for both open and closed porosity) is obtained by the displacement of liquid helium. This measurement requires approximately 5 g of material for a precision of ±0.01 g cm^{-3}. The degree of porosity may be calculated from the helium and bulk densities according to

$$V_p = 1 - \frac{\rho_B}{\rho_{He}}, \tag{7.20}$$

where V_p is the volume fraction of pores, ρ_B the bulk density and ρ_{He} helium density.

7.10 OXIDATION AND OXIDATION PROTECTION

7.10.1 Oxidation protection

There are three issues of concern when designing a protective system: these are particulate inhibitors, coatings and sealants.

Particulate inhibitors

The role of particulate inhibitors is to augment the protection provided by the coating which, due to the differential CTE between substrate and coating, generates microcracks. Enough particulate inhibitor is required to generate rapidly a glassy sealant phase at the tip of the cracks in the coating so that excessive erosion of the carbon underneath does not occur. One of the most catastrophic events to occur to an inhibited system is delamination of the coating from the substrate due to loss of carbon at the substrate/coating interface. The type of sealant generated by the particulate inhibitor, e.g. B or $B_4C \rightarrow B_2O_3$, B/SiC $\rightarrow B_2O_3/SiO_2$, gives different viscosity/wetting glass sealants. Key issues with inhibitors include using a fine powder to achieve an even distribution, good reactivity and choice of processing temperatures to avoid detrimental inhibitor–fibre interfaces, and selections of the optimum loading in relation to fibre and matrix to give the best compromise between oxidative and mechanical performance. Inhibitors always increase the cured ply thickness of the materials, hence reducing some in-plane mechanical properties. There is no 'right' inhibitor; the correct choice will depend on the high-temperature application and operating conditions.

Coatings

Coatings are typically applied by pack cementation where a carbon–carbon component is immersed in a mixture of ceramic powders and subjected to a temperature where vapour phase reaction with the substrates can occur and a diffusion-bonded coating is produced. This method is used on the carbon–carbon parts of the space shuttle. The other technique is CVD, where the coating is deposited via a heterogeneous substrate–gas chemical reaction; CVD is flexible as regards the thickness and stoichiometry of the coating and is in common use.

Again, there is no one 'correct' coating technique. Often multi-layer, multi-technique coating systems are used; multi-layer techniques can minimize the differential CTE problems that plague many coating systems. Different CVD coatings, Si_3N_4 and B_4C/SiC mixed phases, for example, are sometimes used. It is important to mention that coating inspection is an important part of the application process. X-ray non-destructive testing (NDT) techniques are used to check for gross delaminations, and dye penetrometry can be used to investigate the microcrack density of the coating. Unfortunately, most of the information on coatings technology is in classified military reports and are thus inaccessible to the general reader.

Sealants

Sealants are the third method used to provide some oxidation protection for carbon–carbon. Again published literature is hard to find, although there are a number of papers on the LTV space shuttle material describing the TEOS silicate glazes applied to the components. Sealants are typically filled, for example with low-viscosity glasses designed to heal the CTE microcracks in the primary coating.

7.10.2 Testing of oxidation protection systems

This is an extremely complex and difficult issue. The best way to test an oxidation-inhibited substrate is to monitor key properties used as design allowables as a function of time/temperature/gas flow rates/pressures encountered in the real environment of the substrate.

More realistically, the most common test used for screening the oxidation protection system is a weight loss test, either continuously in a thermobalance, or by a series of fixed time exposures at a certain temperature. Absolutely paramount is an awareness of the sensitivity of oxidation rate not just to temperature, but gas flow rates, sample/gas flow geometry, etc. At around 700 °C oxidation kinetics change from being chemically controlled to diffusion dominated. Diffusion-controlled data are extremely difficult to compare unless tests are carried out under identical conditions. For example, static air testing oxidation kinetics at 1000 °C are strongly dependent upon the furnace design in which they are carried out.

Given the need to screen using well-defined test conditions, and test under similar conditions to the desired end use, two further points should be raised. Most people monitor oxidative stability solely in terms of weight loss/gain. Even for lightly loaded materials, however, the key property is mechanical performance at a given time/temperature oxidation exposure. A very commonly used test is a 'stressed' oxidation test in which the substrate is mechanically loaded while being oxidized. Another important issue, particularly when dealing with inhibited materials, is that of humidity. Often, systems for jet engines which looked promising in simple laboratory test time/temperature cycles have failed when exposed to more humid environments due to the absorption of water. Hydration of a borate glass inhibitor and subsequent (near explosive) dehydration of the rehydrated borate on exposure to temperatures above the boric acid dehydration temperature have caused coating spallation. Given a suitable screening test to evaluate potential candidates, testing under conditions (often cyclic) that accurately duplicate the time/temperature/stress/gas pressure and flow rate conditions of service, a number of other factors must be considered. Many, many replications of the test must be performed, as oxidation testing does not duplicate very well due to inconsistencies in substrate and coatings. Oxidation testing may be carried out in a furnace with a static or flowing atmosphere, a flame such as a gas burner test rig or a plasma arc jet which is used to simulate re-entry, for example.

7.11 MEASUREMENT OF THERMAL CONDUCTIVITY

The object of this section is to provide a brief description of the different methods for experimental determination of thermal conductivity. This is by no means a trivial task, and considerable care in experimental design is called for if meaningful date are to be obtained. There are two basic types of thermal conductivity determination experiments; a steady-state method, where the heat flow in and out of the test piece is constantly balanced (as are the temperature differences from one point of the sample to another), and a non-equilibrium method where the test piece is submitted to thermal pulsing – such as a laser flash – and its thermal response monitored. The simplest concept of experiment is the steady-state method for a rod of defined dimensions, with a known heat flow along its length and a known temperature drop. This method is carried out as follows. The thermal conductivity value (K) for the rod material is defined as

$$K = \frac{-ql}{A(T_2 - T_1)}, \tag{7.21}$$

where q is the heat flux across the test piece (W), l the length of test piece (m), A the cross-sectional area of test piece (m^2) and ($T_2 - T_1$) the temperature difference from one end of test piece to the other and has SI units of W m^{-1} K^{-1}. It is worth pointing out that thermal conductivity values of carbons are quoted in a plethora of different units. All values may be converted into SI units using tabulated conversion factors [57]. The simple longitudinal steady-state heat flow experiment has to be very carefully performed if meaningful data are to be obtained. Sidewall heating of the test piece has to be provided at just the right level to prevent radiant heat loss from the test specimen. A whole range of different geometry test pieces can be used, from rods to discs and spheres.

One commonly used technique is to use a comparative method, in either a steady-state and non-equilibrium mode, where the test piece is placed directly adjacent to a test piece of known thermal conductivity. This technique is quite simple to perform, but care must be taken to minimize sources of error – the efficiency of thermal contact to the reference and test specimens must be the same, for example. Considerable variations in conductivity values occur in the literature. This may be due, in part, to the inadequacies of the experimental methods used in determining these values. (Recent texts give a full account of the range of possible techniques of thermal conductivity determination. For more detailed information, the reader is referred to them [57].

REFERENCES

1. Pioneer Gasket Inc., Salt Lake City, Utah, USA.
2. Savage, G. M. (1985) PhD Thesis, Univ. London.
3. Savage, G. M. and Williamson, J., Submitted to *J. Mat. Sci.*
4. Wegman R. F. (1989) Non destructive test methods for structural composites, in *SAMPE Handbook*, **1**, Covina, Ca, USA.
5. Belbin, G. R. (1984) *Proc. I. Mech. E.* **198**, 47.
6. Gibbs, H. H. (1984) *SAMPE J.*, **20**, 37.
7. McAllister, L. E. and Taverna, A. R. (1971) *Proc. Am. Ceram. Soc. 73rd Ann. Mtg.*, Chicago.
8. McAllister, L. E. and Taverna, A. R. (1976) *Proc. Int. Conf. on Comp. Mats*, **1**, *Met. Soc. AIME*, New York, p. 307.
9. Mullen, C. K. and Roy, P. J. (1972) *Proc. 17th Nat. SAMPE Symp.*, p. iii–A–2, AZ, USA.
10. Dietric, H. and McAllister, L. E. (1978) *Proc. Am. Ceram. Soc. 80th Ann. Mtg.*, Detroit.
11. Gray, G. and Savage, G. M. (1989) *Metals and Materials Sept.*, 513.
12. Chard, W., Conaway, M. and Neisz, D. (1974) in *Petroleum Derived Carbons*, **21**, (eds M. L. Deviney and T. M. O'Grady), *ACS Symp. Series*, Ann. Chem. Soc., Washington DC, p. 155.
13. Gray, G., Hunter, A., Payne, R. S. and Savage, G. M. (1990) *Met. Pwdr Rpt.*, **45** (4), 290.
14. Gray, G. and Savage, G. M. (1991) *High Temperature Technology (Butterworths)*, **9**(2), May, 102.

15. ASTM (1987) *ASM Standards and Literature References for Composite Materials*, ASTM, Philadelphia, Pa, USA.
16. Curtis, P. J. (ed.) (1988) *Crag Test Methods for the Measurement of the Engineering Properties of Fibre Reinforced Plastics*, RAE Tech. Rpt 88012, Farnborough.
17. BS 1610 Br. Std Inst., (1989) London.
18. *Student Manual for Strain Gauge Technology*, Bull. 309, Measurements Group, PO Box 27777, Raleigh, N. Carolina, USA.
19. Hysol Division, The Dexter Corporation, Pittsburg, California, USA.
20. Whitney, J. M., Daniel, I. M. and Pipes, R. B. (1984) *Experimental Mechanics of Fibre Reinforced Composite Materials*, rev. edn, Soc. for Exp. Mech., Prentice-Hall, Englewood Cliffs, New Jersey.
21. Pipes, R. B. and Pagano, N. J. (1972) *J. Comp. Mat.*, **4**, 538.
22. Wilkins, D. J., Eisenmann, J. R., Canin, R. A., Margolic, W. S. and Benson, J. W. (1980) *ASTM STP* 775.
23. Delmonte, J. (1981) *Technology of Carbon and Graphite Fibre Composites*, van Nostrand Reinhold, New York.
24. Iosipescu, J. (1967) *J. of Materials*, **2**, 537.
25. Kumosa, M. and Hull, D. (1987) *Int. J. Fracture*, **35**, 83.
26. Charles, J. A. and Crane, F. A. A. (1989) *Selection and Use of Engineering Materials*, Butterworths, Harlow.
27. Gotham, K. V. (1969) *Plastics and Polymers*, **37**, 309.
28. Turner, S. (1983) *Mechanical Testing of Plastics* 2nd edn, George Goodwin.
29. Guess, T. R. and Hoover, W. R. (1973) *J. Comp. Mats.*, **7**, 2.
30. Wu, E. M. (1968) Comp. Mats Workshop Techronic.
31. Outwater, J. O. and Murphy, M. S. (1969) *Proc. 24th Ann. Tech. Conf. Reinforced Plastics*, Comp. Div. Soc. of Plastics Ind., Sect. 11-C, p. 1.
32. Kim, H. C., Yoon, K. J., Pickering, R. and Sherwood, P. J. (1985) *J. Mat. Sci.*, **20**, 3967.
33. Savage, G. M. Submitted to *J. Mat. Sci.*
34. Brown, W. F. Jr (1970) *ASTM STP* 463.
35. Brown, W. F. Jr and Srawley, J. E. (1968) *ASTM STP* 410.
36. Tattersal, H. G. and Tappin, G. (1966) *J. Mat. Sci.*, **1**, 296.
37. Russell, A. J. and Street, K. N. (1982) *Proc. ICCM-IV Bordeaux*, p. 279.
38. Wilkins, D. J. (1982) ASTM STP 775, 168.
39. Amaral, J. E. and Pollock, C. N. (1989) in *Mechanical Testing of Engineering Ceramics at High Temperatures* (eds B. F. Dyson, R. D. Lohr and R. Morrell), Elsevier App. Sci. Pub., London, p. 51.
40. Loveday, M. S., Day, M. F. and Dyson, B. F. (eds) (1982) *Measurement of High Temperature Mechanical Properties of Materials*, HMSO, London.
41. Liu, K. C. and Brinkman, C. R. (1985) *Proc. 23rd Automotive Technology Development Contractors' Meeting*, SAE, Detroit, p. 279.
42. Vaidyanthan, R., Shankar, J. and Avva, V. S. (1987) *Proc. 25th Automotive Technology Development Contractors' Coordination Meeting*, SAE, Detroit, p. 175.
43. Herrmansson, L., Adlerborn, J. and Burstrom, D. ASEA CERAMA AB, company literature.

44. Gouila, R. K. (1983) *J. Mat. Sci.*, **18**, p. 307.
45. Sheshadri, S. G. and Chia, K. Y. (1987) *J. Am. Ceram. Comm.*, **70**(10), 242.
46. Gyekenesi, J. Z. and Herrman, J. H. *NASA Contractor Rpt 180888.*
47. Mandell, J. F., Grade, D. H. and Dannerman, K. A. (1986) *Proc ASTM Symp.*, Phoenix Ariz. USA, 3–4 Nov.
48. Markovic, V., Ragan, S. and Marsh, H. (1984) *J. Mat. Sci.*, **19**, 3287.
49. Forrest, M. A. and Marsh, H. (1981) *Proc. Biennial Conf. on Carbon*, Am. Carbon Soc., Univ. Pennsylvania, Philadelphia.
50. Kowbel, W. and Don, J. (1989) *J. Mat. Sci.*, **24**, 133.
51. Wicks, B. J. (1971) *J. Mat. Sci.*, **16**, 173.
52. Oberlin, A. (1982) *J. Mat. Sci.*, **23**, 7.
53. Johnson, D. J. and Crawford, D. (1973) *J. Mat. Sci.*, **8**, 286.
54. Badami, D. V., Joiner, J. C. and Jones, G. A. (1967) *Nature*, **215**, 386.
55. Weizhou, P., Tianyou, P., Hanmin, Z. and Qias, Yu (1988) *Proc. Int. Conf. Compos. Interfa*, Elsevier, p. 399.
56. Sibilia, J. (ed.) (1988) *A Guide to Materials Characterization and Chemical Analysis*, VCH Publishers, Weinheim, Germany.
57. Eckert, E. R. G. and Irvine, T. F. (eds) (1977) *Progress in Heat and Mass Transfer*, **8** Pergamon Press, Oxford.

The Properties of Carbon–carbon Composites

8

8.1 GENERAL CONSIDERATIONS

The microtexture, that is to say the microstructure and morphology, of the various types of carbon–carbon composite will differ according to the type of raw materials and the processing conditions. Further complication will arise from the use of 'subtle' treatments such as surface modification of the fibres and inclusion of oxidation protection. All such differences in microstructure exert a considerable influence on the properties of the materials.

The architecture of the carbon fibre reinforcement can take many forms: random short fibres, unidirectional (UD) continuous fibres, braided structures, laminated fabrics, orthogonal 3-D weaves (in either Cartesian or cylindrical coordinates) or multi-directional structures may be used. Commercial carbon fibres are manufactured from rayon, acrylic and pitch precursors. The mechanical and other properties of these fibres vary considerably, again depending on raw materials and processing conditions. High-strength fibres are usually obtained from acrylic precursors whereas pitch is preferred for the fibres of highest modulus. The carbon fibre is generally post-treated to improve weaving characteristics and/or to modify the bonding of fibre to matrix.

The matrix may vary from an isotropic, glassy carbon produced from the pyrolysis of a thermoset resin, through to a highly oriented anisotropic graphitic carbon arising from mesophase, developed during the carbonization of pitch. A matrix material may also be formed by chemical vapour deposition (CVD) of carbon from the cracking of hydrocarbons. The structure of the CVD matrix depends very much on the processing conditions. All of the production routes used to manufacture carbon–carbon are slow, inefficient multi-step batch processes. As a result, many finished products tend to be made using a variety of methods. Composites fabricated by the thermoset resin pyrolysis route, for example, are often densified using CVD or a mixture of pitch and a different resin from that used in the initial cycle.

The number of possible combinations of parameters is almost limitless. Rather than provide a complete catalogue of the properties of carbon–carbon, it is intended to present the general trends associated with the different variables. The theme developed, throughout this text, of carbon–carbon as a family of materials whose properties may be tailored to suit a specific application, will thus become apparent.

8.2 MICROSTRUCTURE

8.2.1 CVD-densified composites

The matrix deposited in the CVD process may comprise three different types as discussed in Chapter 3. The first of these, the isotropic type, consists of an optically isotropic morphology of fine particles no more than a few microns in size. The other two matrix types are both optically anisotropic morphologies. The rough laminar structure comprises a combination of layers with strong optical anisotropy, arranged such that the fibres as a whole are surrounded. In the smooth laminar microstructure optically weakly anisotropic layers are oriented so as to surround the fibre uniformly [1]. The principal factors influencing the different microstructures are the concentration, or partial pressure, of the reactant gas and the processing temperature [2]. A simplified representation of the temperature/pressure/microstructure relationships observed in carbon deposition from propane is given in Fig. 8.1 [3–5]. It should be remembered, however, that the mechanisms involved in CVD are extremely complex and as yet poorly understood. Even at the same temperature and pressure, and with the same feedstock gas, the morphologies obtained can be different. The effect of any sizing materials on the fibres can generally be disregarded since deposition is carried out at temperatures in excess of 800 °C at which any such compounds will have been carbonized.

Figures 8.2 and 8.3 are optical micrographs of CVD brake materials produced by the French company SEP for use in aircraft and Formula 1 racing cars. The microstructure in Fig. 3.2 illustrates an important characteristic of CVD-densified material, in that CVD is extremely good at filling small pores, but generally poor at filling large ones.

Figure 8.3 shows a detail of this microstructure in Fig. 8.2. The matrix is optically anisotropic and is highly reflective. Around the fibres the matrix exhibits extinction crosses indicative of the presence of circumferentially stacked, highly graphitic material that has been deposited around individual carbon fibres. The circular cross-sections, characteristic of PAN-based fibres, can be clearly seen. The irregular extinction crosses (blue, yellow and

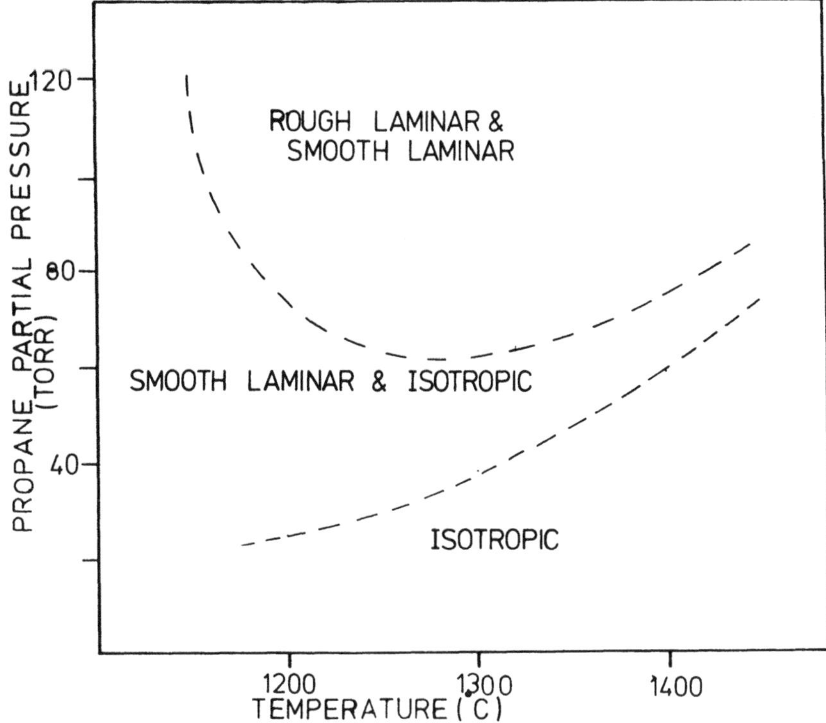

Fig. 8.1 Relationship of CVD matrix microstructure to processing temperature and propane gas partial pressure.

purple under polarized light) are typical of the rough laminar microstructure [6].

8.2.2 Thermoset resin derived material

The microstructure of a composite produced using the thermoset resin pyrolysis route will vary dramatically according to the heat treatment temperature (HTT). When carbonized by itself, a thermoset resin will form a glassy isotropic carbon [7]. X-ray analysis reveals no evidence of the formation of graphite. The material is optically isotropic, very hard, and has a low porosity and permeability. When heat-treated in the form of a composite with carbon fibres on the other hand, graphitic material is observed in the matrix in the region of the interface with the fibres [8]. This effect is most marked in the case of a furan precursor, in which optical anisotropy is seen even at HTTs below 1000 °C. Should the temperature of processing exceed 2200 °C, the domains of anisotropy gradually change to graphite. An HTT of 2800 °C or greater will result in a completely graphitic matrix.

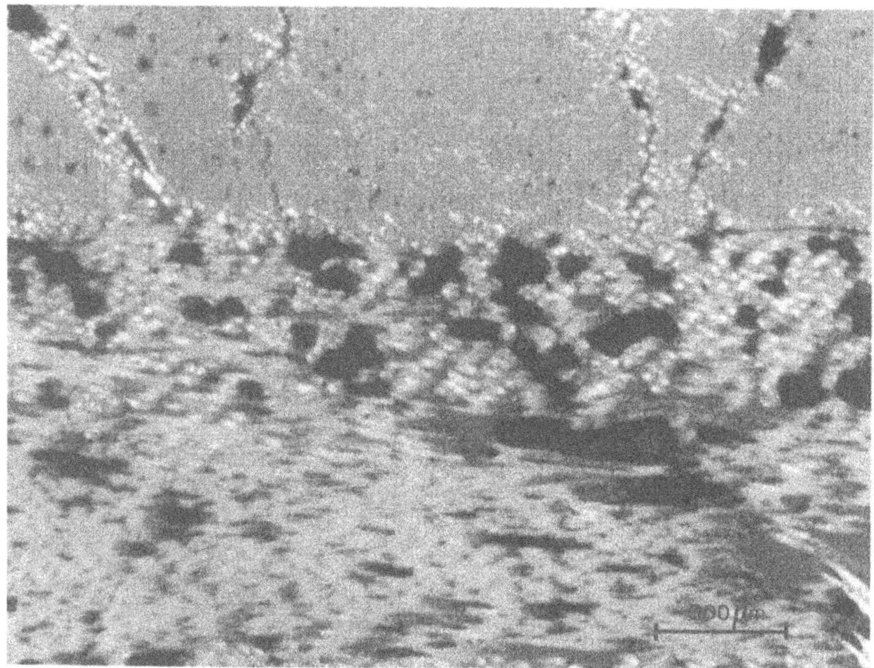

Fig. 8.2 Optical micrograph of SEP brake material.

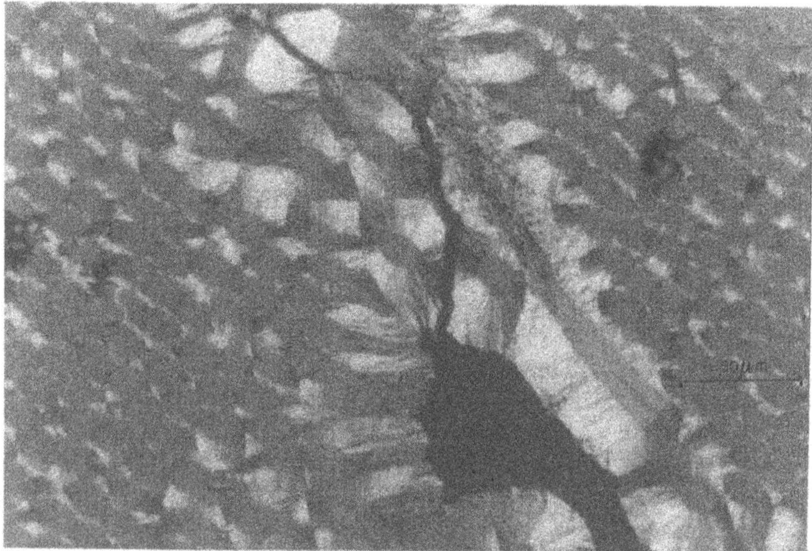

Fig. 8.3 Detailed microstructure of SEP brake material showing rough lamellar matrix morphology.

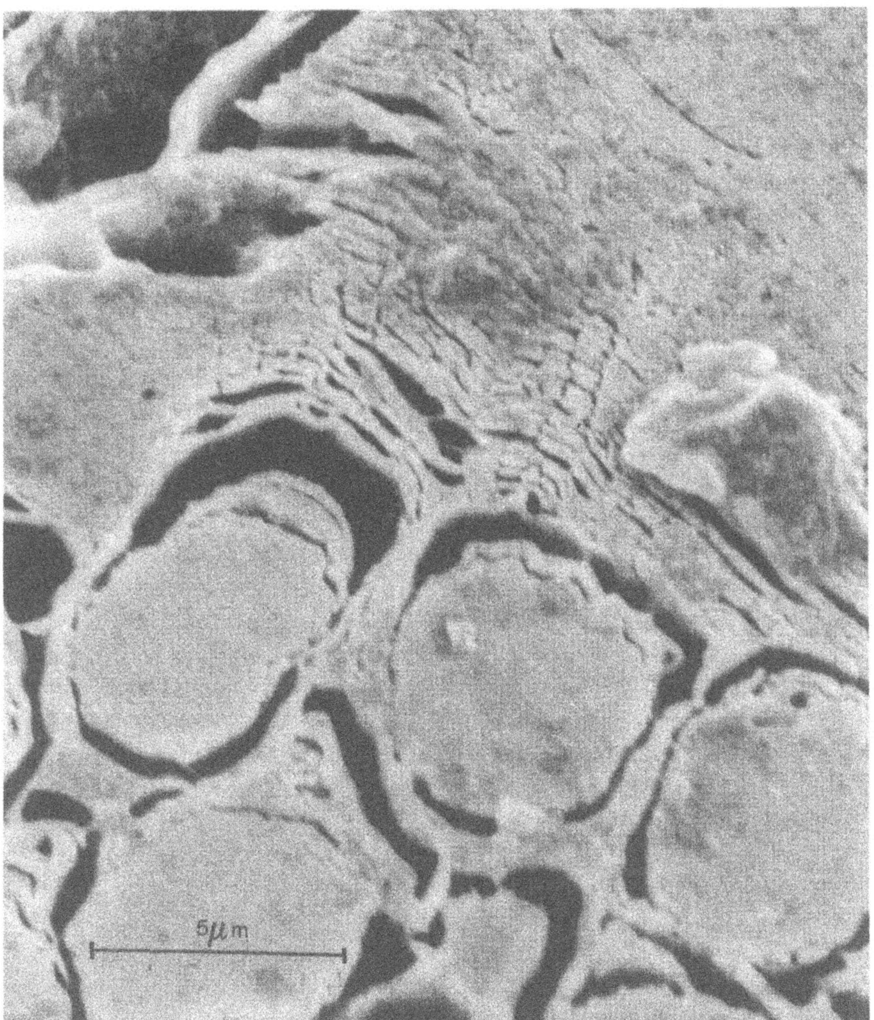

Fig. 8.4 Scanning electron micrograph of a graphitized carbon–carbon composite, acid etched to show the lamellar structure of the matrix in the vicinity of the fibre/matrix interface.

Research suggests that the graphite microstructure in thermoset-derived matrices is oriented in such a way that the graphite planes encircle the fibres [9]. Figure 8.4 is a scanning electron micrograph of a polished surface perpendicular to the fibre axis which has been subjected to acid oxidation. It appears that the graphite layers are divided into a number of domains and, as a whole, envelop the fibre. The graphitization of what are essentially 'non-graphitizing carbons' is believed to be due to the high stresses set up between fibre and matrix as a result of thermal expansion mismatch at high temperatures. The process is described as 'stress-graphitization' [10,11]. The temperature and rate at which the matrix transformation proceeds depend very much on the type of reinforcing fibre. The higher the modulus of the fibre the lower will be the graphitization temperature range and the quicker the rate of change. A number of studies of stress-graphitization are reported in the open literature [12–14] but the mechanism remains poorly understood.

Very little work has been published relating to the fibre surface treatments which are so important in the polymer composite industry [15]. Fitzer *et al.* reported oxidizing the surface of the fibres, using nitric acid, to result in an increase in the strength of a carbonized composite [16]. Their results indicated an optimum level of pre-treatment since excessive oxidation caused a lowering of strength. The majority of sizing agents carbonize at temperatures lower than that of the matrix precursor with a very poor carbon yield. Their presence is therefore considered deleterious to the properties of the carbon–carbon composite. All weaving lubricants and sizing materials are generally removed, by thermal treatment, prior to processing.

The microstructure of ex-thermoset composites is dominated by large-scale porosity and shrinkage cracks. The porosity results from the volatilization of small molecules and heteroatoms during pyrolysis (Fig. 8.5). Carbonization is generally performed at temperatures up to 1000 °C even though the matrix will retain some percentage of residual hydrogen. The hydrogen content diminishes to below 0.1% with further heat treatment to between 1350 and 1400 °C. Heat treatment above 2000 °C results in significant shrinkage of the carbon matrix as it gradually transforms into a graphite structure. The use of intermediate graphitization cycles is thus observed to improve the efficiency of subsequent impregnation and recarbonization processes by increased access to porosity. Even when fully graphitized composites are not required, it is often standard practice to graphitize the composite after the first carbonization as an aid to sub-sequent densification.

Figure 8.6 shows a polarized light micrograph of a densified carbon–carbon composite produced from a thermoset resin (phenolic) reinforced with intermediate modulus (IM) fibres (Hercules AS4). Following carbonization the sample was heat treated to 2700 °C to open up the porosity and then densified four times with a mixture of pitch and furan resin. Reimpregnation, even after four iterations, has failed to densify fully the composite as may be witnessed by the fact that the reimpregnant carbon tends to line the pores rather than fill them in. A 'matrix rich' region in the same composite is shown in Fig. 8.7.

Optical anisotropy in the phenolic-based matrix can be clearly discerned in the bottom left-hand corner of the micrograph. The graphitic carbon can be seen grading continuously into an isotropic matrix as the distance from the fibre increases. One may thus conclude that the microstructure of the matrix derived from phenolic resins is strongly affected by the presence of fibres and the associated thermal stresses at the interfaces.

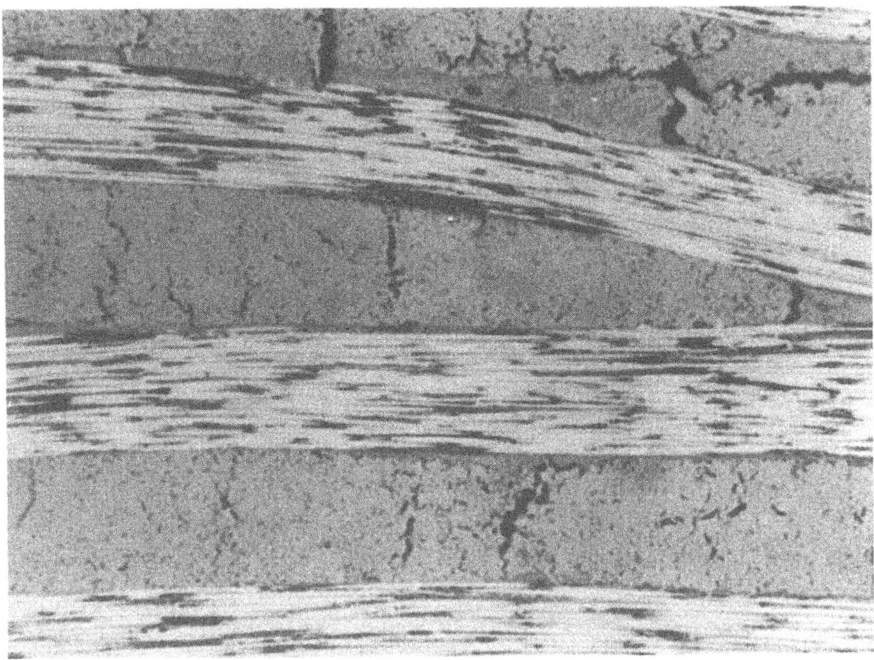

Fig. 8.5 Electron micrograph of densified carbon–carbon composite made from phenolic precursor, showing evidence of microcracking and porosity.

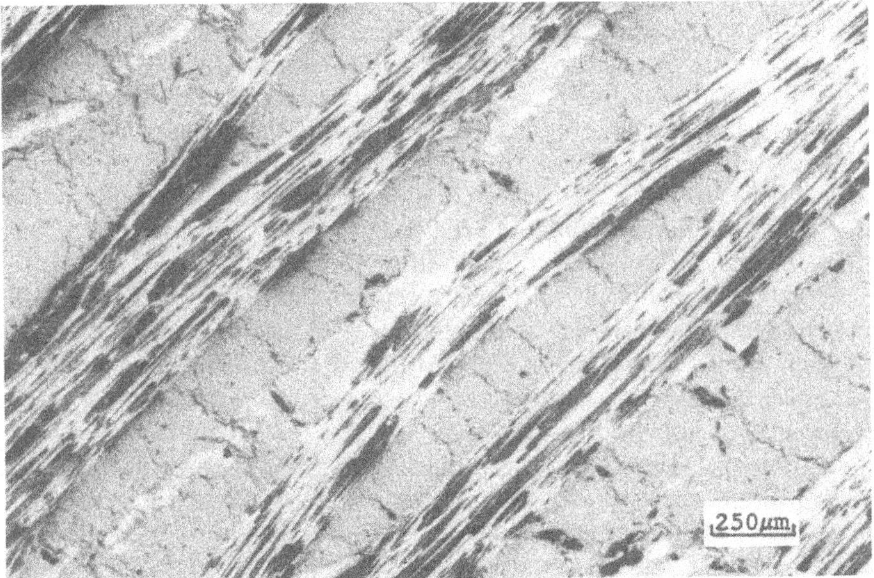

Fig. 8.6 Polarized optical micrograph of densified composite showing optical anisotropy resulting from graphitization and lined but open microcrackes and pores illustrating inefficient reimpregnation.

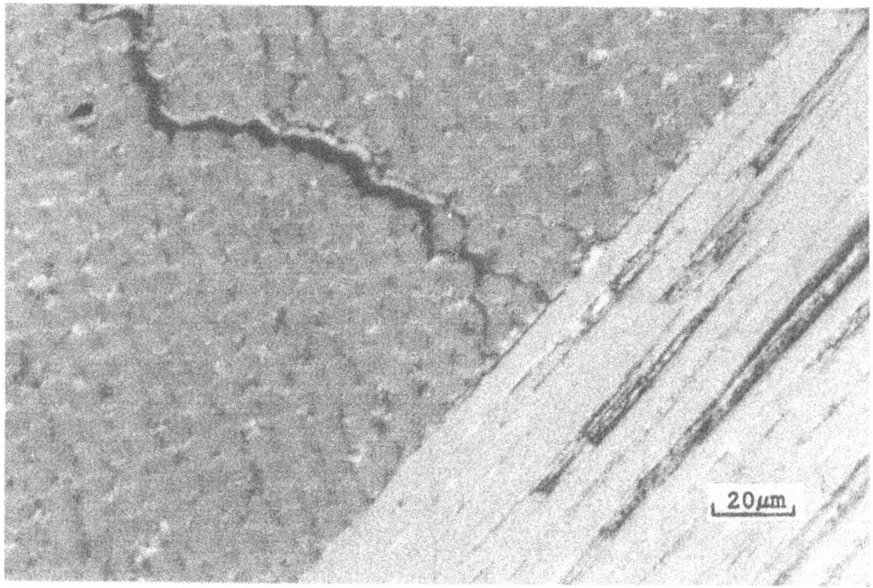

Fig. 8.7 Polarized light micrograph illustrating stress graphitization in phenolic-based matrix.

8.2.3 Thermoplastic-based composites

The microstructure of pitch matrix materials pyrolysed at pressures below 10 MPa is hardly affected by the fibre type. Graphite planes are oriented parallel to the fibre axis at the interface, becoming more and more disordered with distance from the fibre. In Chapter 5 it was shown how the carbon yield from pitch can be substantially increased by the application of pressure during pyrolysis. The suppression of mesophase incorporation occurs at pressures of above 30 MPa such that the matrix as a whole is isotropic, except in very close vicinity to the fibres where the graphite planes are oriented parallel to the fibre axis [17]. A completely isotropic matrix may be obtained if fibres with a thin (0.5 μm) CVD carbon layer are applied to the fibres prior to impregnation with pitch [18,19]. It is considered that the CVD treatment impairs the wetting ability of the pitch. Pitch, however, is never a straightforward and predictable material, examples having been reported in which mesophase incorporation is not suppressed at high pressures [20].

Ambient pressure carbonization of pitch matrix composites results in large-scale open porosity of complex geometry. Pressure carbonization, on the other hand, results in many closed pores which tend to be spherical in shape and of uniform size distribution. Further methods of increasing the carbon yield of pitch involve oxidation (Chapter 5) and the addition of sulphur [21,22]. Both processes essentially serve to cross-link and 'pseudo-thermoset' the pitch in order to prevent the loss of lower molecular weight

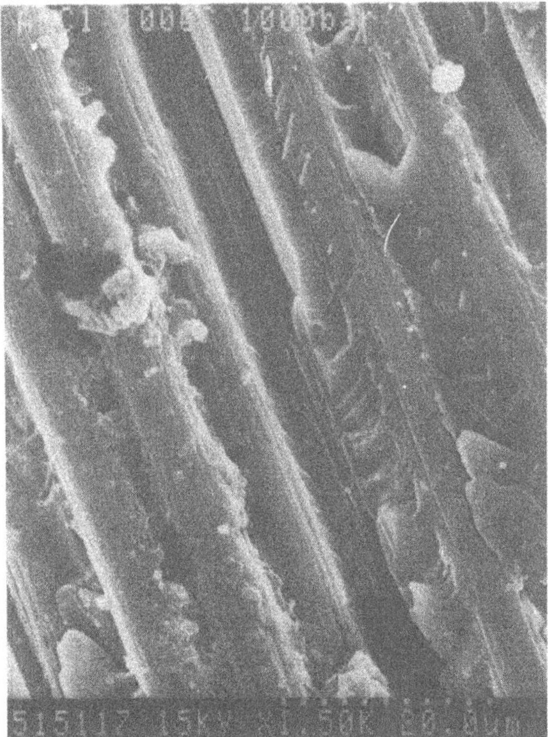

Fig. 8.8 Carbon–carbon composite with PEEK-derived matrix showing unique 'polymeric-type' morphology.

compounds as volatiles. A hard/rigid carbon microstructure akin to a glassy carbon tends to result.

The use of engineering thermoplastic polymers, such as PEEK, as matrix precursors is a relatively recent development [23,24]. The microstructures obtained tend to be of the glassy, isotropic type, but with a unique morphology very similar to that of the polymer whence it was derived (Fig. 8.8). Microstructural [25] and electron diffraction [26] evidence suggests that the matrix may be stress graphitized, especially in the presence of ultra-high modulus pitch-based fibres. The materials may be considered as intermediate between those obtained from thermosets and that from pitch. Carbonization at high pressures (≈100 MPa) results in a low porosity and a high density (Fig. 8.9).

8.3 INTERFACES IN CARBON–CARBON COMPOSITES

The nature of the interfaces in carbon–carbon composites and their influences on thermomechanical properties is extremely complex. The type of bonding across the interfaces is not well understood [27]. A number of

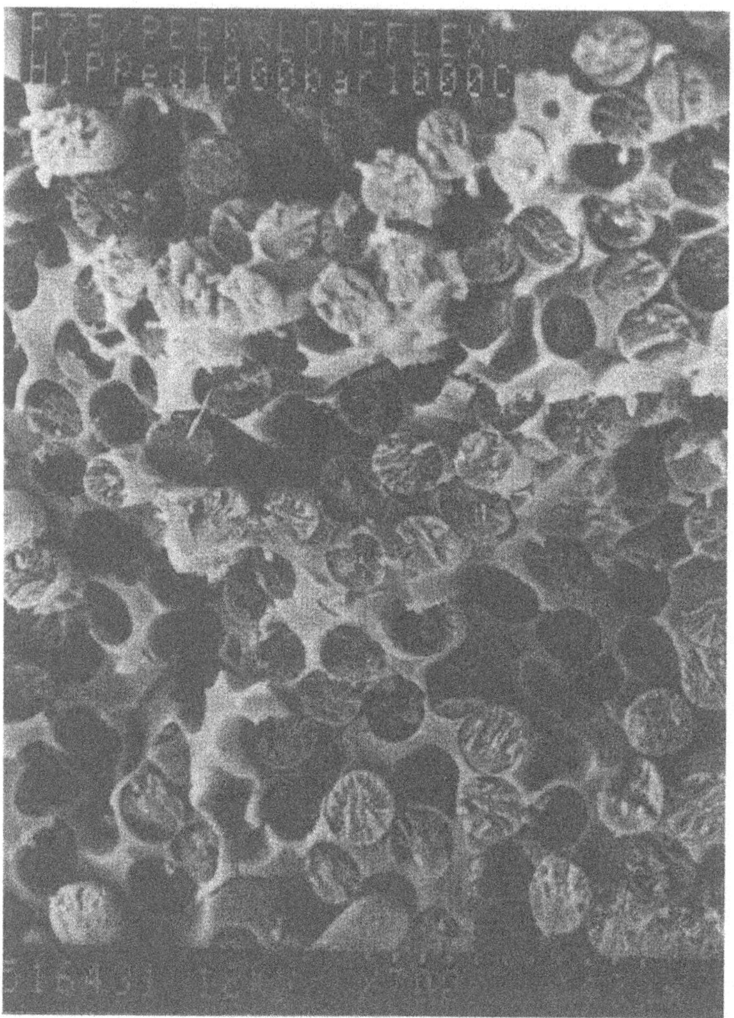

Fig. 8.9 High-density carbon–carbon made from P75 fibres in PEEK matrix.

different types of interface may exist in carbon–carbon: between fibres and matrix, within fibre bundles, between fibre bundles and between the different microstructures which may exist within the matrix. The bonds formed may be strong or weak chemical links, mechanical interlocking and friction couplings. The nature of the interfaces in a particular composite depends, as do so many other properties, on the choice of raw materials and processing conditions.

The possible orientations of the matrix at the interface with the fibres are shown in Fig. 8.10 [28]. The isotropic form is generally seen in the

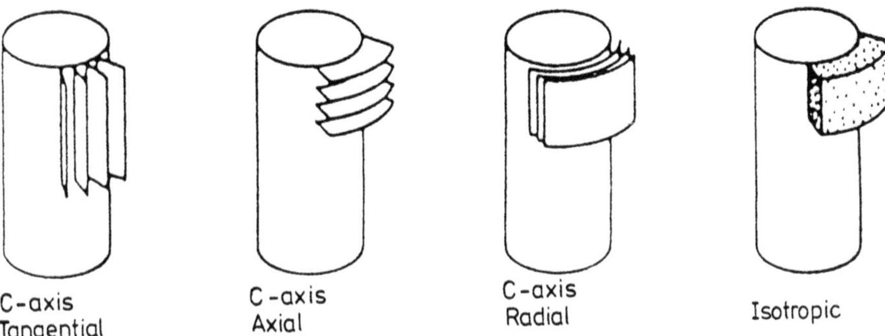

Fig. 8.10 Elementary possibilities for the orientation of the carbon matrix about a carbon fibre [28].

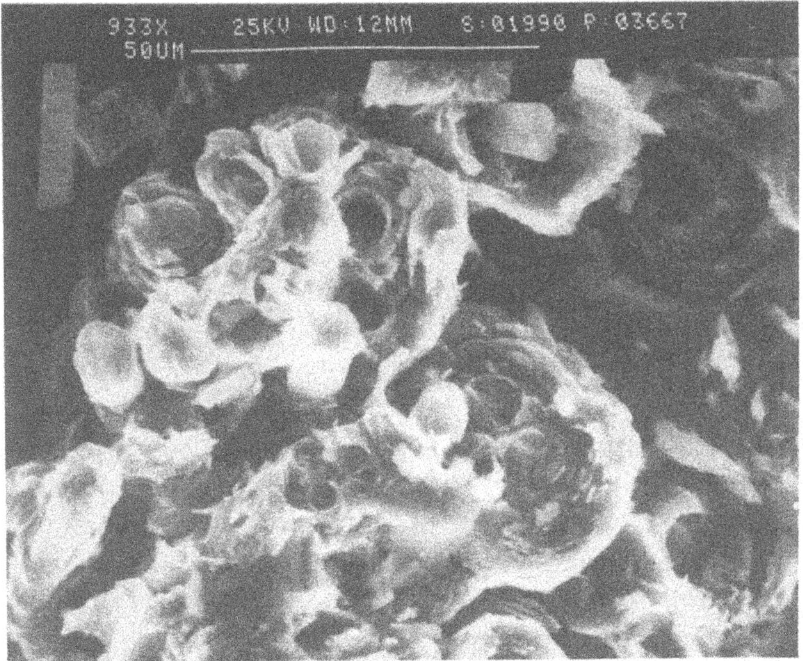

Fig. 8.11 Strong fibre matrix bonding in CVD-densified carbon–carbon.

matrix of ex-thermosetting resin materials although it is sometimes observed in CVD matrix materials [29]. Of the laminar forms, the *c*-axis radial (sheath) and *c*-axis axial (transversely aligned) are the most common. The fibre/matrix bond in CVD-formed composites is extremely strong, having been progressively built up atom by atom (Fig. 8.11). In the case of liquid matrix precursors, strong bonds may form when the fibre substrate has active carbon sites available to which the matrix can attach. The majority

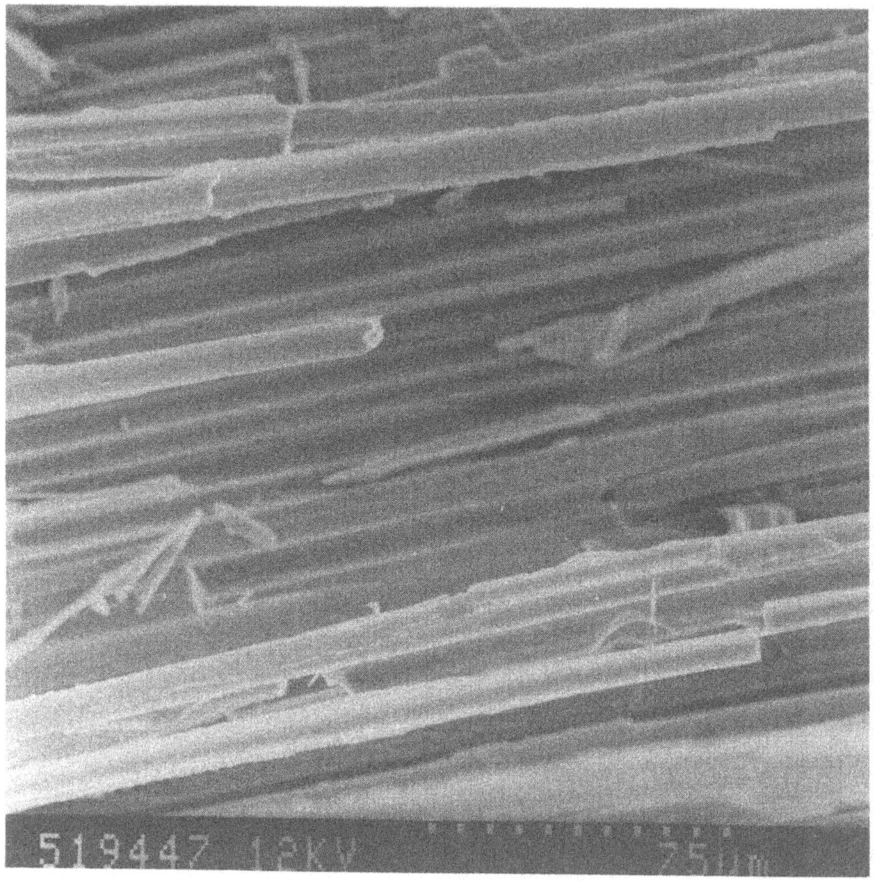

Fig. 8.12 Moderately strong interface around radial texture pitch-based ultra-high modulus fibres.

of PAN-based intermediate or high modulus (HM) fibres possess a skin-core structure with a sheath-like HM skin [30]. One would, therefore, only expect weak fibre–matrix bonds for this material. Ultra-high modulus pitch-based fibres, on the other hand, exhibit a radial texture with an active surface to which the matrix may bond well [31]. Evidence of this can be seen in Fig. 8.12, showing the relatively strong interface which is formed in P75/PEEK-based composites. A pitch matrix, however, tends to form a sheath around such fibres despite the postulated availability of edges to which the mesophase ought to have formed covalent bonds [32]. It is discrepancies such as this which highlight the severe gaps in our understanding of carbon–carbon.

The evidence presented in the literature shows a predominance of weak fibre–matrix bonding in carbon–carbon composites produced from liquid precursors. Interfacial bonding in CVD materials, on the other hand, is quite strong, the material being limited by the remaining large-scale porosity that is inherent to the process. The exact nature of the fibre/matrix interface in liquid pyrolysed composites remains a subject of some debate. In many respects, the outcome of the discussion is of little concern since if, as is believed, the interfacial bonds throughout the matrix are weak, the predominant failure mode will be a cohesive fracture of the matrix. The effects of such a phenomenon would be closely analogous to adhesive failure at the interface.

Unidirectional carbon–carbon composites exhibit low transverse flexure and shear strengths as a result of their weak matrix and interfaces. Their usefulness in structural applications is thus very limited. Nevertheless, they are extensively studied when researching the fundamental properties of carbon–carbon (Chapter 7). Carbon–carbon components thus tend to be multi-directionally reinforced; 2-D fabric laminates or spatially reinforced n-D structures are employed in order to increase damage resistance and reduce the risk of delamination in service.

The fibre to matrix bonding depends on the stresses developed between the two components as a result of matrix shrinkage during processing and by differences in thermal expansion. Increases in the degree of chemical bonding or intimate contact (mechanical interlocking) have been shown to manifest themselves in high tensile and compressive strengths in the fibre direction.

If the interface becomes stronger than the matrix itself, then the load transfer behaviour between fibre and matrix will depend on the mechanical properties of the matrix. Poor interfacial bonding will, conversely, result in failure at the interface. This has been demonstrated in tensile fractures showing fibres pulled out of composites without any discernible matrix adhering to their surface [33]. It is possible to engineer the nature of the interface by the selection of fibre type, the matrix and processing conditions [25]. A 'moderately strong' bond (transverse flexure strength ≈16 MPa) is preferred. Too weak an interface impairs stress transfer from matrix to fibres, whereas too strong a bond may promote damage to fibres during heat treatment and brittle fracture behaviour.

The thermal expansion behaviour of carbon–carbon is strongly dependent on the extent of fibre/matrix interaction. The CTE of highly aligned carbon fibres between 0 and 500 °C is negative. In contrast, a glassy carbon matrix will possess a CTE which is both positive and isotropic and is approximately four times the absolute value of that of the fibres. As a result, there will be a thermal mismatch between fibres and matrix which must, by definition, be resisted by the interface [34]. There is evidence, therefore, that the coupling between fibre and matrix may be altered from one of a chemical nature in the precursor into something akin to a 'frictional' type of bond.

In Section 8.2 it was shown how microscopy reveals a myriad of cracks running in a variety of directions within a carbon–carbon composite. This three-dimensional crack network arises due to inefficient penetration in CVD materials, and owing to the shrinkage which occurs from heteroatom elimination and differences in CTE between fibres and matrix in the case of liquid precursors. It is this network of cracks which permits the further densification of partially processed materials. The crack network enlarges the surface area available to oxidative attack [35]. The cracks do, however, aid the infiltration of the composite with protective compounds. Coatings and inhibitors may thus be anchored to the composite by mechanical interlocking, allowing the material to act as a reservoir for substances which soften at high temperatures and produce a protective barrier (Chapter 6) and also increasing the coating–substrate adhesion/spallation resistance.

Additional microcracking in service is a consequence of the thermal stress cracks, weak matrix and interfaces and the porosity present in all carbon–carbon composites. The resulting effects on thermomechanical properties include inefficient utilization of fibre properties, unpredictable anomalies in thermal expansion behaviour and highly non-linear responses to shear stresses [36]. The understanding of interfacial behaviour in carbon–carbon is paramount to the development of composites with attractive and predictable properties. Much of the behaviour of carbon–carbon may be explained, in part, by assuming that load transfer across interfaces occurs primarily by a friction mechanism. That is to say, much of the thermal and mechanical behaviour of carbon–carbon may be modelled assuming debonded interfaces which are only able to transmit stress in compression across the interface [36].

8.4 MECHANICAL PROPERTIES

The structural properties of continuous fibre 'advanced' composites are generally controlled by the properties, volume fraction and geometry of the fibres. Carbon–carbon composites, however, are by nature very complex as a result of the physical and chemical changes, and interactions that occur during processing. Pyrolysis of an organic precursor, for example, to form a carbon matrix involves a 50% reduction in volume. Shrinkage of this magnitude can create severe damage to the composite by the build-up of large process-related stresses [37]. The transformation from a carbon fibre/organic interface to a carbon/carbon fibre interfacial bond will vary considerably, dependent on materials and process variables. Extended heat treatments, process-induced stresses, and fibre–matrix interactions are very likely to affect adversely the primary properties of the fibre reinforcement [38,39]. Severe thermal stresses, due to thermal expansion mismatches between fibres and matrix, and between different matrix structures, will result in a time-dependent deterioration of composite mechanical properties during heat treatment and service cycles.

A complete description of how composition and processing influence the properties of carbon–carbon is beyond the current level of knowledge and understanding. The data available are only capable of qualitatively indicating the trends that have been observed. A number of these trends will be reviewed in an attempt to illustrate the factors and principles that influence the mechanical behaviour of carbon–carbon.

In keeping with all fibre-reinforced materials, the mechanical properties of a carbon–carbon composite show marked anisotropy as a result of the anisotropy in the fibres. For the case of UD fibres, the maximum stress theory states that the tensile strength of the fibre will be manifest when the angle (θ) between the applied load and fibre axis is no more than 4°. If θ exceeds 4° the strength (σ_{11}) will be governed by the shear strength (τ) and may be expressed as

$$\sigma_{11} = \tau / \sin\theta \cos\theta. \tag{8.1}$$

Should θ exceed 24°, the strength approaches that of the matrix. The relationship between the angle at which a tensile load is applied and the strength of a UD carbon–carbon composite is shown in Fig. 8.13 [40]. The composite's tensile strength (σ_c) in the fibre axis of a 1-D material can be expressed according to a simple rule of mixture:

$$\sigma_c = \sigma_f V_f + \sigma_m (1 - V_f), \tag{8.2}$$

where σ_f is the fibre strength, σ_m the matrix strength and V_f the volume fraction of fibres. The strength of the matrix can generally be neglected when compared to the strength of the fibres such that the strength of a composite is proportional to the fibre content.

It is very difficult, in practical applications, to realize the theoretical strength of a composite due to twisting and distortion of the fibres, variations in fibre orientation, and stress concentrations associated with the method of test. Furthermore, should the fibre content exceed 70%, uniform impregnation by the matrix becomes impossible, resulting in a fall in strength. The large-scale, irregular porosity, poor fibre/matrix interface, and inherent brittleness of the system have the result that the carbon–carbon never attains its theoretical strength, values of 50–60% being typical. As a general rule of thumb, it is usually assumed that a balanced woven fabric reinforced material (or 0/90° UD lay-up) will have roughly half the strength of a UD material and a quasi-isotropic laminate one-third its strength.

The strength of carbon–carbon composite materials does not generally follow the simple law of mixtures relationship. The weak interfacial bonding results in an inefficient transfer of applied loads on to the fibre reinforcement, such that theoretical strengths are seldom if ever, achieved. Mechanical properties of carbon–carbon tend to vary between 10 and 60%

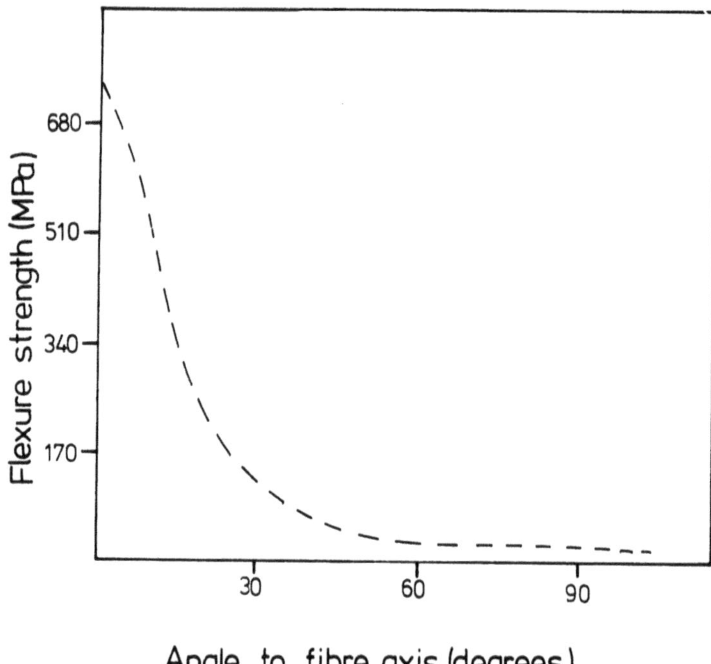

Fig. 8.13 Effect of angle between fibre axis and the load directions on the strength of a carbon–carbon composite.

lower than the values calculated from the rule of mixtures. It is possible to increase the strength of the fibre/matrix interface but, as previously discussed, this manifests itself in extreme brittleness of the composite.

8.4.1 Influence of the fibre microstructure and architecture on mechanical properties

The mechanical properties of carbon–carbon are very much dominated by the properties of the fibres which depend on precursor type and processing conditions (Chapter 2). Microstructural variations in shape, porosity and cross-section can occur within filaments. Variations in fibre architecture on the thermomechanical properties of carbon–carbon are ill-understood, but it is postulated that they may result in a heterogeneous distribution within the composite. Such a distribution may result in understresses in some fibres while overstressing others at adjacent locations [41]. The key factor in optimum fibre utilization is good alignment with respect to the axis of the load [42]. Stress–strain curves in 2- and 3-D composites show reductions in fibre properties and behave non-linearly as a consequence of misalignment [43] due to the 'crimping' which occurs as a result of the weaving process. The fibres are not aligned with the principal stress axis and their effective strengths are thus reduced.

A study has been carried out by Manocha and Bahl [44] on the effect of fibre type and weave pattern on the mechanical properties of 2-D carbon–carbon. They were able to demonstrate clearly how the choice of reinforcement is paramount to the performance of the composite. The basic properties of the fibre were found to influence the ultimate properties of the composites. The fibre surface affects the fibre/matrix interaction, which in turn adopts a significant role in the densification process.

High modulus fibres are observed to lead to better densified composites than high strength or IM fibres. The weave pattern also contributes to the properties of the 2-D materials. An eight-harness satin weave was found to produce superior composites to those employing a plain weave geometry. The eight-harness weave exhibits fewer 'kinks' in its structure and thus a greater utilization of fibre properties. Furthermore, the kinked plain-weave pattern results in a carbon–carbon composite with heterogeneous matrix pockets.

8.4.2 CVD material

One of the major virtues of carbon–carbon is the non-brittle fracture behaviour exhibited, especially in the case of continuous fibre reinforcement. Schmidt [45] and Cristina [46] have reported a psuedo-plastic failure behaviour under tensile loading. This phenomenon is attributed to matrix cracking and fibre movement to accommodate the applied load. Similar behaviour has been observed under flexural loading [47]. It has been discussed how the fracture behaviour is strongly influenced by the type of interface established between fibre and matrix. The mechanical properties of material formed by the CVD method are strongly influenced by the matrix structure, which in turn depends on the deposition conditions employed. The dependence of the flexural strength of carbon-felt/CVD carbon composites on the matrix has been demonstrated by Pierson and Northrop [48]. A similar result was also observed by Delhaes *et al.* for a carbon-fabric/CVD material [49].

In Chapter 3 it was shown how the structure of the pyrolytic carbon matrix exhibits a systematic variation with changes in the deposition temperature and gas composition. A wide range of structural parameters, such as density and crystallite size, as well as microstructures, is possible. These variations in matrix structure, as one would expect, influence the mechanical properties. The mechanical properties of CVD carbon–carbon composites will be affected by the variations in elastic properties of the matrix which are due to the difference in crystalline perfection and to the presence of defects.

Oh and Lee [50] carried out a series of densification experiments using a reactant gas concentration of 10% propane at temperatures between 1100 and 1440 °C with a carbon fabric. In material processed at 1100 °C, the matrix was observed to be well infiltrated between, and well bonded to, the fibres. By contrast, a great number of pores are observed in

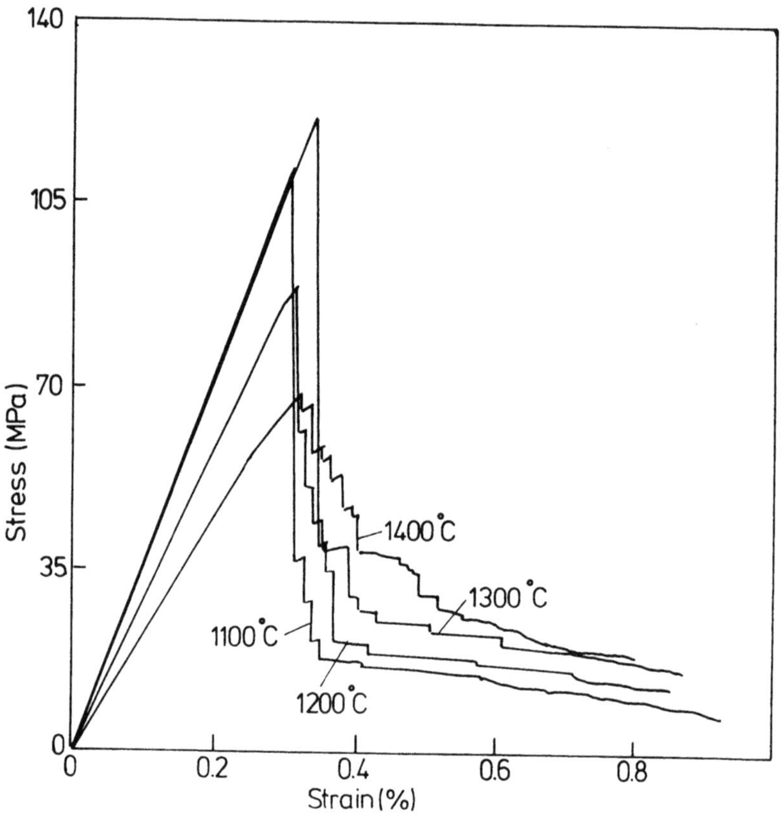

Fig. 8.14 Stress–strain behaviour of CVD carbon–carbon composites prepared from 10% propane at various temperatures [50].

composites made at 1400 °C, the matrix being loosely bound to the fibres. The increased porosity at high temperatures is due to the blockage of the pore channels to reactant gases, which is caused by the deposition of carbon on to the outer fibres before those in the interior are completely infiltrated. The bulk density decreased from 1.79 to 1.37 g cm^{-3} with increasing deposition temperature [51]. This is a consequence of a decrease in matrix density from 1.99 to 1.55 g cm^{-3} and the greater porosity. A decrease in the optical activity is also observed, corresponding to a microstructural change from smooth laminar to isotropic, the arrangement of planar carbon atoms in the matrix being better at lower deposition temperatures [50].

The variation in the stress–strain curves (in flexure) with respect to the deposition temperature is shown in Fig. 8.14 [51]. The stress–strain behaviour of the material deposited at 1100 °C is very linear elastic, representing a near catastrophic failure mode. Increasing the temperature of deposition leads progressively to a stepwise failure behaviour. Both strength and modulus decrease with increasing processing temperature (Fig. 8.15)

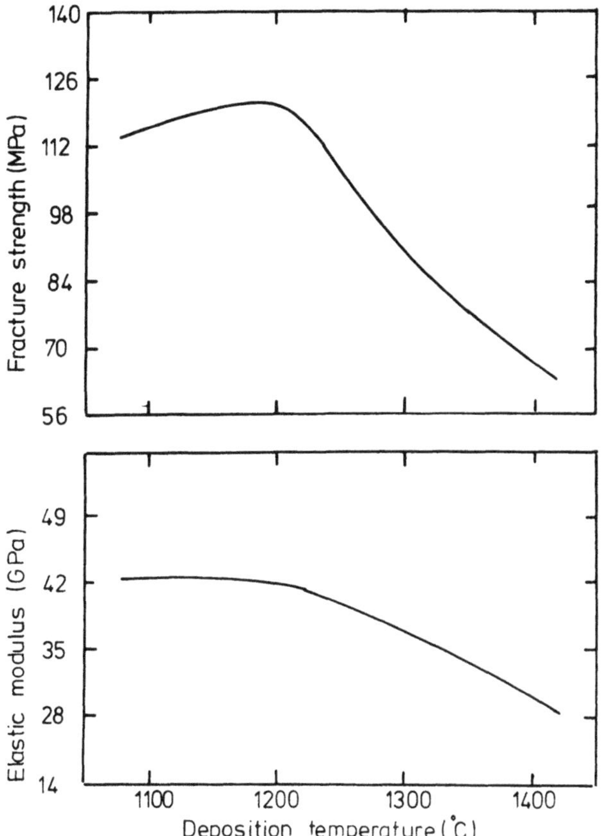

Fig. 8.15 Effect of processing temperature on the mechanical properties of 2-D CVD carbon–carbon using a 10% concentration propane feedstock [50].

due, in part, to the reduced bulk density of the composite and also the increased flow/pore size [52]. The strain to failure, on the other hand, remains almost constant. This observation was made for both fabric and felt reinforcement. The strain to failure of the composite is dominated by that of the matrix, from which one may conclude that the different matrix microstructures have roughly identical failure strains [53].

Microscopical observations on fabric-reinforced material reveal two failure modes: fracture perpendicular to the plane of the reinforcement and delamination. Composites formed at 1100 °C possess a smooth fracture surface with no evidence of fibre pull-out. Material produced at 1400 °C, by contrast, fails predominantly by delamination. Pyrolytic carbon deposited at high temperatures would tend to be more chemically active, having edge atoms with unsaturated C–C bonds as a result of a lower degree of preferred orientation. One would therefore expect a stronger fibre/matrix interface in composites densified at 1400 °C rather than those produced at 1100 °C. Experimental observations, however, reveal the converse to be

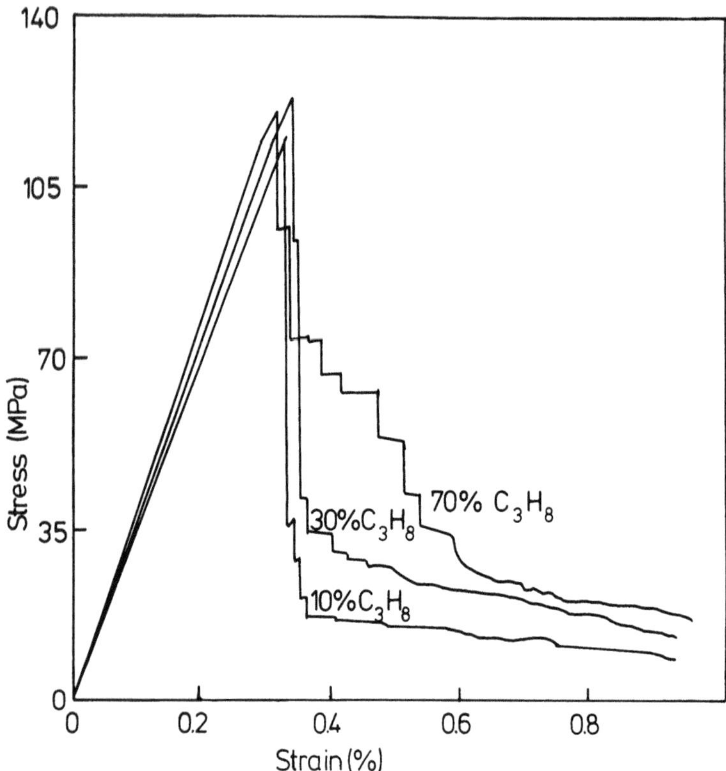

Fig. 8.16 Stress–strain behaviour of CVD carbon–carbon composites with respect to reactant (propane) gas concentration [51].

true. This suggests, then, that the bond between fibres and matrix in CVD composites is one of mechanical interlockings rather than a chemical bond. The strong interface at 1100 °C is most likely due to a compact infiltration of matrix into the pores between fibres. Cracks may propagate through such a microstructure with ease. Fibres may fracture in the primary crack plane, resulting in a catastrophic failure mode devoid of delamination or fibre pull-out. The loosely bound fibre/matrix interface is much less efficient at load transfer but can serve to arrest cracks. Such a material will fail in a stepwise process with fibre pull-out and delamination.

Increasing the concentration of the reactant gas at constant temperature results in an increased optical activity in the microstructure of the carbon–carbon [51]. A higher composite bulk density is observed, caused mainly by the increased matrix density, which in turn results from improved arrangement of planar carbon molecules. Stress–strain curves change from a catastrophic fracture to a stepwise failure with increasing hydrocarbon concentration. The results of Oh and Lee, using propane as the reactant, are shown in Fig. 8.16. The strength and modulus do not, however, vary

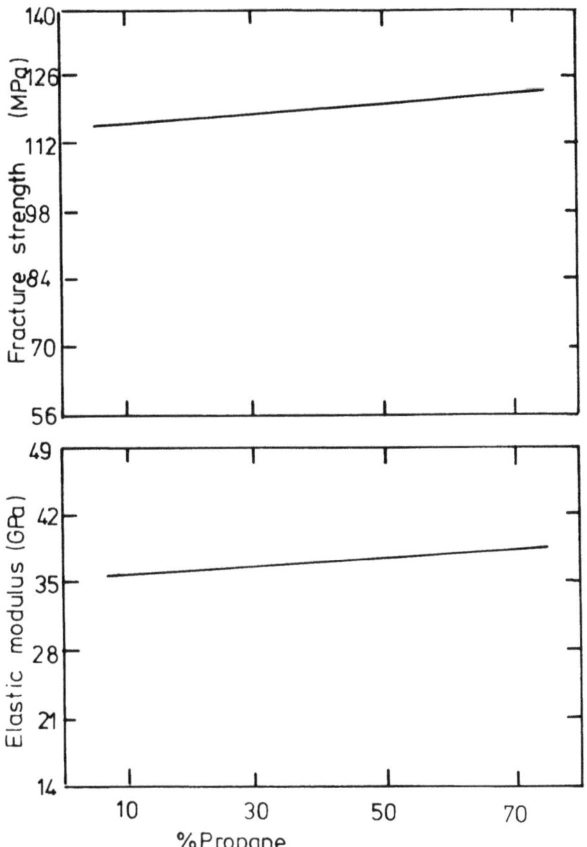

Fig. 8.17 Effect of reactant gas concentration on the mechanical properties of CVD carbon–carbon densified at 1100 °C [51].

significantly but, rather, are slightly increased with increasing reactant concentration (Fig. 8.17). A higher concentration of propane in the system was shown to yield a higher degree of preferred orientation of the matrix with respect to the fibre surface, resulting in a weaker bond between fibre and matrix, and hence a pseudo-plastic fracture behaviour.

To summarize, the lower mechanical properties of an isotropic matrix deposited at higher temperatures are due to its lower density and the formation of closed porosity. A high-density smooth laminar matrix produces composites with high strength and stiffness but brittle fracture behaviour. Rough laminar matrix material shows the highest degree of preferred orientation. It is associated with severe microcracking and has a lower modulus and strength than the smooth laminar structure in spite of their similar densities. Increasing the reactant gas concentration results in a slight improvement in mechanical properties and a pseudo-plastic rather than linear-elastic fracture behaviour. The stepwise failure is believed to

Fracture stress (MPa)
0
35
70
105
140
1100
1200
1300
1400
Deposition temperature (°C)
10
20
30
40
50
60
70
% Propane
35
70
105
140

(a)

Elastic modulus (GPa)
0
7
14
21
28
35
1100
1200
1300
1400
Deposition temperature (°C)
10
20
30
40
50
60
70
% Propane
7
14
21
28
35

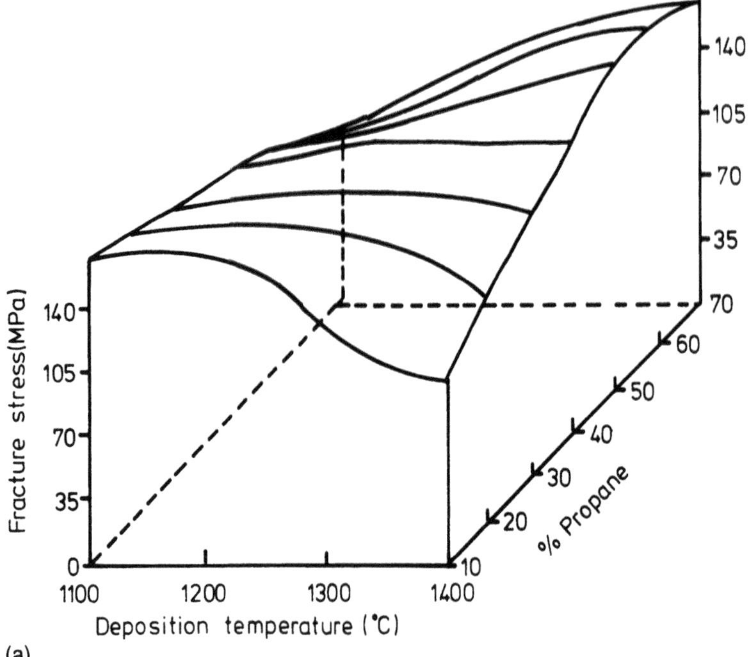

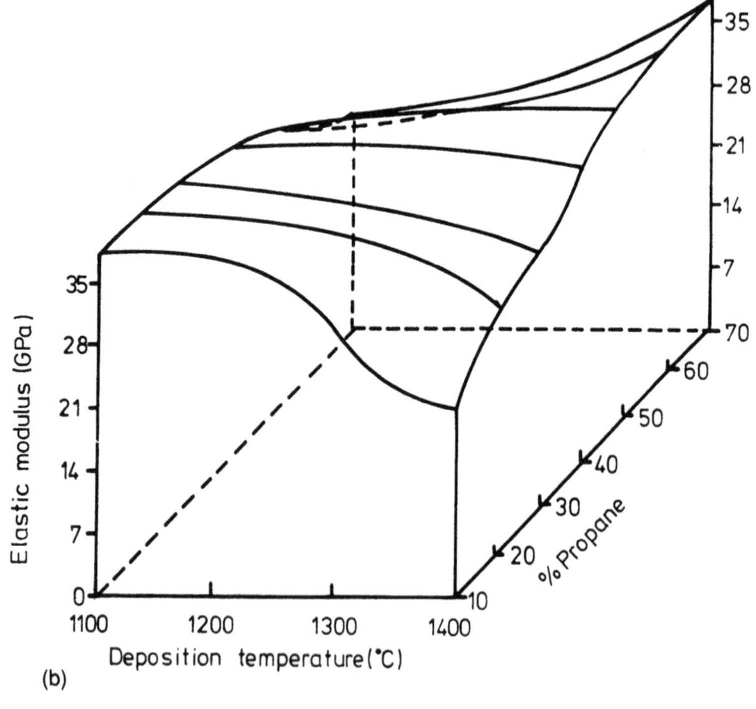

(b)

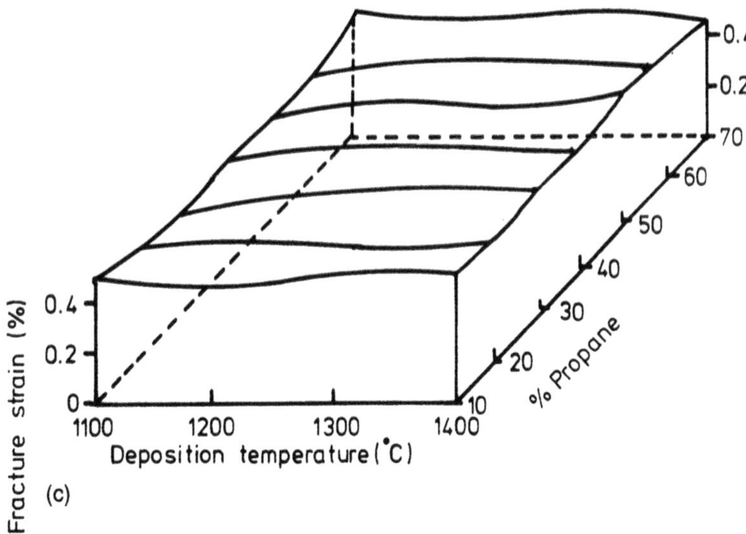

Fig. 8.18 3-D surface plots showing the complex relationships between processing conditions and mechanical properties of CVD carbon–carbon [51].

be due to a loosely bound interface between fibre and matrix. Figure 8.18 shows some 3-D surface plots illustrating the complex relationships between processing conditions and mechanical properties of CVD carbon–carbon.

8.4.3 Thermoset-derived carbon–carbon

The first requirement for a suitable thermoset resin matrix precursor is a high carbon yield which must be achievable under simple pyrolysis conditions. A second requirement is that the shrinkage of the matrix on carbonization should not damage the carbon fibres. Finally, the carbon matrix formed by pyrolysis of the precursor should contain open rather than closed porosity. Only under these preconditions can further improvement of density and mechanical properties be achieved by subsequent impregnation steps. Cross-linked aromatic resins with high molecular weights, such as polyimides, polyphenylene and acetylene-terminated aromatics, achieve a high initial strength after only one carbonization. This is a consequence of a high carbon yield and isotropic shrinkage without damage to the fibre. There is a tendency, however, to form closed porosity such that the mechanical properties achieved after one processing cycle are not significantly improved by further post-impregnation steps.

A second group of matrix precursors, which includes phenolic and furans, forms porous carbon matrices with low composite mechanical strength after the first carbonization. The high accessible porosity allows strong increases in mechanical properties by repeated impregnation/recarbonization cycles.

The highest mechanical properties are achieved by using HM PAN or mesophase pitch-based fibres, preferably without surface treatment. If not surface treated, HM fibres show very low adhesion to the resin. The bulk cross-sectional shrinkage on pyrolysis is therefore very low. In the case of IM and isotropic fibres, especially those which have been surface treated, the first carbonization causes a bulk shrinkage as a result of good fibre/matrix adhesion. The shrinkage of the resin is resisted by the low compressibility of the fibres. As a result, a high degree of compressive pre-stress is built up in the fibres which reduces the strength of the composite.

The major influence on the mechanical properties of carbon–carbon produced from thermosetting resin precursors is the HTT [54]. At an HTT of below 2000 °C, the strength is low and brittle fracture the predominant failure mode. The matrix so formed is a brittle, isotropic glassy carbon in which propagating cracks readily sever the fibres without changing direction. At HTTs above approximately 2400 °C a graphite component begins to develop at the fibre/matrix interface. An apparent pseudo-plastic, 'ductile' fracture behaviour is the consequence. The strength and modulus increase up to an HTT of around 2750 °C.

Processing at 2800 °C and above tends to reduce the strength due to thermal damage to the fibres [9]. A graphical representation of the effect of HTT on the flexural strength of furan-based composites is shown in Fig. 8.19 [54]. The ductile behaviour of the graphitized composites may be illustrated by considering the energy of fracture [55]. At an HTT of 1000 °C this is measured at 0.3 kJ m^{-2} whereas at 2700 °C it is around 4 kJ m^{-2}, more than an order of magnitude higher.

Exotic high carbon yield resins are seldom used as matrix precursors in commercial operations due to their high cost, limited commercial availability and the formation of closed porosity. The thermosetting resins which are employed generally yield between 50 and 60% by weight of carbon on pyrolysis. A single impregnation carbonization treatment of a 2-D fabric reinforced composite would typically lead to a material with a low bulk density of 1.3–1.4 g cm^{-3}. The impregnation and carbonization thus require to be repeated several times in order to raise the densities to 1.7–1.8 g cm^{-3}. As previously discussed (Chapter 4), the density increase only follows the 1/2 power of the number of impregnation cycles due to blockage of the pores by the impregnant. Above a certain number of times, the density will hardly increase, such that four to six cycles are generally taken to be the optimum. Graphitization results in a shrinkage of the matrix, thus making the porosity more accessible, and is therefore routinely used as an aid to densification.

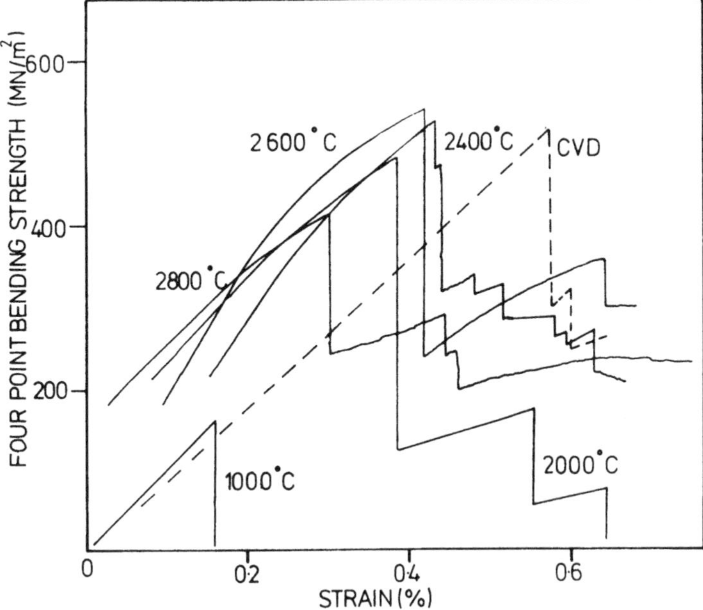

Fig. 8.19 Fracture behaviour of a carbon–carbon composite made from a furan precursor, with respect to HTT.

8.4.4 Mechanical behaviour of carbon–carbon with thermoplastic precursor matrices

The carbon yield of a mesophase pitch is increased from around 35% at ambient pressure to 80% by the application of 100 bar of inert gas pressure [56]. A gas pressure of 1000 bar may effect yields of up to 90% [57]. No systematic results are available in the open literature on the effect of high and ultra high pressures on the final mechanical properties of carbon–carbon. Generally, only an increase in bulk density up to 2 g cm^{-3} and a more rapid approach to 'desirable' mechanical properties is reported [58].

The properties of pitch-based carbon–carbon are not only controlled by the carbon yield of the matrix precursor, but also by the microstructure. Pitch carbonized under relatively low pressures results in a well graphitizable matrix carbon with a sheath structure parallel to the fibre surface and thus in anisotropic matrix properties [59]. A transverse oriented (TO) matrix structure has been observed when high pressures (≈1000 bar) have been applied during carbonization [60]. Such TO structures tend to produce a more isotropic matrix as is found in some CVD material. Preferred orientation of the carbon matrix has a significant influence on the mechanical bulk properties of the composite. This preferred orientation can be influenced by the heating rate during carbonization and the addition of

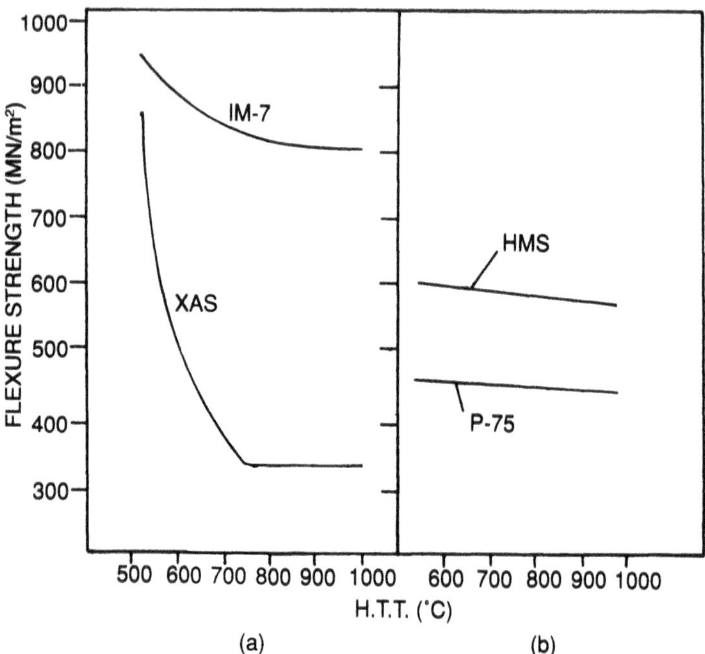

Fig. 8.20 Effect of HTT on the strength of pitch matrix carbon–carbon composites: (a) IM fibres; (b) HM fibres.

cross-linking agents such as sulphur to improve carbon yield. In the best cases (low heating rates ≈20 °C h^{-1} and no sulphur additions), complete matrix orientation parallel with the fibre surfaces results in a strong contribution of the matrix to the composite's modulus of the order of that of the fibres.

A carbon–carbon composite with a pitch-based matrix shows a pseudo-plastic fracture behaviour but the strength shows the reverse pattern of dependence on processing temperature to resin-based materials. Strength is observed to fall as the HTT is increased [61]. Figure 8.20 shows a typical strength/HTT relationship for pitch matrix composites with both IM and HM fibres. Figure 8.21 shows the increase in flexural strength and bulk density achieved as a function of the densification cycles for composites prepared from pitch [62]. As a general rule of thumb, four densification cycles generally increase the flexural strength of a pitch-based carbon–carbon composite by a factor of three to four times. A similar result is observed in thermoset-derived material [63]. A flexural strength of around 1000 MPa can be achieved after four impregnation/carbonization cycles. Four impregnation/graphitization cycles yield a strength of 700 MPa.

Microscopy shows that cracks between fibre and matrix and within the matrix are present in fully densified composites. The extent of matrix and interfacial cracking depends on the increasing anisotropy of the matrix

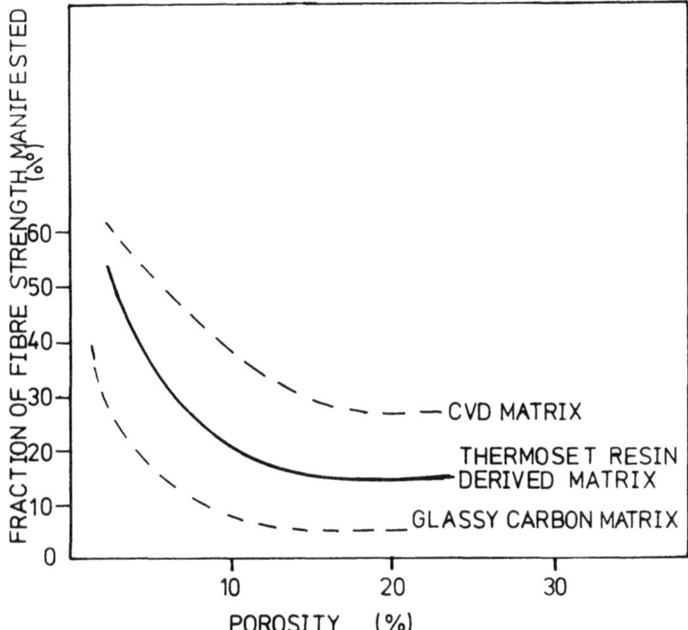

Fig. 8.21 The percentage of fibre strength manifested according to the percentage of open porosity in a carbon–carbon composite [66].

microstructure. In the case of a high degree of anisotropy, thermal contraction during cooling from the graphitization temperature matches that of the fibres and the oriented matrix. The shrinkage stresses resulting from a more isotropic matrix, on the other hand, cause crack formation. The bonding between fibre and matrix is very weak. This mechanical interlocking can be improved by repeated impregnation. Repeated graphitization cycles increase the size of the lens-shaped porosity between fibres and matrix. Graphitization thus aids densification but at the price of a lower 'ultimate' composite strength.

Recent studies by the author and co-workers [64] have concentrated on the production of carbon–carbon from high carbon yield thermoplastic resin precursors such as PEEK. Flexural strengths of close on 1000 MPa were obtained from a single carbonization cycle (IM fibre reinforcement). That is to say, the mechanical properties of the thermoplastic-based systems were equal or superior to those obtained from thermoset materials after four cycles. The thermoplastic precursor systems may well be viable as structural materials since their high raw materials costs are offset by lower processing costs. Of special interest are the composites made using ultra-high modulus pitch-based fibres. An optimum bond strength between fibres and matrix produced a flexural strength close to that of the polymer system whence it was derived, while an improved modulus was obtained due

Table 8.1 Mechanical properties of carbon–carbon derived from thermoplastics compared with those of their precursors

Property	*AS4/PEEK*	*P75/PEEK*	*AS4/PEEK (carbonized)*	*P75/PEEK (carbonized)*
Flexural strength (MPa)	1800	728	900	600
Flexural modulus (GPa)	150	280	190	300
Transverse flexural strength (MPa)	137	51.8	10	16

to the contribution of the matrix. A summary of the properties of the thermoplastic-based materials is given in Table 8.1 Ultra-high modulus materials of this type may well have a market niche in space structures which are lightly loaded, stiffness-critical and where high premiums are paid to save weight.

8.4.5 Porosity

One of the major factors determining the strength of carbon–carbon is the density of the material. The density may be considered as a description of the composites' porosity or void content. The strength of a ceramic material may be expressed in terms of its porosity by a simple empirical formula [65]:

$$\sigma = \sigma_0 \, e^{-BP} \tag{8.3}$$

where σ_0 is the strength of the material with zero porosity, B a constant and P the porosity.

A similar relationship has been observed in carbon–carbon composites as illustrated in Fig. 8.21 [66]. The upper curve shows a CVD material, the central curve a graphitized thermoset and the lower curve a carbonized thermoset with a glassy matrix. The change in strength due to differences in the type of matrix is thus clearly demonstrated. In flexural testing, cracks develop from the outer, tensile surface. In other words, the open pore content governs the strength, whereas in a tensile test a uniform stress field is generated such that the overall porosity will be the important factor. Microscopy shows that, even after four or more densification cycles, the composites still contain a large degree of porosity. Despite the inefficiency of the reimpregnation process, large increases in strength are obtained. This can be explained in part by a reduction in the porosity, but also because the reimpregnation tends to smooth the geometry and reduce the size of the voids thus lessening stress concentration.

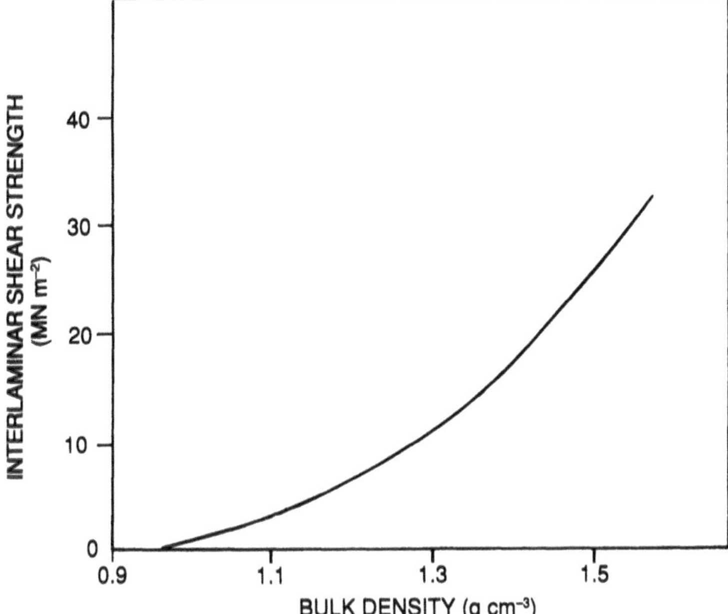

Fig. 8.22 Effect of increased composite density on the interlaminar shear strength of a pitch-based carbon–carbon composite [56].

8.4.6 Compression and shear properties

There are very few instances in which a carbon–carbon composite is used as a compression member [67]. The property which controls the compressive response is the interlaminar shear strength. The relationship between density and ILS of a pitch-based composite is shown in Fig. 8.22 [56]. The interlaminar shear strength increases with density in the same way as the flexural and tensile strengths. There is, however, an ultimate value of ILS of approximately 30 MPa. It is clear then that the shear strength of the matrix governs the shear strength of the composite. The same is true for a CVD-type matrix.

8.4.7 Effect of heat treatment temperature on the modulus of carbon–carbon

The modulus, like the strength of carbon–carbon, is dependent on the matrix type and HTT. Consider, for example, a UD material fabricated from a thermoset precursor as shown in Fig. 8.23. Increase in HTT causes a rapid increase in the modulus in the fibre axial direction. The specific modulus, along that axis, may reach a value three times that of steel for an HTT of ≈2800 °C [68]. In contrast, the modulus perpendicular to the fibre axis falls, albeit not so sharply as the increase in the parallel direction,

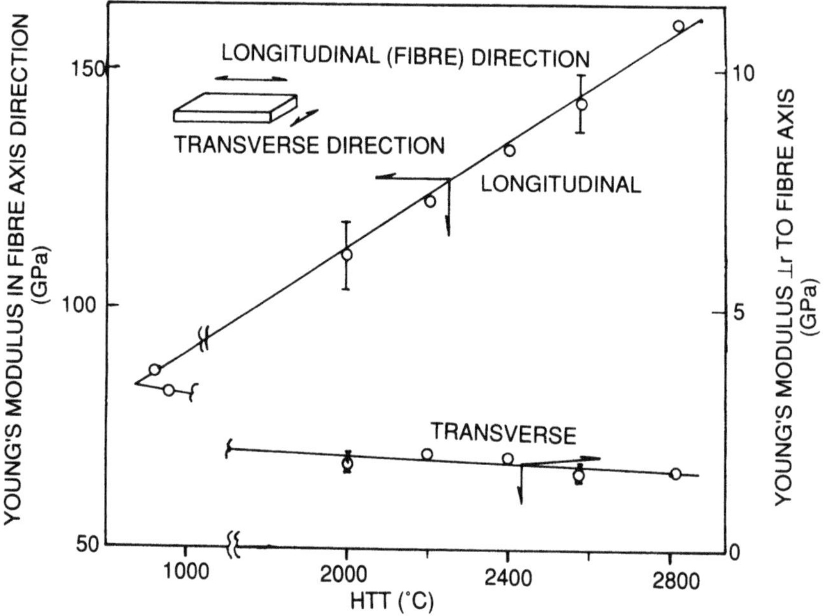

Fig. 8.23 Relationship between Young's modulus and HTT for a UD carbon fibre thermoset (furan) composite [68].

with increased HTT. These effects are due to the morphological changes in the matrix, the development of graphite crystallites and increase in the degree of preferred orientation.

8.4.8 Mechanical properties at high temperatures

One of the major assets of a carbon–carbon composite is that of retention of mechanical properties at high temperatures, as shown in Fig. 8.24 [69,70]. It can be seen from Fig. 8.24 that up to 1500 °C, there is a 10–20% increase in strength and modulus compared to ambient temperatures.

8.4.9 Fatigue properties

Very little work is presented in the open literature concerning the fatigue performance of carbon–carbon. Fitzer and Heym [71] showed a lifetime of 10^7 cycles at a stress of 40% of the static bending strength (Fig. 8.25). A Woeler diagram presented in the literature of Schunk [72] suggests the material shows a fatigue limit analogous to that of steel. An infinite fatigue life is predicted at 45–50% of the static strength (Fig. 8.26) under a cyclic, sinusoidal loading. The fatigue endurance properties of carbon–carbon appear, therefore, to be outstanding.

A question arises, however, as to whether the pore opening and closing

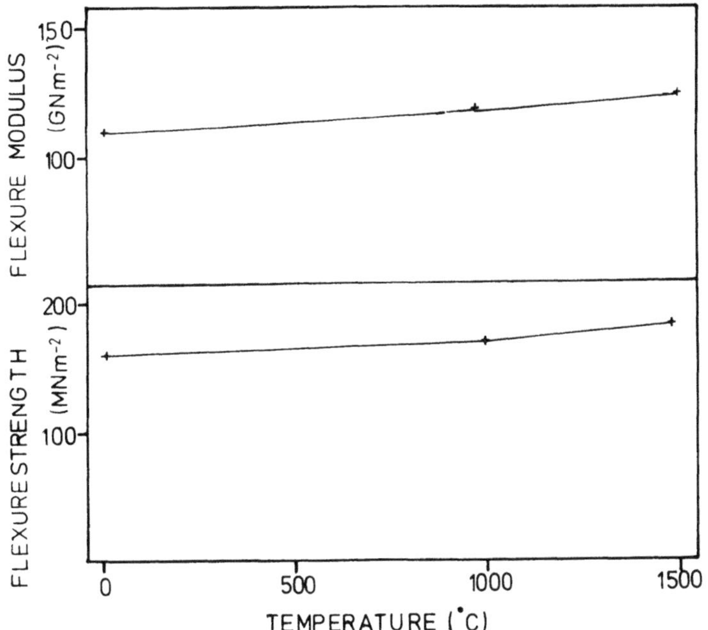

Fig. 8.24 Change in flexure strength and modulus of a carbon–carbon composite made from a thermoset precursor with respect to operating temperature [69].

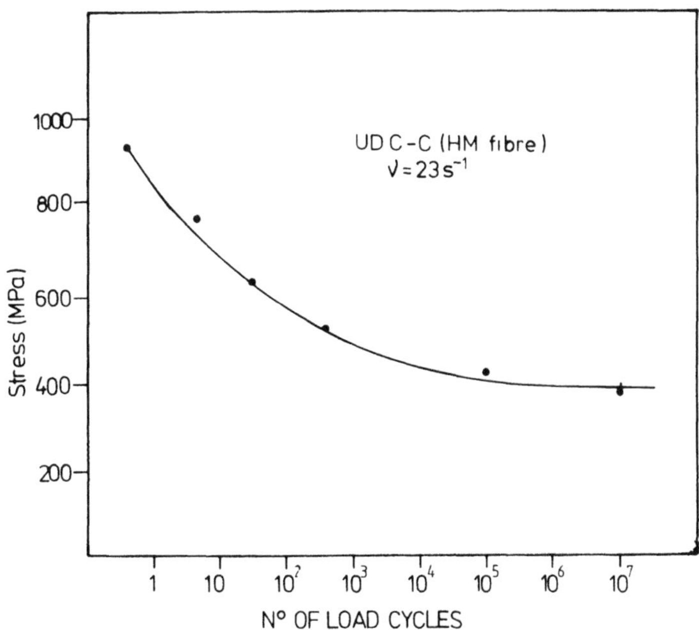

Fig. 8.25 Fatigue curve for a carbon–carbon composite [71].

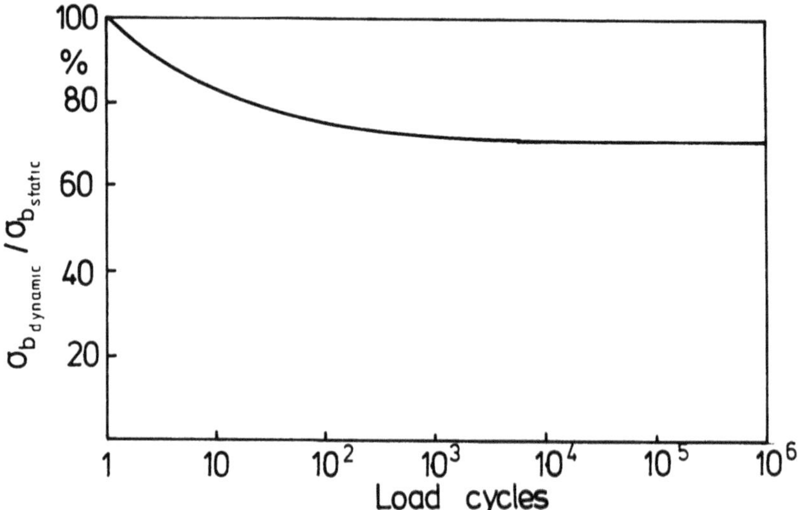

Fig. 8.26 Fatigue behaviour of carbon–carbon composites under alternating flexural load, suggesting a 'fatigue limit' as observed in ferrous metals.

mechanism, which must occur under fatigue loading, will continue indefinitely. The pore volume is increased during long time cycling. A probability exists that in local regions, the fracture strength of the matrix will be exceeded. The result will be the fracturing of the matrix in those areas and its loss as dust from the composite. This phenomenon, for obvious reasons, is known as 'dusting-out'. Dusting-out has been observed at high temperatures (≈1400 °C) in centrifugal loading operations (fan blades). There is therefore a severe limitation on the long time application of carbon–carbon under high alternating stresses.

8.4.10 Fracture toughness

Materials such as carbon–carbon which have internal flaws or voids as a consequence of fabrication processes must be evaluated for fracture toughness prior to consideration for a specific application. The toughness is interpreted as the ability of a material to resist crack propagation from existing voids. Very few data have been reported to date on the fracture toughness of carbon–carbon, due in the main to the difficulty in making accurate and meaningful measurements.

The various fracture toughness parameters are found to depend strongly on the type of carbon fibre used and the orientation of the initial crack with respect to the fibre reinforcement structure. In a two-dimensional, fabric-reinforced composite severe crack blunting and delamination are observed when crack propagation is perpendicular to the fibres [73]. K_{Ic}

values of around 7.6 MN $m^{-3/2}$ have been calculated in this orientation [63,68]. Such results are, however, invalid since linear elastic fracture mechanics is not applicable to pseudo-plastic failure. When crack propagation occurs collinear with the initial crack, such as between the layers of a 2-D material, fracture toughness parameters can be measured – 1- and 2-D laminates are particularly susceptible to this mode of failure. The interlaminar fracture toughness of a phenolic-derived material was found to be of the order of 0.11 kJ m^{-2}. This value is very low, around half that measured for even the most brittle epoxy composites [63]. The extremely brittle nature of the matrix and the presence of voids and shrinkage cracks may result in failure of components during processing. Cracks, etc. formed during the first pyrolysis cycle may grow as a result of thermal loading of subsequent cycles such that the materials are extremely unpredictable. A 2-D composite may thus fail by delamination without warning during the processing.

The interlaminar fracture toughness of 2-D ex-thermoset carbon–carbon may be increased by more than a factor of 2 by the introduction of a dispersed, third phase of particulate graphite. The graphite reduces the shrinkage of the matrix during carbonization, while increasing the G_{Ic} by a crack deflection and blunting mechanism. Optimization of filler content and particle size have resulted in G_{Ic}, values as high as 0.25 kJ m^{-2} being measured [63]. From a practical standpoint, 2-D composites containing graphite fillers were less prone to delamination during processing (in fact, in laboratory preparation this type of failure was totally eliminated) and easier to machine when finished.

As shown in Fig. 8.27, three-dimensional reinforcement leads to a pseudo-plastic fracture behaviour. The formation of stepwise cracks, as seen in the stress–strain diagram, results from a high degree of fibre pull-out during fracture and makes for an exceptionally tough material [72].

8.5 THERMAL PROPERTIES

Next to mechanical properties, the most important characteristics of a carbon–carbon composite are thermal conduction and thermal expansion. The number of papers in the literature relating to these is only exceeded by those on mechanical properties.

8.5.1 Thermal conductivity

Theory of thermal conductivity

Two mechanisms exist for thermal conduction in a solid; the first is via electron charge cloud drift (cf. electrical conductivity) and the second is

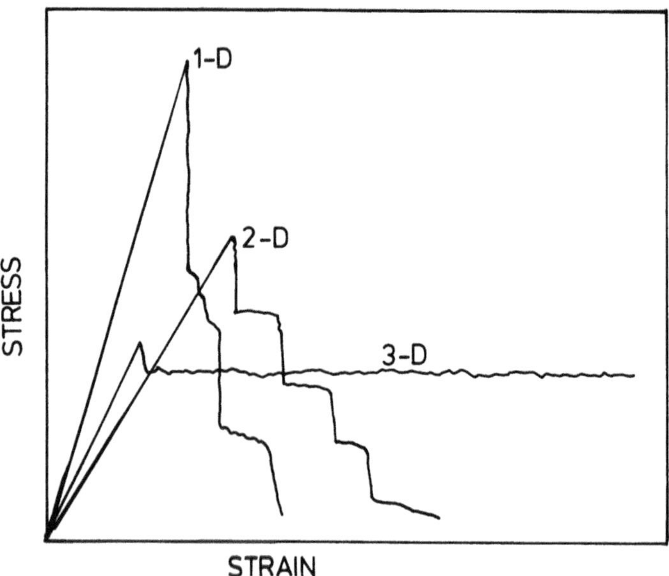

Fig. 8.27 Stress–strain behaviour of carbon–carbon showing increased 'pseudo-plasticity' with dimensions of reinforcement.

by lattice vibrations, known as phonons. The relative contributions from these two mechanisms have great implications for both the size and temperature dependence of thermal conductivity. The method for assessing the relative contributions of electronic and phonon mechanisms of thermal conductivity is by calculation of the Wiedmann–Franz ratio as a function of temperature [74]:

$$\text{Wiedmann–Franz ratio} = \frac{\text{Thermal conductivity} \times \text{electrical resistivity}}{\text{Temperature}} (\text{in } \mathrm{V}^2\,\mathrm{K}^{-2}). \tag{8.4}$$

A thermal conductivity dominated predominantly by the electronic component is signified by a near-constant Wiedmann–Franz ratio. Carbon-based materials have wildly varying Wiedmann–Franz ratios, indicative of a phonon-based thermal conductivity mechanism. The phonon mechanism is believed to be responsible for 99% of the thermal conductivity [75].

The thermal conductivity of a phonon-dominated material can be described by the Debye equation [76]:

$$K = bCvl, \tag{8.5}$$

where K is the thermal conductivity, b a constant, C the specific heat per unit volume, v the phonon velocity and l the mean free path of phonons.

Phonons are scattered by two mechanisms; collision with other phonons and collision with lattice defects, crystal boundaries or pores. The two contributions to phonon scattering can thus be expressed as

$$\frac{1}{l} = \frac{1}{l_e} + \frac{1}{l_d}, \tag{8.6}$$

where l is the mean free path of phonons, l_e the phonon–phonon scattering path length and l_d the spacing of inhomogeneities, defects grain boundaries, etc.

The diversity of different carbon structures gives rise to very different contributions from the two phonon-scattering mechanisms. For highly crystalline materials, such as natural graphite, there are few defects and grain boundaries, so the contribution from l_d is relatively small. Two factors influence the temperature dependence of the thermal conductivity of these highly crystalline materials. The phonon–phonon scattering path length, l_e, which dominates the overall phonon scattering, is inversely proportional to temperature. From intermediate to high temperatures, the thermal conductivity thus tends to decrease. At low temperatures, on the other hand, the specific heat rapidly decreases to zero despite an increase in the phonon–phonon scattering path length which aids thermal conduction. For less crystalline carbon materials inhomogeneity-induced phonon scattering is much more important than simple phonon–phonon scattering. This manifests itself in the relative temperature invariance of thermal conductivity, as shown in Fig. 8.28 [77].

Thermal conductivity in carbon–carbon composites

Carbon–carbon composites present the opportunity to 'tailor' thermophysical properties into carbon materials. The complexity of possible options can be gauged by brief consideration of the different fibre/matrix/processing permutations possible:

1. fibres – rayon, PAN, isotropic and mesophase pitch (plus variations in heat treatment);
2. matrix – Thermoset pyrolysis, CVD and thermoplastic/pitch pyrolysis;
3. fibre architecture – Felt, UD, fabrics and 3-D structures.

With such a range of options, a very wide range of thermal conductivities is possible.

CVD matrices

As previously discussed, a range of different carbon microstructures is possible from the CVD process, each of which has considerable influence over the thermomechanical properties of the composite. Liebermann and Pierson at the Sandia Laboratories have made several investigations into

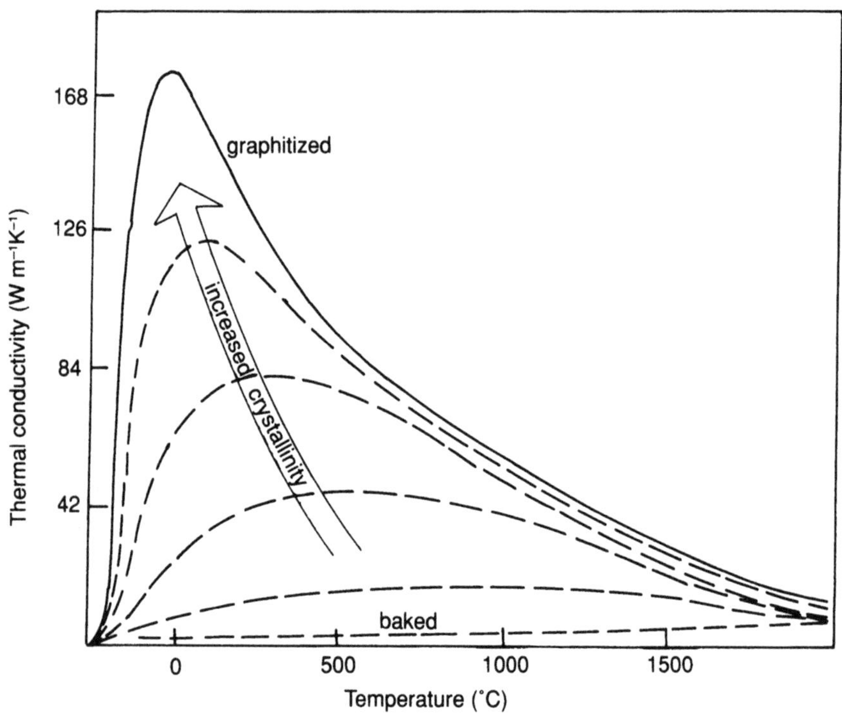

Fig. 8.28 The temperature dependence of thermal conductivity for crystalline and non-crystalline carbon material [77].

the CVD process/property relationships for different carbon–carbon composite materials [78]. Using a thermal gradient infiltration technique, they prepared a range of composites, based on a 0.1 g cm^{-3} felt preform, from CH_4/H_2 at 1250–1700 °C. Each composite was then given a final heat treatment at 3000 °C. The low density/volume fraction of the fibre preform allowed the assumption that the thermal conductivity property of the composite were matrix dominated.

The correlation of matrix microstructure with the thermophysical properties is relatively straightforward, as shown in Table 8.2. The rough laminar microstructure is clearly a good deal more 'graphitizable' than the smooth laminar and isotropic microstructures, and thus exhibits the highest thermal conductivity. A whole range of other studies validate Liebermann's theories on CVD matrix microstructure/property relationships. Curlee and Liebermann reported the thermal conductivity values for a filament-wound, CVD-densified tube where either CH_4 or C_6H_6 was used as the hydrocarbon feedstock [79]. While both matrices possessed the same smooth laminar microstructure, the composites prepared from benzene were found to have 80–120% higher thermal conductivity than the methane-derived matrix samples. This is again attributable to the

Table 8.2 CVD matrix microstructure/thermal conductivity relationships

Matrix microstructure	*Crystallite size* (Å)	*Thermal conductivity at 350 °C* ($W\ m^{-1}\ K^{-1}$)
Smooth laminar	125	25
Rough laminar	385	96
Isotropic	90	25

relative graphitizability of the matrix. Curlee and Northrop prepared CVD-impregnated, filament-wound hoops from CH_4/H_2 mixtures at identifiable temperatures and pressures but with a different C/H feed ratio [80]. They produced both laminar and isotropic matrix microstructures. The laminar matrix material was found to possess roughly twice as great a value of thermal conductivity as the isotropic matrix composite (54 versus 22 W $m^{-1}\ K^{-1}$ at 350 °C).

Thermoset-derived glassy carbon matrices

In these carbon–carbon systems, the fibres will dominate the thermal conductivity of the composite due to the fundamentally low thermal conductivity of glassy carbon. Fitzer and Heym [71] prepared UD composites from HM fibres and a phenolic precursor. Three reimpregnation–carbonization–graphitization cycles resulted in composites having a thermal conductivity of 40 W $m^{-1}\ K^{-1}$ (at room temperature) parallel to the fibres and 3 W $m^{-1}\ K^{-1}$ in the perpendicular direction. Their results were consistent with a 'rule of mixtures' basis, albeit a little on the low side. The surprisingly low transverse thermal conductivity probably results from the presence of porosity which will reduce the thermal conductivity.

McAllister and Lachmann have presented a range of thermal conductivity data for different carbon–carbon materials [81]. A lack of clear information on processing makes it difficult to discern, however, which factors influence the thermal conductivity. They present data for an early 3-D composite, based upon a 2-D graphitized rayon fabric pierced with HM PAN-derived fibres (T-50). The pseudo-3-D fibre structure was processed using repeated phenolic impregnation/pyrolysis/graphitization cycles to a final density of 1.65 g cm^{-3}. Thermal conductivity for this material is presented in Table 8.3 along with comparable data for a related Avco material called MOD-3 [82].

The Avco material was based on an unspecified woven carbon yarn and phenolic resin densified to approximately 1.6 g cm^{-3}. The two materials have almost identical thermal conductivities. Both materials show a relatively isotropic thermal conductivity when compared with the Fitzer/Heym UD

Table 8.3 Thermal conductivity of 3-D phenolic derived carbon–carbon materials

Thermal conductivity ($W\ m^{-1}\ K^{-1}$)	*McAllister–Lachmann*		*Avco*	*MOD-3*
	X–Y-plane	*Z-direction*	*X–Y*	*Z*
At 250 °C	830	55.4	83.0	55.3
At 2500 °C	27.7	24.5	29.4	24.2

material, a consequence of the 3-D fibre array. The authors of the Avco paper, however, point out that the thermal conductivities of these 3-D carbon–carbons are considerably lower than that of a premium grade aerospace graphite, such as the ATS grade.

Two conflicting reports have been published on the thermal conductivity of 2-D phenolic-derived carbon–carbon materials. Laramee *et al.* [83] (from Morton Thiokol) describe the thermophysical properties of 2-D carbon–carbon, from T-50 PAN fabric and phenolic processed to a density of 1.5 g cm^{-3}, proposed as a candidate material for rocket nozzle construction. In-plane conductivities in the range of 93 W m^{-1} K^{-1} (800 °C) were reported.

The high thermal conductivity value perpendicular to the ply direction – in comparison with Fitzer's previously discussed value of 3 W m^{-1} K^{-1} for transverse conductivity – seems a bit surprising. Curry, *et al.* have presented thermal conductivity data for the structural carbon–carbon used for the leading edge parts of the space shuttle [84]. The material is based on plain weave graphitized fabric/phenolic with three subsequent furan reimpregnation–pyrolysis steps, and then given a final SiC–$Si0_2$ oxidation protection coating. The thermal conductivity values quoted for this material are extremely low. At room temperature, the values are 5 W m^{-1} K^{-1} parallel to the plies and 3 W m^{-1} K^{-1} perpendicular. Oddly, the values of conductivity rise as the temperature increases, up to 14.5 and 6 W m^{-1} K^{-1} (parallel and perpendicular) at 1650 °C. This is very different behaviour from the other thermoset-derived carbon–carbons and leads one to question how much influence the SiC/$Si0_2$ coating plays on the thermal conductivity values.

Pitch-derived carbon matrices

Pitch-derived carbon matrix–carbon fibre composites are generally processed via the HIPIC technique (hot isostatic pressure impregnation and carbonization) of pitch into a dry carbon fibre preform. Repeated HIPIC cycles are required to build the composite matrix up to an acceptably high density/low porosity for deployment in severe ablative environments. Two sets of thermal conductivity data have been published on 3-D HIPIC-processed composites.

Table 8.4 Thermal conductivity of HIPIC-processed 3-D carbon–carbon materials

Thermal conductivity ($W\ m^{-1}\ K^{-1}$)	*Battelle 3-D carbon–carbon*		*Morton Thiokol 3-D carbon–carbon*	
	X–Y	*Z*	*X*	*Y*
At 30 °C	149	246	202	171
At 800 °C	—	—	91	80
At 1600 °C	44	60	—	—

Workers at the Battelle Institute [81] (the originators of the HIP process) fabricated a 3-D weave from T-50 fibre, with a 2–2–6 fibre distribution (*X–Y–Z*), and then used four HIPIC cycles with a coal-tar pitch impregnant to produce a composite of density 1.9 g cm^{-3}. The other documented 3-D composite is a body optimized for ring strength – a 9–2–9 *X–Y–Z* preform based on T-50 PAN fibre, HIPIC processed to a density of 1.9 g cm^{-3}; this material was described by workers at Morton Thiokol [83]. The thermal conductivity data are presented in Table 8.4.

The values are higher than those shown in Table 8.3 for phenolic-derived materials of similar fibre architecture. They reflect the more graphitic, thermally conducting nature of the HIPIC-derived matrix. It is worth noting, however, that the thermal conductivity of these materials is, at best, only a marginal improvement on those displayed by conventional synthetic graphites, despite their far superior mechanical properties.

Effect of heat treatment temperatures on thermal conductivity

Figure 8.29 shows the relation between the thermal conductivity and the HTT of a carbon–carbon composite where a thermosetting resin was used a the resin precursor, and Fig. 8.30 the relationship between HTT and L_a. High temperature treated composites based on pitch or on a thermosetting resin will have a high thermal conductivity [86].

Thermal conductivity at high temperatures

There are a number of reports in the literature concerning the high-temperature thermal conductivity of carbon–carbon [87,88]. Figure 8.31 shows the relationship between the measurement temperature and the thermal conductivity of a CVD-type composite material [87]. As the temperature is increased, so is the amount of phonon scatter due to vibration of the crystal lattice, and the thermal conductivity falls [89].

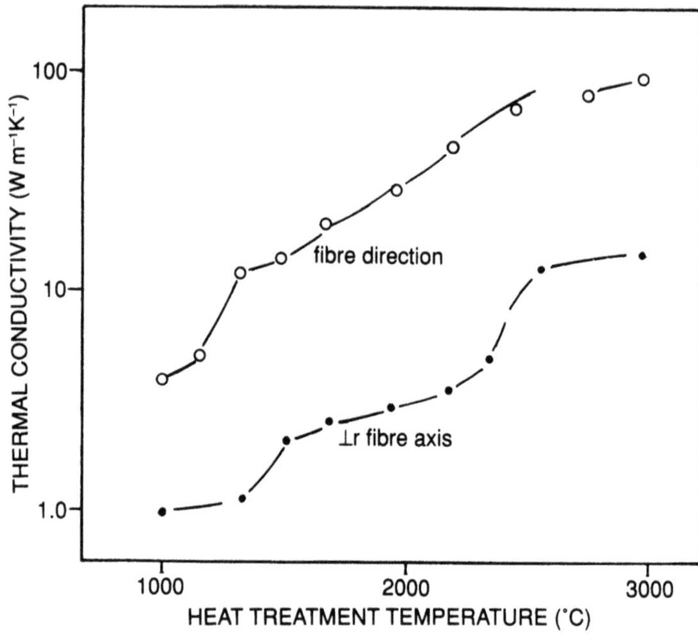

Fig. 8.29 Relationship between thermal conductivity and HTT for a thermoset resin derived composite [85].

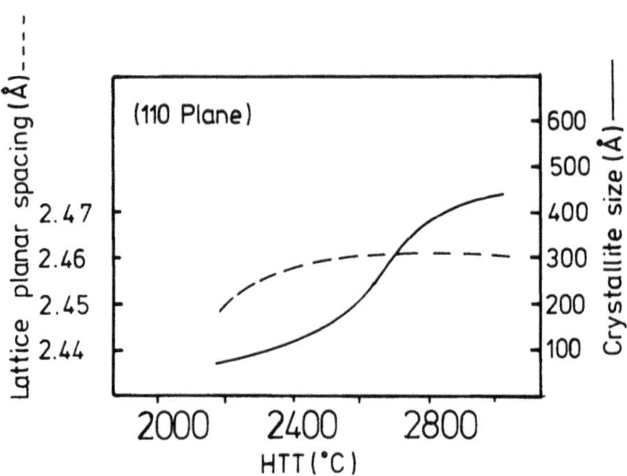

Fig. 8.30 Relation between X-ray pattern and the HTT for a furan resin matrix carbon–carbon composite [85].

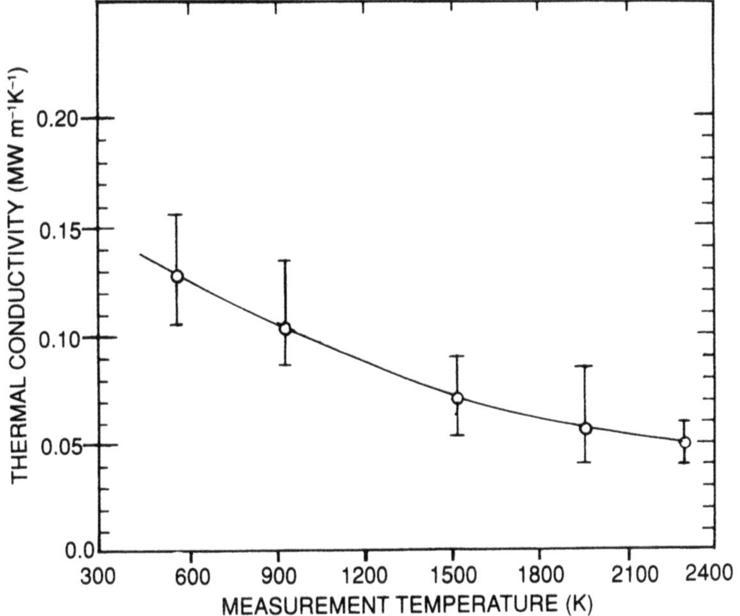

Fig. 8.31 High-temperature thermal conductivity of a CVD carbon–carbon material [87].

8.5.2 Thermal expansion

Thermal expansion is an important property when using materials at high temperatures. The thermal expansion in the fibre axis direction is chiefly governed by the thermal expansion of the fibre. That is to say, thermal expansion of a composite material in which a highly oriented PAN fibre is used resembles the expansion behaviour in the planar direction of graphite Fig. 8.32 [87,88].

8.6 ELECTROMAGNETIC PROPERTIES

When a thermoset-based carbon–carbon is heat treated to above 2800 °C, three-dimensional crystallite growth takes place such that L_{112} becomes very large. The electrical conductivity, as shown in Fig. 8.33, increases correspondingly [85]. The development of a three-dimensional structure is seldom, if ever, observed in carbon fibres. The electrical conductivity of a carbon–carbon composite is primarily governed by the density and the matrix microstructure, including porosity.

Magneto-resistance measurements have been used to investigate the graphitization process [90,91]. Tanaka *et al.*, [90] for example, report that the magneto-resistance of thermoset-derived composites changes from negative to positive values above an HTT of ≈2400 °C. The activation

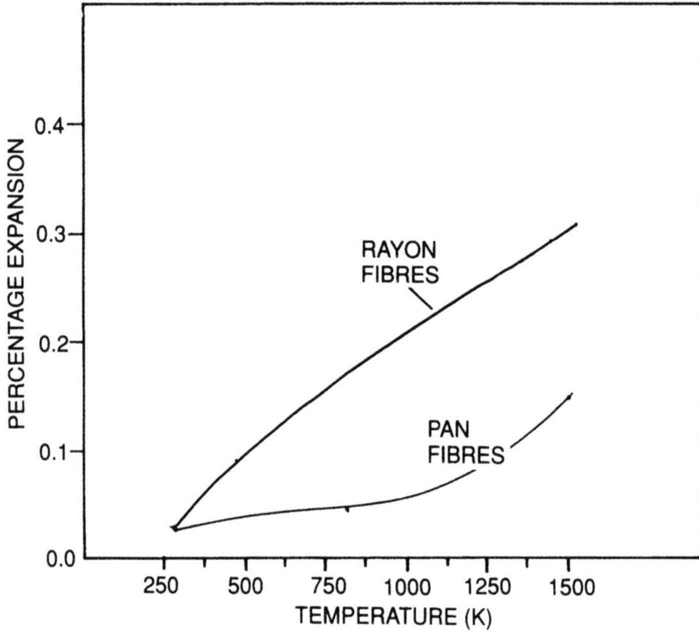

Fig. 8.32 Thermal expansion of a CVD carbon–carbon [87].

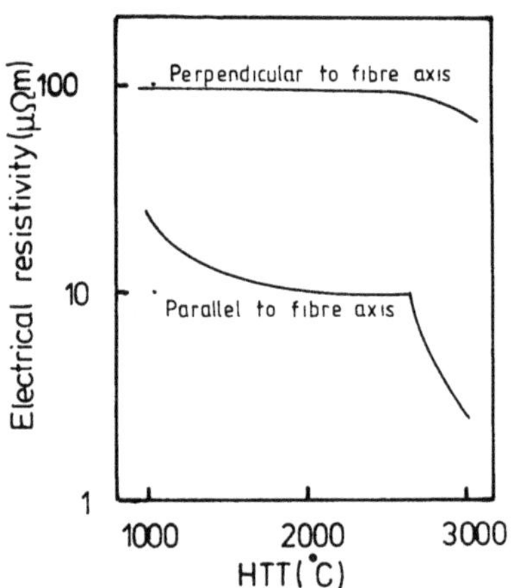

Fig. 8.33 Relationship between electrical resistance and HTT for a thermoset resin carbon–carbon [85].

energy obtained from the dependence on temperature was of the order of 1000 kJ mol^{-1}. They thus concluded that the rate-determining step in the graphitization of the matrix is the healing of defects. The three types of CVD microstructure show different magneto-resistances, especially following high-temperature heat treatment. It is shown that the rough laminar type is close to being a readily graphitized carbon [91].

REFERENCES

1. Liebermann, M. L. and Pierson, H. O. (1973) *Ext. Abst. 11th Biennial Conf. on Carbon*, p. 314.
2. Kotlensky, W. V. (1973) in *Chemistry and Physics of Carbon*, **9** (ed. P. L. Walker Jr), Marcell Dekker, New York, p. 174.
3. Kimura, S., Yasuda, E., Takase, N. and Kasuya, S. (1981) *High Temp. High Press*, **13**, 193.
4. Devillard, J., Berthier, J., and Clary, J. (1976) *Ext. Abst. 2nd Int. Carbon Conf.*, p. 489.
5. Ehrburger, P., Lahaye, J. and Bourgeois, C. (1981) *Carbon*, **19**, 1.
6. Pierson, H. O. and Liebermann, M. L. (1975) *Carbon*, **13**, 159.
7. Savage, G. M. (1985) PhD thesis, Univ. London.
8. Yamada, R. (1968) *The Science of New Industrial Materials* (eds Takei and Kawajima) Tokyo, p. 265.
9. Kimura, S., Yasuda, E., Tanaka, H. and Yamada, R. (1975) *J. Ceram. Soc. Jap.*, **83**, 122.
10. Yasuda, E., Kimura, S. and Shibusa, Y. (1980) *Trans. Jap Soc. Comp. Mat.*, **6**, 14.
11. Kimura, S. Tanabe, Y., Takase, N. and Yasuda, E. (1981) *J. Chem. Soc. Jap. (pure Chem. Sect.)*, **9**, 1474.
12. Saxena, R. R. and Bragg, R. H. (1978) *Carbon*, **16**, 373.
13. Marchand, A., Lespade, P. and Conzi, M. (1981) *Ext. Abst. 15th Biennial Conf. on Carbon*, Am. Chem Soc., Washington, DC, p. 282.
14. Tanaka, H., Kaburagi, Y. and Kimura, S. (1978) *J. Mat. Sci.*, **13**, 2555.
15. Marks, B. S., Mauri, R. E. and Bradshaw, W. G. (1975) *Ext. Abst. 12th Biennial, Conf. on Carbon*, Am. Chem Soc., Washington, DC, p. 337.
16. Fitzer, E., Geigle, K. H. and Huettner, W. (1979) *Ext. Abst. 14th Biennial. Conf. on Carbon*, Am. Chem. Soc., Washington, DC, p. 236.
17. Fitzer, E. and Huettner, W. (1981) *J. App. Phys.*, **14**, 347.
18. Gebhardt, J. J. (1979) *Ext. Abst. 14th Biennial. Conf. on Carbon*, Am. Chem. Soc., Washington, DC, p. 234.
19. Marsh, H. and Forrest, M. (1981) *Ext. Abst. 15th Biennial. Conf. on Carbon*, Am. Chem. Soc., Washington, DC, p. 270.
20. Tanaka, H., Mariyama, S. Yasuda, E. and Kimura, S. (1976) *Tanso (Carbon)*, no. 86, 107.
21. Fitzer, E., Heym, M. and Rhee, B. (1976) *Ext. Abst. 2nd Int. Carbon Conf.*, p. 508.

22. Fitzer, E., Huettner, W. and Manocha, L. (1979) *Ext. Abst. 14th Biennial Conf. on Carbon*, Am. Chem. Soc., Washington, DC, p. 240.
23. Gray, G., Hunter, A. and Savage, G. M. (1990) *Proc. Int. Iso. Press. Conf.*, Stratford-upon-Avon, 5–7 Nov, paper 25.
24. Gray, G. and Savage, G. M. (1991) *High Temp. Processing*, special edition, Butterworths, Oct. Oxford.
25. Gray, G. and Savage, G. M. (1992) submitted to *J. Mat. Sci.*
26. Payne, R., von Bradsky, G., Gray, G. and Savage, G. M. (1990) *Proc. Conf. Adv. in Elec. Mic.*, Seattle, p. 1024.
27. Peebles, L. H., Meyer, R. A. and Jortner, J. (1988) *Proc. 2nd Int. Conf. on Comp. Interfaces*, ICCI II, Cleveland, Ohio, USA, 13–17 June.
28. Jortner, J. (1976) *Proc. Army Symp. Solid Mech.*, US Army Mat. α Mech. Res. Ctr., AMMRC MS 76–2, Sept. p. 81.
29. Stouer, E. R., D'Andrea, J. F., Bolinger, P. N. and Gebhardt, J. J. (1977) *Ext. Abst. 13th Biennial Conf. on Carbon*, p. 166.
30. Diefendorf, R. J. and Tokarsky, F. M. (1971) *AFML-TR-72-133*, Parts I and II, US Air-force, Oct.
31. Gray, G. and Savage, G. M. (1992) submitted to *J. Mat. Sci.*
32. Zimmer, J. E. and White, J. L. (1983) *Carbon*, **21**, 323.
33. Meyer, R. A. and Gyetvay, S. R. (1986) in *Petroleum Derived Carbons* (eds J. D. Bacha, J. N. Newson and J. L. White), ACS Symp. 303, Am. Chem. Soc., Washington, DC.
34. Sheaffer, P. M. (1987) *Ext. Abs. 18th Biennial Conf. on Carbon*, p. 20.
35. McKee, D. W. (1987) *Carbon*, **25**, 551.
36. Jortner, J. (1986) *Carbon*, **24**, 603.
37. Fitzer, E. and Barger, A. (1971) *Proc. Int. Conf. on Carbon Fibres, their Composites and Applications*, London, Paper no. 36.
38. McAllister, L. E. and Taverna, A. R. (1971) *Proc. Am. Ceram. Soc. 73rd Ann. Mtg.*, Chicago.
39. Adams, D. F. (1976) *Mat. Sci. Eng.*, **23**, 55.
40. Hill, J., Thomas, C. R. and Walker, E. J. (1973) *Ext. Abst. 11th Biennial Conf. on Carbon*, Am. Chem. Soc., Washington, DC, p. 278.
41. Meyer, R. A. (1986) *Proc. Carbon 1986 Conf.*, Baden-Baden, Germany.
42. Evangelides, J. S. (1976) *Proc. Army. Symp. on Solid Mech.*, Sept., US Army, p. 98.
43. Jortner, J. *J R & E Rpt no. 8514*, p. 39.
44. Manocha, L. M. and Bahl, O. P. (1988) *Carbon*, **26** (1) 13.
45. Schmidt, D. L. (1972) *SAMPE J.*, May/June, 9.
46. Cristina, V. D. (1971) *Proc. AIAA Thermophys. Conf.*, Tullahoma, Tennessee, USA, April, Paper no. 71–416.
47. McDonald, J. E. (1971) *Mech. Eng.*, Feb., 21.
48. Pierson, O. H. and Northrop, D. J., (1975) *J. Comp. Mat.*, **9**, 118.
49. Delhaes, P., Trinquescoste, M. Pacault, A., Goma, J., Oberlin, A. and Thebault, J. (1984) *J. de Chem. Phys*, **81**, 809.
50. Oh, S. M. and Lee, J. Y. (1988) *Carbon*, **26**(6), 769.
51. Oh, S. M. and Lee, J. Y. (1989) *Carbon*, **27**(3), 423.
52. Marinkovic, S. and Dimitrijevic, S. (1985) *Carbon,* **23**, 691.

53. Parmee, A. C. (1972) *Carbon*, **10**, 333.
54. Adams, D. F. (1974) *J. Comp. Mats*, **8**, 320.
55. Zhao, J. X., Bradt, R. C. and Walker, P. L. Jr (1981) *Ext. Abst. 15th Biennial Conf. on Carbon*, Am. Chem. Soc., Washington, DC, p. 274.
56. Fitzer, E. and Terwiesch, B. (1973) *Carbon*, **11**, 570.
57. Chard, W. and Keck, H. (1977) *Proc. 13th Biennial Conf. on Carbon*, Soc., Washington, DC, p. 212.
58. Dietrich, H. and McAllister, E. (1978) *Proc. 8th Ann. Mtg Exp. Am. Ceram. Soc.*, Detroit.
59. Terwiesch, B. (1972) PhD thesis, Univ. Karlsruhe.
60. Evangelides, J. S. (1977) *Proc. 13th Biennial Conf. on Carbon*, Am. Chem. Soc., Washington, DC, p. 76.
61. Fitzer, E., Heym, M. and Karlisch, K. (1974) *Ext. Abst. 4th Lon. Int. Carbon and Graphite Conf.*, p. 39.
62. Fitzer, E. Huettner, W. and Manocha, L. M. (1980) *Carbon*, **18**, 291.
63. Savage, G. M. (1992) submitted to *J. Mat. Sci.*
64. Norton-Berry, P., Savage, G. M. and Steel, M. L. (1989) European Patent app. W35016/EP.
65. Ryshkewitch, E. (1953) *J. Am. Ceram. Soc.*, **36**, 65.
66. Kimura, S., Kasuya, S. and Yasuda, E. (1978) *Proc. Int. Symp. of Factors in Dens. and Sintering of Ceramics*, Japan, p. 229.
67. Hill, J. (1975). *Ext. Abst. 12th Biennial Conf. on Carbon*, Pittsburg, p. 287.
68. Yasuda, E. Tanaka, H. and Kimura, S. (1980) *Tanso (Carbon)*, no. 100, 3.
69. Kimura, S. Yasuda, E. and Tanabe, Y. (1982) *Proc. Int. Conf. Comp. Mats*, **IV**, p. 1601.
70. Fitzer, E. and Heym, M. (1977) *Ext Abst. 13th Biennial Conf. on Carbon*, p. 128.
71. Fitzer, E. and Heym, M. (1980) *Kunstofftechnik*, 85.
72. *Carbon Fibre Reinforced Carbon*, Schunk promotional brochure.
73. Kim, H. C., Yoon, K. J., Pickering, R. and Sherwood, R. J. (1985) *J. Mat. Sci.*, **20**, 3967.
74. Kelly, B. T., (1969) *Chem. Phys. Carbon*, **5**, 119.
75. Carbon and artificial graphite, in *Kirk-Othmer Dictionary of Chemical Technology*, 3rd edn, **4**, Wiley, New York.
76. Taylor, R., Gilchrist, K. E. and Poston, L. J. (1968) *Carbon*, **6**, 573.
77. Bokros, J. C., (1969). *Chem. Phys. Carbon*, **5**, 1.
78. Liebermann, M. L. and Pierson, H. O. (1973) *Proc. 11th Biennial Carbon Conf.*, p. 314.
79. Curlee, R. M. and Liebermann, M. L. (1973) *Proc. 11th Biennial Carbon Conf.*, Am. Chem Soc., Washington, DC, p. 280.
80. Curlee, R.M. and Northrop, D. A. (1973) *Proc. 11th Biennial Carbon Conf.*, Am. Chem. Soc., Washington, DC, p. 318.
81. McAllister, L. E. and Lachmann, W. (1983) in *Fabrication of Composites* (*Handbook of Composites*, **4**) (eds A. Kelly, and S. T. Millieko), North-Holland, p. 109.
82. Rolincik, P. G. (1973) *Proc. 11th Biennial Conf. on Carbon*, Am. Chem. Soc., Washington, DC, p. 343.

83. Laramee, P., Lamere, G., Prescott, B., Mitchell, R., Sottosanti, P. and Dahle, D. (1975) *Proc. 12th Biennial Conf. on Carbon*, Am. Chem. Soc., Washington, DC p. 74.
84. Curry, D. M., Scott, H. C. and Webster, C. N. (1979) *Proc. 24th SAMPE Symp.*, p. 1524.
85. Kimura, S., Yasuda, E. and Tanabe, Y. (1982) *Proc. Int. Symp. on Carbon in Japan*, p. 410.
86. Curlee, R. H. and Northrop, D. A. (1973) *Ext. Abst. 11th Biennial Conf. on Carbon*, Am. Chem. Soc., Washington, DC, p. 318.
87. Pierson, H. O., Northrop, D. A. and Smatana, J. F. (1973) *Ext. Abst. 11th Biennial Conf. on Carbon*, Am. Chem. Soc., Washington DC, p. 275.
88. Granoff, B. and Apodaca, P. (1973) *Ext Abst. 11th Biennial Conf. on Carbon*, Am. Chem. Soc., Washington, DC, p. 273.
89. Lamieq, P. Mace, J. and Perez, B. (1980) *Ext. Abst. 15th Biennial Conf. on Carbon*, Am. Chem. Soc., Washington, DC, p. 528.
90. Tanaka, H., Kaburagi, Y. and Kimura, S. (1978) *J. Mat. Sci.*, **13**, 2555.
91. Kimura, S., Hagis, N., Yasuda, E. Tanabe, Y., Hishiyama, S. and Kaburage, Y. (1983) *J. Ceram. Soc. Jap.*, **91**, 121.

Applications of Carbon–carbon Composites

9

9.1 BRAKES AND CLUTCHES

9.1.1 Introduction

Something like 63% by volume of the carbon–carbon produced in the world is used in aircraft braking systems. Carbon–carbon brake materials were originally developed by the Super Temp Division of B. F. Goodrich Inc. in the USA. Their process was licensed by Dunlop in the UK. Dunlop were the first company to manufacture and fit carbon–carbon composite brakes into regular service.

Trials were carried out in 1973 on a VC10 aircraft followed a year later by standard fitment to Concorde SST [1]. At the time when they were first introduced, the cost of the brakes was around £550 kg^{-1}. As a result, their use could only be considered economically viable on supersonic transports and high-performance military aircraft. Advances in technology have now reduced the cost of carbon–carbon suitable for brakes to around £100–£150 kg^{-1}.

It is now commercially advantageous to employ carbon–carbon brakes on civil subsonic aircraft. Furthermore, the use of carbon–carbon has been exploited or postulated for a number of land vehicles such as racing cars, high-speed trains and even main battle tanks (MBTs).

Today, probably the largest producer of carbon brakes is Hitco, based in California in the USA, who are owned by the UK's BP Corporation. Other major 'players' are the French company SEP, along with Bendix (Allied Signal) and Arcraft Braking Systems in the USA. Dunlop themselves still retain a market share, although not on the same scale as their competitors.

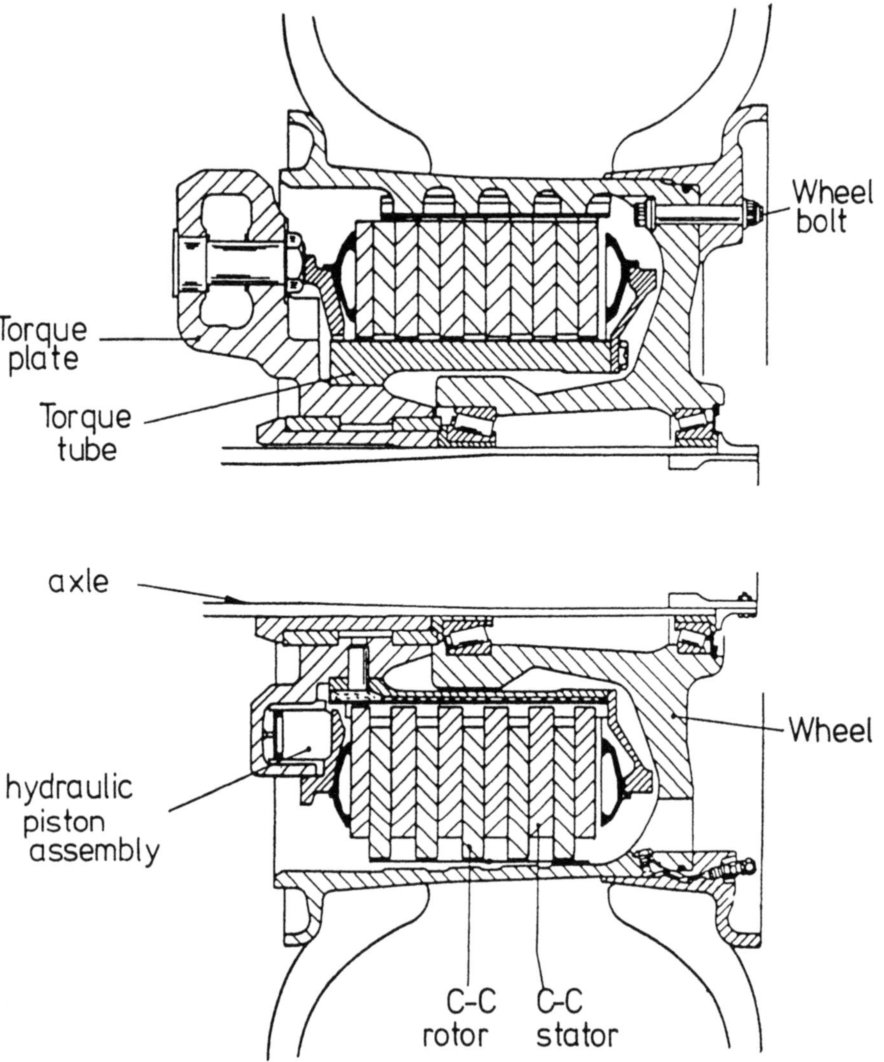

Fig. 9.1 Schematic diagram of a carbon–carbon aircraft disc brake system.

9.1.2 Aircraft brakes

Aircraft brakes are multiple disc brakes of rotors sandwiched between stators (Figs 9.1 and 9.2). The rotors are driven by the wheel and the stators held stationary by the brake structure. The braking action is performed by pressing the discs together using some form of hydraulic actuator system. The brake discs are required to provide the frictional torque which stops the aircraft, serve as a heat sink to absorb the heat generated during the braking action (of the order of several hundred MJ) and also act as a structural component. Friction between the rotating and stationary discs

Fig. 9.2 Sectioned carbon–carbon brake from a commercial airliner (Airbus A320).

causes them to heat up to around 500 °C with surface temperatures reaching as high as 2000 °C as the kinetic energy of the aircraft is absorbed [2]. Consequently, the materials used in this environment must exhibit a good thermal shock resistance. The high thermal conductivity and very low coefficient of thermal expansion of carbon–carbon make it an ideal choice.

A number of performance parameters are important in aircraft brake design. These include peak torque, oxidation and stability. Such variables are controlled by engineering design, composition and processing conditions of the friction material. The two primary advantages of carbon–carbon are heat capacity (2.5 times greater than that of steel) and high strength at elevated temperatures (twice that of steel). The result is a 40% saving in weight compared to metal brakes and a doubling of their service life when measured as landings per overhaul (LPO). A wide-bodied airliner such as the Airbus A320 would expect to complete something like 2500 LPO when fitted with carbon brakes as opposed to around 1500 using a metal equivalent.

9.1.3 Materials and processing

Three distinct types of carbon–carbon composite are currently used in braking systems (Fig. 9.3). These are carbon fabric laminates, semi-random

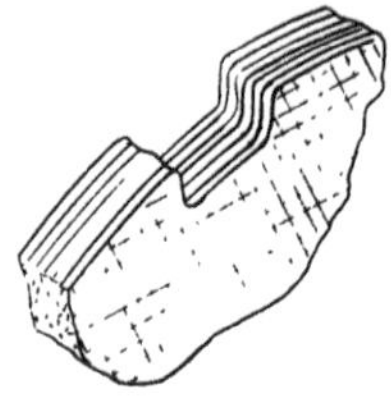

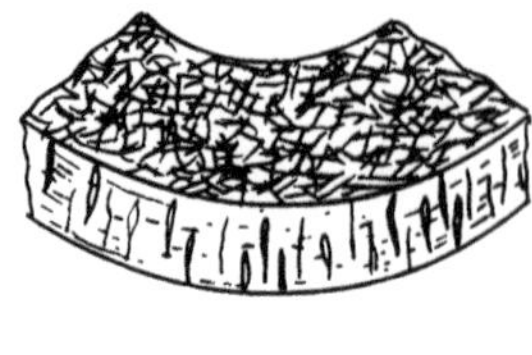

Chopped fibre semi-random orientation.

Laminated matte.

Fig. 9.3 The three types of carbon fibre reinforcement used in carbon–carbon composite brakes.

chopped carbon fibres (Hitco) and laminated carbon fibre felts with cross-ply reinforcement (SEP). The matrix consists exclusively of pyrolytic carbon or a combination of pyrolytic and glassy carbons. The pyrolytic carbon matrix is obtained via a chemical vapour deposition (CVD) process. The glassy carbon is produced from the carbonization of a high char yield resin, generally a phenolic. The resin is used in the consolidation of carbon fibres in the shape of a disc or in the densification of the porous disc by resin impregnation in the final stages of processing. A flow diagram showing a typical carbon–carbon brake manufacturing process is shown in Fig. 9.4.

Carbon–carbon composite brake discs must be protected from oxidation which occurs from repeated excursions to high temperatures during braking. Oxidation protection is achieved by the application of a 'trowelable' glass-forming penetrant to block the active sites on non-rubbing surfaces or by adding an inhibitor during fabrication.

9.1.4 Frictional properties and braking mechanisms

Brakes and clutches are devices for transferring rotating energy and may thus be considered together [3].

A schematic dynamic representation of a brake or clutch is shown in Fig. 9.5. Two inertias I_1 and I_2, travelling with angular velocities of ω_1 and ω_2 respectively, are brought to the same speed by engaging the clutch. In the case of a brake, one of the angular velocities will generally be zero. The two elements are rotating at different speeds thus resulting in slippage.

Energy dissipation during actuation causes a rise in temperature. When undertaking analysis of the performance of a device of this type, the factors of concern are:

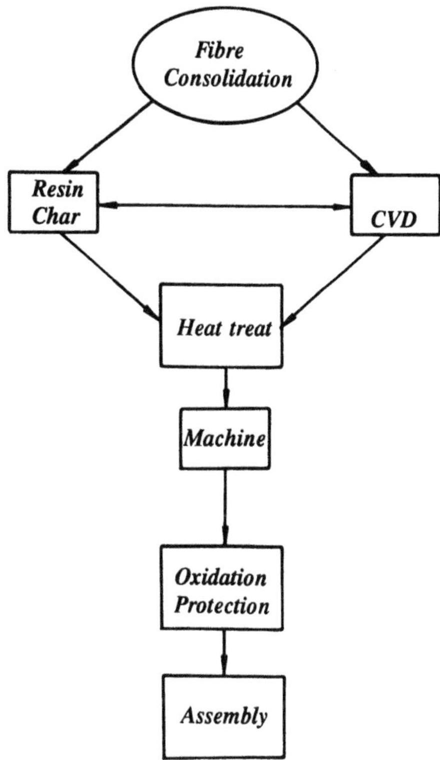

Fig. 9.4 Typical process flow chart for the manufacture of carbon–carbon composite brake discs.

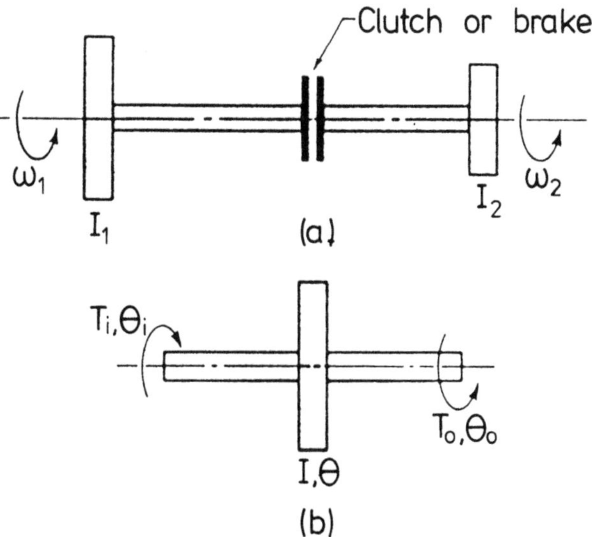

Fig. 9.5 Dynamic representation of a clutch or brake.

Fig. 9.6 Carbon disc clutch system from a Formula 1 racing car (courtesy AP Racing).

1. the actuating force;
2. the torque transmitted;
3. the energy loss;
4. the rise in temperature.

The torque transmitted must be studied individually for each geometric coefficient of friction (μ) and the geometry and *modus operandi* of the device. Temperature rise, on the other hand, is related to the energy loss which may be studied independently of the type of apparatus because the only geometry of concern is that of the heat-dissipating surfaces.

The disc–disc brake systems used on aircraft and the clutches of Formula 1 racing cars (Fig. 9.6) are examples of what is known as an 'axial clutch', that is to say that the mating frictional members are moved in a direction parallel to the shaft. Advantages of the disc system include the freedom from centrifugal effects, a large frictional area installed into a small space, a more effective heat dissipation from rubbing surfaces and a uniform distribution of applied pressure.

The capacity of a disc brake system can be analysed in terms of the

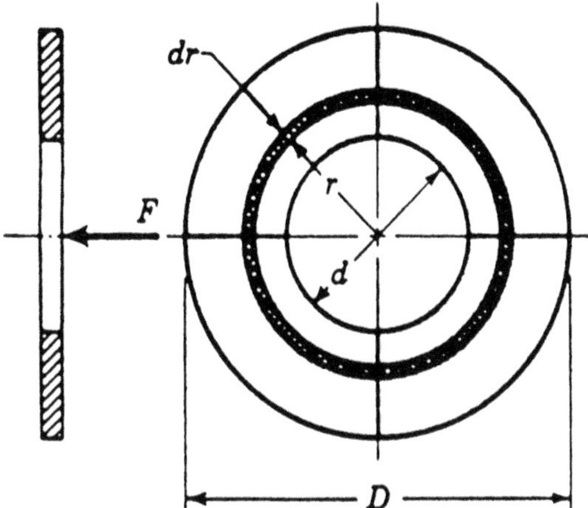

Fig. 9.7 Disc friction member.

material used and its dimensions. A friction disc of outside diameter D and internal diameter d is shown in Fig. 9.7. Two possible methods exist to calculate the axial force F necessary to produce a torque T and pressure p. The greatest wear on a rigid disc will initially arise in the outer areas, in which the work of friction is greatest. Once a degree of wear has taken place, the pressure distribution will change, permitting the wear to be uniform. The other solution is to employ devices such as springs, for example, to maintain a uniform pressure over the braking area.

Uniform wear

Uniform wear occurs once the discs have worn down to a point such that the greatest pressure occurs at $r = d/2$. By letting the maximum pressure be denoted p_a, we may derive the condition in which the same amount of work is done at radius r as is done at radius $d/2$:

$$p_r = p_a \frac{d}{2}, \quad \text{i.e. } p = p_a \frac{d}{2r}. \tag{9.1}$$

Now considering Fig. 9.6, we have an element of area of radius r and thickness dr. The area of this element, then, is $2\pi r\,\mathrm{d}r$. The normal force acting upon the element will be

$$\mathrm{d}F = 2\pi p r \mathrm{d}r. \tag{9.2}$$

The total normal (actuating) force acting on the disc may be calculated by integrating equation (9.2) between the limits $d/2$ and $D/2$:

$$F = \int_{d/2}^{D/2} 2\pi pr \, \mathrm{d}r = \frac{\pi p_a d}{2} \int_{d/2}^{D/2} \mathrm{d}r = \frac{\pi p_a d}{2}(D - d). \tag{9.3}$$

The torque is obtained by integrating the product of the frictional force and the radius:

$$T = \int_{d/2}^{D/2} 2\pi\mu pr^2 \mathrm{d}r = \pi\mu p_a d \int_{d/2}^{D/2} r \mathrm{d}r = \frac{\pi\mu p_a d}{8}(D^2 - d^2). \tag{9.4}$$

A more convenient expression of the torque may be obtained by substituting the value of F calculated from equation (9.3):

$$T = \frac{F\mu}{4}(D + d). \tag{9.5}$$

When designing a disc brake or clutch system, equation (9.3) is used to calculate the required actuating force per friction surface pair for the selected maximum pressure p_a. Equation (9.5) is then used to obtain the torque capacity per friction surface.

Uniform pressure

If uniform pressure can be assumed over the area of the disc, the actuating force is simply the product of the pressure and the area, i.e.

$$F = \frac{\pi p_a}{4}(D^2 - d^2). \tag{9.6}$$

The torque is found, as before, by integrating the product of the frictional force and the radius:

$$T = 2\pi\mu p \int_{d/2}^{D/2} r^2 \mathrm{d}r = \frac{2\pi\mu p}{24}(D^3 - d^3) \tag{9.7}$$

since $p = p_a$

$$T = \frac{F\mu}{3} \frac{D^3 - d^3}{D^2 - d^2}. \tag{9.8}$$

In both solutions the torque has been calculated for a single pair of mating surfaces. In a complete braking system, this value must therefore be multiplied by the number of surfaces in contact.

Energy dissipation

When the rotating components of a vehicle or machine are slowed or brought to a halt by means of a brake, their kinetic energy must be absorbed by the brake. The energy is manifested in the form of heat within the brake. Similarly, when the members of a machine which are initially at rest are brought up to speed, slippage must occur in the clutch until the driven members attain the same speed as the driver. It is this process of slippage by which kinetic energy is absorbed and appears in the appliance as heat.

The analysis in the previous two subsections showed how the torque capacity of a brake depends upon the coefficient of friction of the material and the safe, normally applied pressure. The nature of the braking load may be such however that, if the calculated torque value is permitted to develop, the device may be destroyed by its own generated heat. The performance of a brake or clutch is therefore limited by the friction characteristics of a material and its ability to dissipate heat. Should heat be generated quicker than it is dissipated, a temperature rise problem will result. It is apposite therefore to consider the amount of heat generated during the braking or clutch operation.

In order to model the events during a simple braking or clutch action, it is necessary to refer back to Fig. 9.5. Two inertias I_1 and I_2 possess angular velocities ω_1 and ω_2 respectively. Application of the clutch causes both angular velocities to change and eventually become equal. Assuming the two shafts to be rigid and a constant torque, the equations of motion relating the two inertias may be written:

$$I_1 \frac{d^2\theta_1}{dt^2} = -T \tag{9.9}$$

$$I_2 \frac{d^2\theta_2}{dt^2} = T, \tag{9.10}$$

where $d^2\theta_1/dt^2$ and $d^2\theta_2/dt^2$ are the angular accelerations of I_1 and I_2 respectively. T is the clutch torque.

The angular velocities at any time t may be found by integrating (9.9) and (9.10):

$$\frac{d\theta_1}{dt} = -\frac{T}{I_1}t + \omega_1 \tag{9.11}$$

$$\frac{d\theta_2}{dt} = \frac{T}{I_2}t + \omega_2. \tag{9.12}$$

The difference in velocities, $d\bar{\theta}/dt$, is known as the relative velocity:

$$\frac{d\theta_1}{dt} - \frac{d\theta_2}{dt} = \frac{d\bar{\theta}}{dt} = \frac{T}{I_1}t + \omega_1 - \left(\frac{T}{I_2}t + \omega_2\right)$$

$$= \omega_1 - \omega_2 - T\left(\frac{I_1 + I_2}{I_1 I_2}\right)t. \tag{9.13}$$

The clutching process is complete when $d\theta_1/dt = d\theta_2/dt$, i.e. when $d\bar{\theta}/dt = 0$. The time required to complete the operation, t, is therefore defined as

$$t = \frac{I_1 I_2 (\omega_1 - \omega_2)}{T(I_1 + I_2)}. \tag{9.14}$$

We may thus conclude that the time required to execute the braking or clutching operation is directly proportional to the velocity difference and indirectly proportional to the torque generated.

Throughout the analysis, the clutch torque is assumed to be constant. Using equation (9.13) therefore, the rate of energy dissipation (dE/dt) may be found:

$$\left(\frac{dE}{dt}\right) = T\frac{d\bar{\theta}}{dt} = T\left[\omega_1 - \omega_2 - T\left(\frac{I_1 + I_2}{I_1 I_2}\right)t\right]. \tag{9.15}$$

It is clear then that the greatest dissipation of energy occurs at the start of the action, i.e. when $t = 0$. The total energy dissipated may be found by integrating equation (9.15) between $t = 0$, and $t = t_1$.

$$E = \int_0^{t_1} \frac{dE}{dt} dt = T\int_0^{t_1}\left[\omega_1 - \omega_2 - T\left(\frac{I_1 + I_2}{I_1 I_2}\right)t\right]dt$$

$$= \frac{I_1 I_2 (\omega_1 - \omega_2)^2}{2(I_1 + I_2)}. \tag{9.16}$$

The energy dissipated is thus proportional to the velocity difference but independent of the torque.

Temperature effects

The temperature rise (ΔT) resulting from the braking action may be approximated using classical thermodynamics

$$\Delta T = \frac{E}{Cm}, \tag{9.17}$$

where C is the specific heat (J $kg^{-1}K^{-1}$), m the mass of device (kg) and E the total energy dissipated during braking operation (J). An equation of this form may be used qualitatively to explain events during the operation of the brake or clutch. There are, however, far too many variables involved for any useful or accurate quantitative information to be generated. The major advantage of a temperature rise analysis is the identification of the most important design parameters by experimental examination.

9.1.5 Experimental and in-service analysis of brake performance

Carbon–carbon friction materials are tested on a dynamometer using a ring-on-ring configuration. On a research scale it is performed with a small

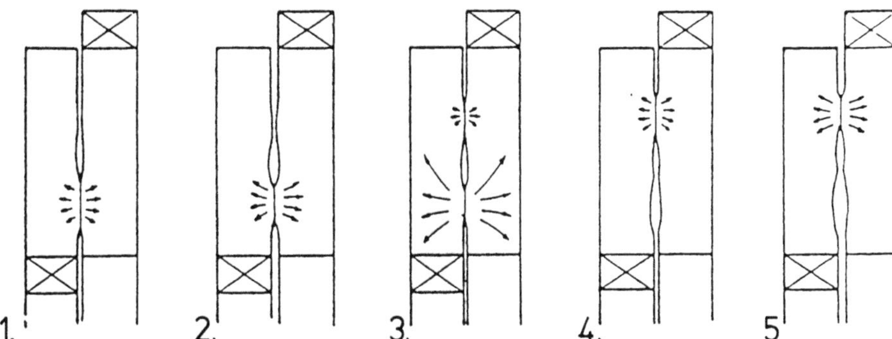

Fig. 9.8 Bands of contact formed during braking as a result of thermo-elastic and wear effects.

annular ring (around 5 cm OD × 3 cm ID) mounted on a rotating shaft and pressing against a similar, but stationary, ring positioned on the same axis. Full-scale testing is accomplished using a complete brake assembly mounted on a wheel with tyre. In both cases, test conditions are chosen to simulate in-service operation. The critical parameters are; relative sliding speed at the friction surface, pressure on the friction surface, kinetic energy loading per unit mass and per unit friction surface area and the rate of energy dissipation. The performance of the carbon–carbon brake is also considerably influenced by the ambient conditions, that is to say, the brake temperature and humidity. The final contributory factor is prior test history which establishes the condition of the friction surface.

The average coefficient of the friction (μ) in static or low-speed sliding configuration is generally low. When the brake is hot, however, μ can be very high, reaching as much as 0.6. Under wet conditions, μ is generally very low (around 0.1) due to the lubricating effect of absorbed water. The thermal loading on a disc is extremely severe and has a controlling influence on the brake design.

Frictional contact of brake components involves thermo-elastic and wear phenomena. The sliding surfaces initially contact one another at local points only. The work done at those points results in local expansion. Rotation of one of the surfaces causes the formation of a narrow annular band of contact. Further work and expansion take place at the radius until heat flow, wear at the band of contact and changes in mechanical loading in adjacent discs cause a new band of contact to become established. The work at the first band diminishes, the material cools locally and, because wear has taken place, that particular band of contact is unlikely to be re-established for some time. This mechanism, illustrated in Fig. 9.8, results in a very high rate of work per band of contact such that transient temperatures at the contact surface may exceed 2000 °C.

The micro-mechanism of braking involves the formation of a friction film from ground and compacted wear debris (Fig. 9.9). Repeated sliding

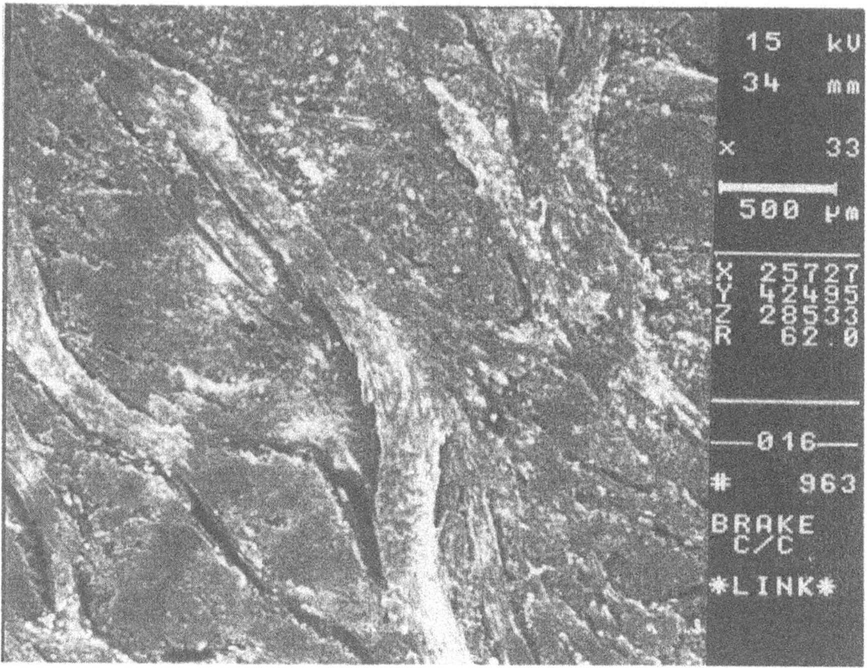

Fig. 9.9 Scanning electron micrograph of a fully developed friction surface in a carbon–carbon brake.

causes a delamination of part of the friction film resulting in the formation of new wear debris by abrasion of the substrate. The majority of this debris is compacted again, regenerating the friction film, while a small amount is lost to the atmosphere, contributing to wear. A schematic diagram of the wear process is shown in Fig. 9.10.

9.1.6 Racing car brakes

Carbon–carbon brakes were introduced into Formula 1 motor racing in the early 1980s by the Brabham team. Today, brakes of this type are universally used in the sport. In contrast to aircraft brakes, those employed in Formula 1 are of the disc and pad type, having evolved from the metal brakes they replaced. The brakes are operated by hydraulic callipers pressing the carbon–carbon pads against the ventilated discs (Figs 9.11 and 9.12) [4]. Disc systems are used, however, in the clutches of Formula 1 vehicles. The operation of the carbon–carbon brakes depends on their environment and on cooling. It is important that the geometry of the upright and cooling assembly complies with specifications concerning the disc assembly (Fig. 9.13). The essential requirement is to provide adequate passages for cooling air within the uprights, airflow being directed into the

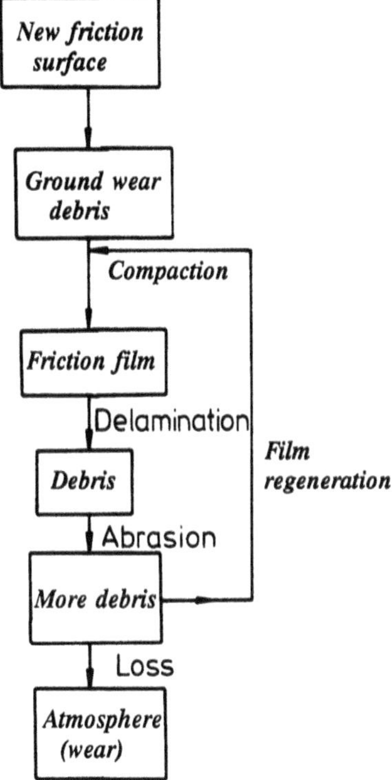

Fig. 9.10 Schematic flow diagram of the wear processes for carbon–carbon aircraft brakes.

vent holes in the disc. A cross-section of at least 70 cm^2 of air flow is generally required for each disc.

Referring to Fig. 9.11, a floating disc to bell assembly is necessary to avoid problems due to thermal expansion mismatch. The bells are made of titanium or a high-temperature aluminium alloy (e.g. alloy 2618) with a surface treatment to improve hardness and resistance to elevated temperatures. Dimensioning of contact surfaces between the sleeves and the bell is critical to prevent fretting. The brakes have an operation window of between 400 and 600 °C. Operating at too low a temperature results in poor performance due to a low coefficient of friction, whereas usage above 600 °C runs the risk of excessive oxidation.

It is important the driver run in and warm up his brakes correctly to develop an optimum friction surface and attain the operating window. During a practice lap he must brake firmly several times, but not during acceleration as this would cause glazing of the discs and pads caused by an absence of carbon dust on the rubbing surfaces. During pit stops the temperature of the brakes must be kept above 350 °C and is monitored constantly using an optical pyrometer to enable the determination of correct operation, brake balancing and temperature control on the callipers to prevent 'vapour lock' [4].

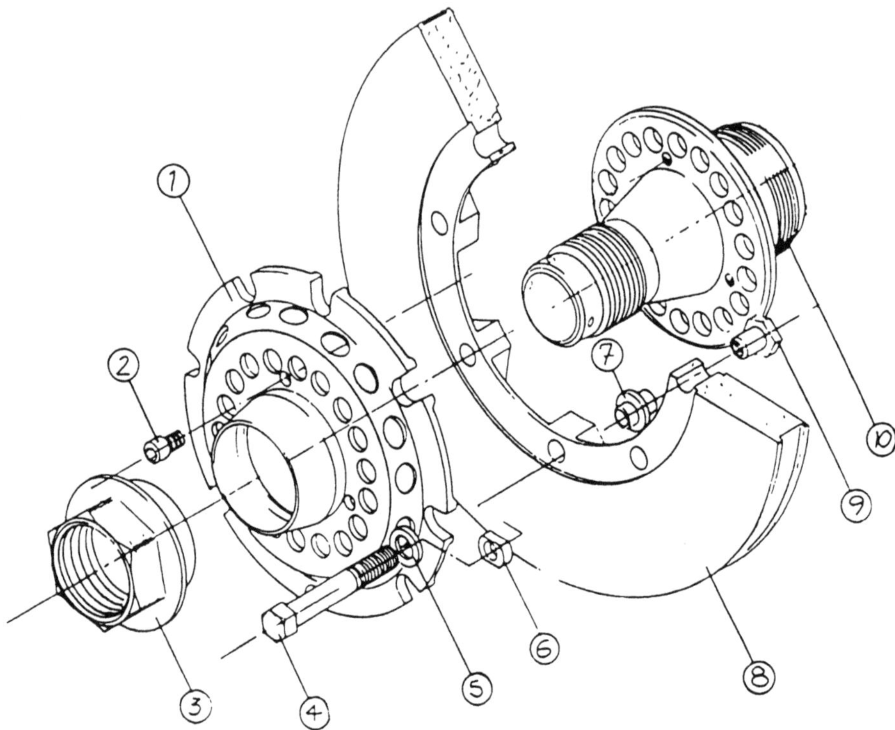

Fig. 9.11 Exploded diagram of Formula 1 carbon–carbon brake system: 1, disc bell; 2, bell-retaining screw; 3, extended wheel nut; 4, disc/disc bell bolt; 5, retaining washer; 6, drive lug; 7, mounting bush; 8, carbon–carbon disc; 9, mounting nut; 10, axle.

Ventilation of the discs plays an important role in cooling to avoid their mass temperature exceeding 600 °C. Location of the ventilation intakes is designed during aerodynamic studies of the vehicle in the wind tunnel. Brake ducts are placed in the zone where air flow is greatest. Determination of potential wear is carried out by regular measurement of disc and pad thicknesses by race engineers. Wear readings are taken using a precision (1/100 mm) micrometer. Measurements are taken at the centre of the friction surface of the disc and at both ends of each pad (Fig. 9.14).

Calculation of wear on discs and pads is carried out as follows. For measurement of maximum disc and pad wear at front and rear (D), let T_i = initial thickness and T_f = final thickness; then

$$D = T_i - T_f. \tag{9.18}$$

If the car travels X laps, in which the lap during which the car leaves the pit, plus the lap during which the car comes back to the pit is considered as one full lap, then wear per lap is

Fig. 9.12 Carbon–carbon brake fitted to a Formula 1 racing car.

$$W = D/X. \tag{9.19}$$

Average thickness of brake element is

$$T = \frac{T_i + T_f}{2}. \tag{9.20}$$

The potential of a new rubbing element, under identical conditions of use, can then be determined on a wear curve provided by the manufacturer (Fig. 9.15).

Two manufacturers provide products to the Formula 1 circuit, SEP and Hitco. The discs and pads used are generally made from the centre sections removed from aircraft brake discs to allow the fitting of the axle and are thus a good use of what might otherwise be scrap material – if anything as expensive as carbon–carbon could ever be thought of as such! Although nominally the same material, their characteristics are very different due to their different structures and processing routes.

Hitco brakes are made from phenolic resin, impregnated carbon fibres chopped into 38 mm lengths (known as 'matchsticks') which are compression moulded into discs. The resin is carbonized followed by resin reimpregnation and CVD densification to around 1.75 g cm^{-3}. SEP brakes, on the other hand, are a CVD-densified carbon fibre felt of roughly the same density. The performance characteristics of each can be explained using electron micrographs of pads tested on a Formula 1 car at a European

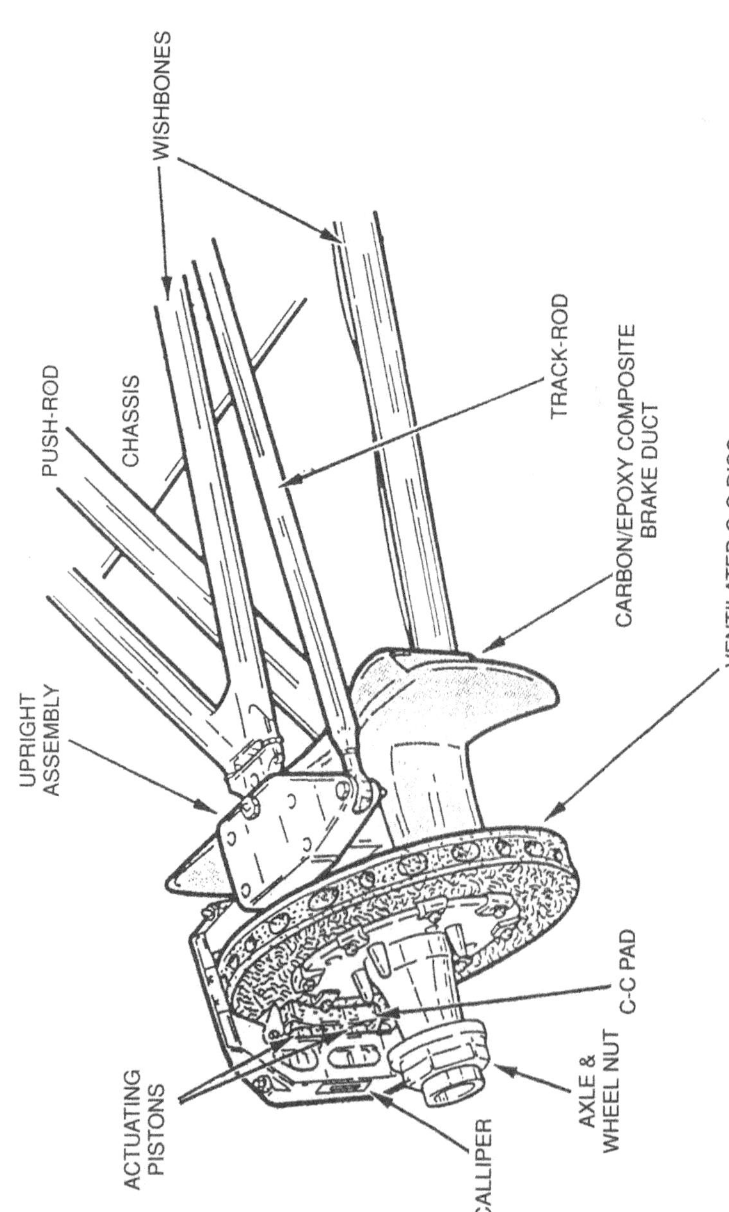

Fig. 9.13 Configuration of brakes and ventilation ducts on a Formula 1 racing car.

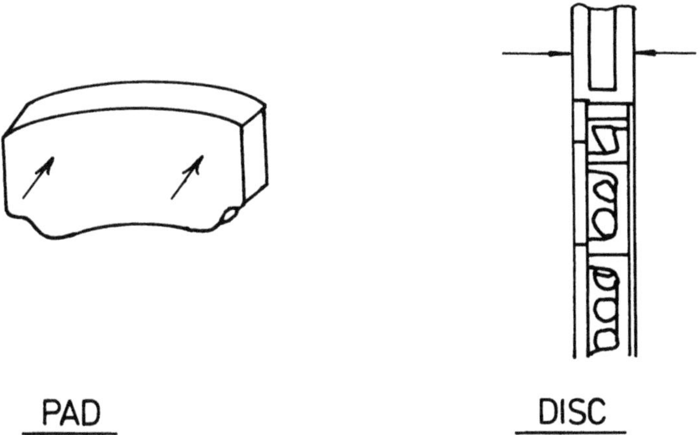

Fig. 9.14 Measurment points for calculation of carbon–carbon brake wear.

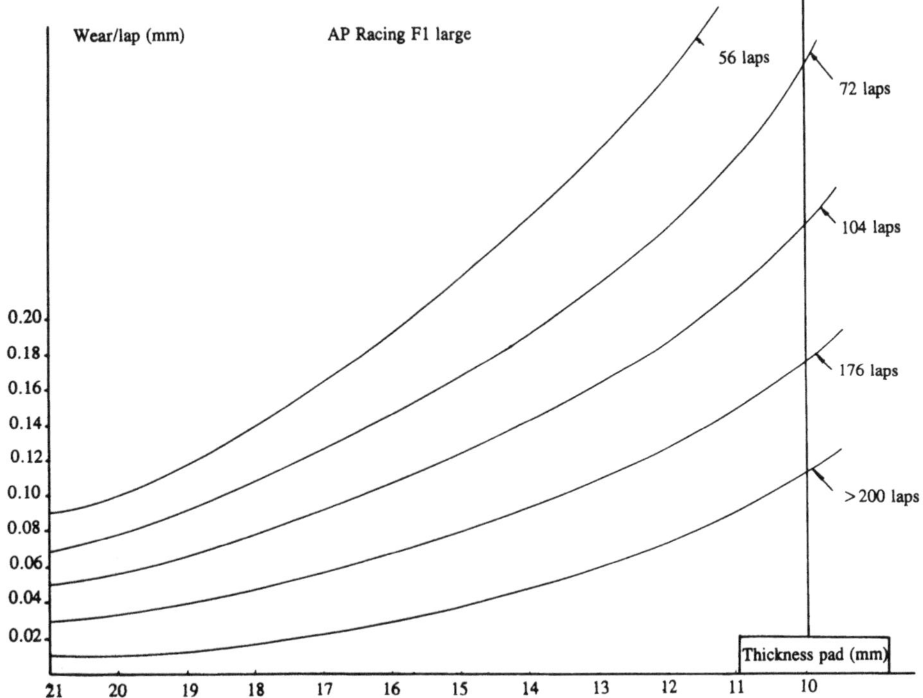

Fig. 9.15 Example of manufacturers' brake wear curves. (Courtesy of AP Racing).

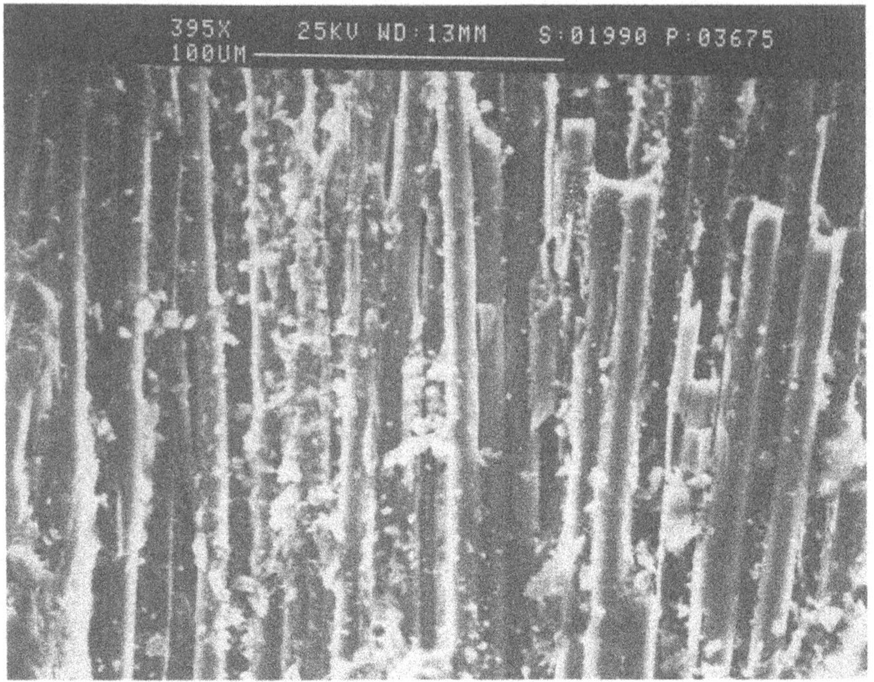

Fig. 9.16 Scanning electron micrograph of powdery isotropic carbon debris on the friction surface of a pad produced by resin pyrolysis.

circuit. As a prelude, it is interesting to note the driver's comments on the behaviour of each type of material and to correlate these with the microscopical observations. Resin based brakes heat up very quickly and perform excellently during the first couple of laps. Braking becomes increasingly difficult, however, with increasing time. Wear of the discs and pads is relatively slow. CVD brakes, on the other hand, are more difficult to heat up, requiring a great deal of care during the initial laps. Once up to temperature, performance remains fairly constant unless they become overheated. Wear is significantly faster than their competitors. Both types of brake were observed to become 'spongy' if overheated.

The friction surface of the thermoset resin based brakes heats up relatively quickly because of the essentially two-dimensional nature of the fibre reinforcement. Thermal conductivity is far greater along the fibres than through the glassy carbon matrix, thus the dissipation of heat energy away from the rubbing surface is relatively inefficient. The matrix is very brittle and only weakly attached to the fibre reinforcement. The friable isotropic carbon forms a characteristic 'powdery' debris which is subsequently compacted into the surface (Fig. 9.16). The resulting glazed

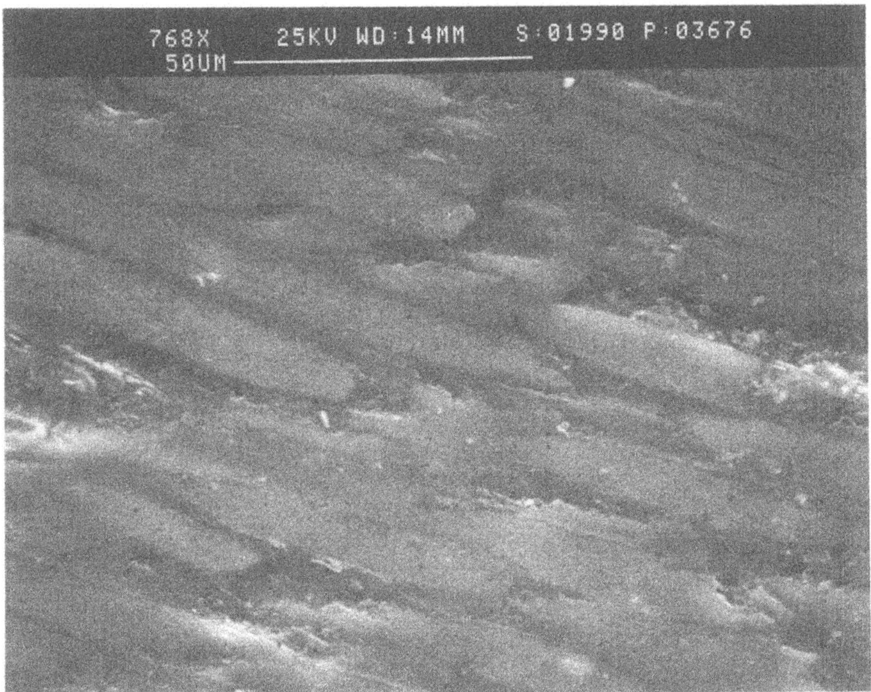

Fig. 9.17 Scanning electron micrograph of glazed friction surface of a resin based carbon–carbon brake pad.

surface thus loses a degree of breaking efficiency, but as a direct consequence, wear is reduced (Fig. 9.17).

Micrographs of an CVD brake's friction surface (Figs 9.18 and 9.19) illustrate the characteristics of a material densified by this technique. The surface is dominated by large pores as one would expect, since CVD is extremely efficient at filling small pores but very poor at filling large ones. The bond between fibres and matrix is much stronger than a resin-densified material so that debris is much larger and more irregularly shaped, producing a much rougher friction surface (Fig. 9.20). The observation that the CVD materials take longer to attain operating temperature can be explained by the microstructure of the carbon felt reinforcement and the insulating effects of the porosity. Although essentially lamellar in construction, the felt possesses a 'pseudo-three-dimensional' appearance such that fibres are available to conduct more heat away from the rubbing surface as it is being heated during operation. The maintenance of a rougher friction surface topography accounts for the more constant braking behaviour but also increases the wear. Overheating of the brakes results in oxidation. Oxidative attack takes place preferentially in the most thermodynamically unstable areas. In the case of carbon–carbon, as we have

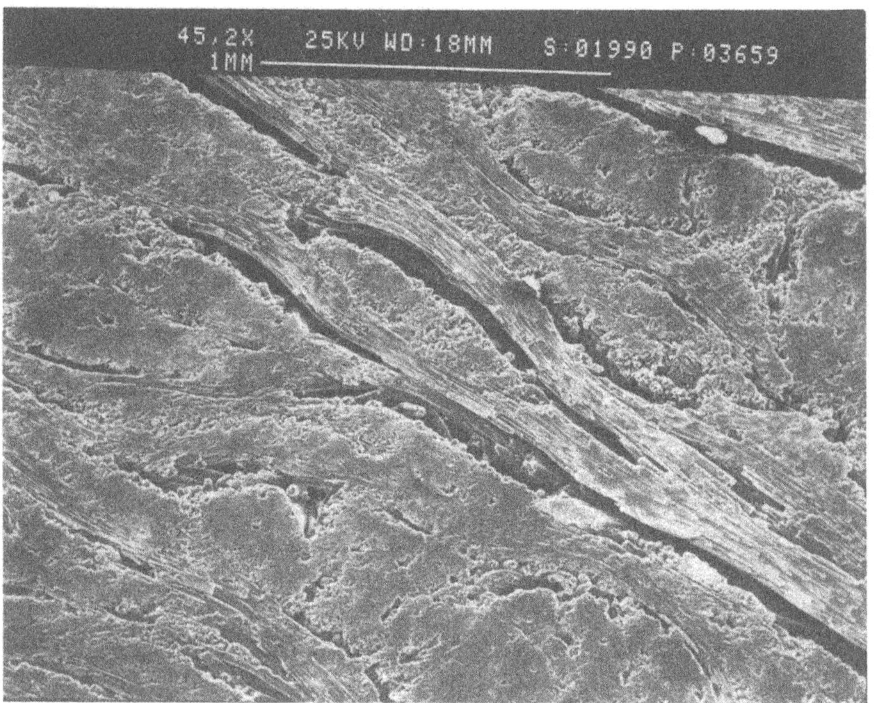

Fig. 9.18 Electron micrograph of CVD braking surface showing large-scale porosity.

seen in Chapter 6, this is in the matrix. Figure 9.21 shows an oxidized area at the corner of one of the pads, close to the rubbing face (i.e. the hottest part). Preferential attack of the matrix can clearly be seen. Degradation of the material in this way means that the support to the fibres is relaxed and stress transfer impeded, hence explaining the driver's observation that the brakes become 'spongy' as the material 'collapses' under the applied pressure. Oxidation protection is attempted using a trowelable 'paint' which does not penetrate the network of pores particularly well. Furthermore, the rubbing surfaces cannot be protected as this would impair their performance. Oxidation kinetics are reduced by ventilation due to the reduction in temperature. It should be remembered, however, that ventilation increases the partial pressure of oxygen available for reaction and the aerodynamics forces the oxygen into the pore structure in the interior of the material, eating it away from within. A number of teams prefer the CVD variety of brake, although they can result in severe problems of wear on some circuits. Other teams, in contrast, choose the thermoset resin derived brakes, sacrificing a little in performance to give improved longevity. It is indeed not uncommon for a mixture of both types of material to be used in the quest for the optimum set up.

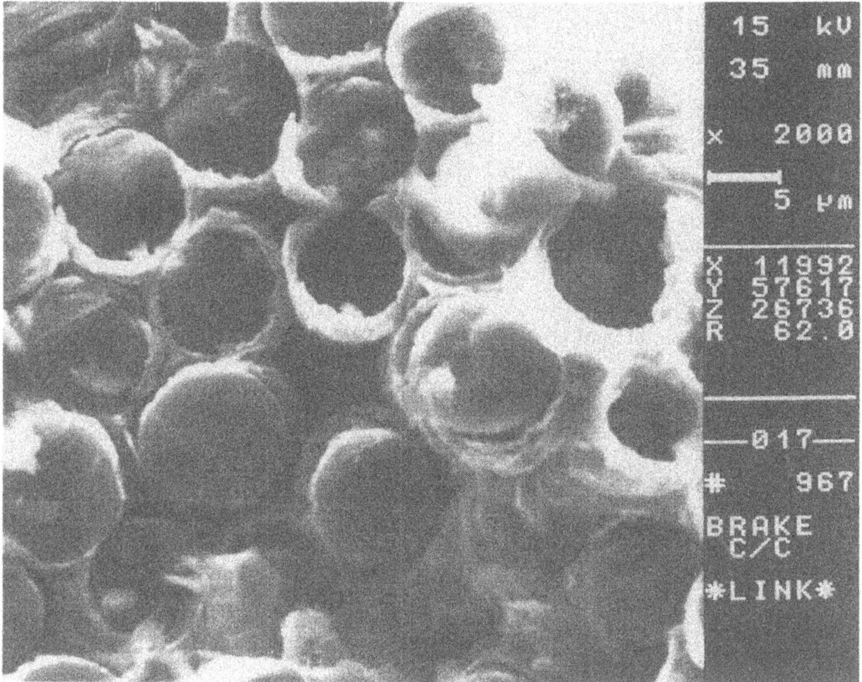

Fig. 9.19 Strong fibre matrix interface in CVD-densified carbon–carbon brakes.

9.1.7 Other braking applications

The major advantages of carbon–carbon braking materials are their light weight, excellent thermomechanical performance and inertness. Disadvantages are those inherent to all carbon–carbon applications, cost and oxidation. Aircraft and racing cars use carbon–carbon friction materials to obtain improved performance at lower weight than a metal alternative. The low density of carbon–carbon coupled with the tendency towards lower costs suggests possible applications in other forms of transport or machinery. The French have already tested such systems on their TGV Atlantique high-speed passenger trains. The thermomechanical stability of carbon–carbon allows it to absorb much more energy over a short period of time than a metal or organic brake. The use of carbon–carbon in emergency situations is a very important consideration. One could imagine the development of emergency brakes in high-momentum situations such as stopping trains and large road vehicles over short distances or preventing injury and damage due to the failure of lifting gear in mine shafts for example. Carbon–carbon brakes and clutches employed in military vehicles (mainly in Japan) improve airportability or allow a more

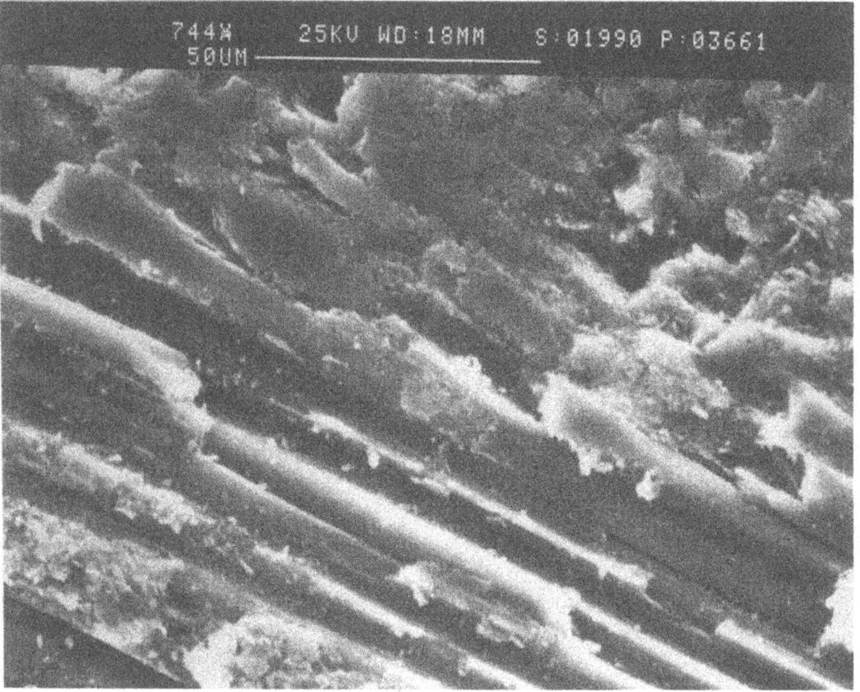

Fig. 9.20 Rougher friction surface developed in CVD brakes.

efficient distribution of the vehicle's weight into the area where it is most required, that of armoured protection. The cost of carbon–carbon would generally be considered prohibitive for use in domestic vehicles. Work has been carried out, with a degree of success, developing motor car brakes operating carbon pads on cast-iron discs [5]. It is difficult to foresee the introduction of carbon–carbon into the brakes and clutches of urban motor transport unless demanded by pollution control legislation, carbon–carbon particles being relatively harmless by comparison with the asbestos and organic materials presently used.

9.1.8 Design and optimization of carbon–carbon friction members

The firbre orientation within a carbon–carbon brake disc is determined by heat flow and structural requirements. Figure 9.22 illustrates the four criteria for preferred fibre orientation:

1. to conduct heat away from the friction surface and into the body of the disc;
2. to conduct heat radially;
3. to withstand shear forces at the drive slot;
4. to minimize the effects of the drive slot.

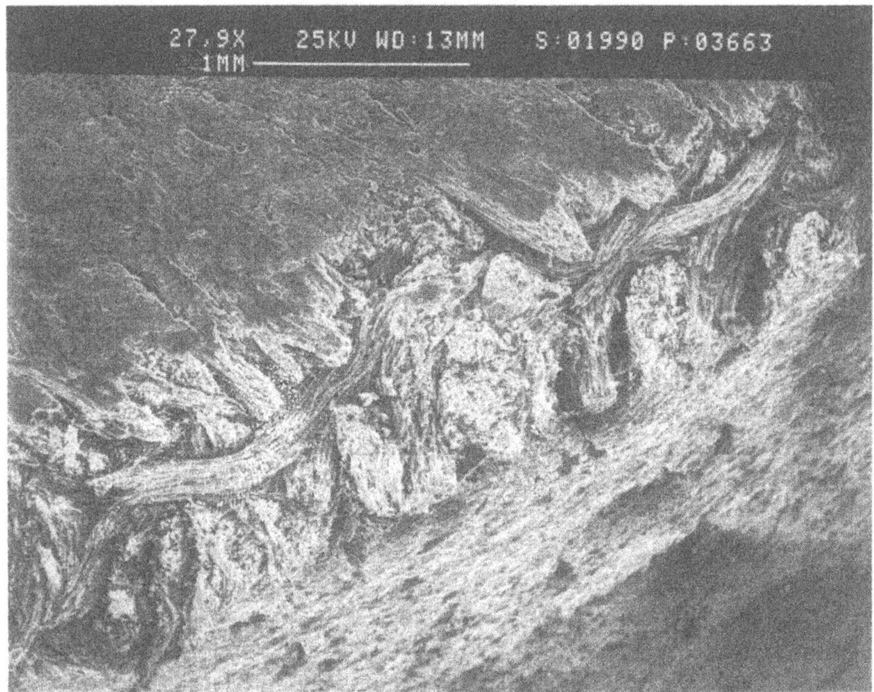

Fig. 9.21 Oxidation of carbon–carbon brakes occurs preferentially in the matrix phase.

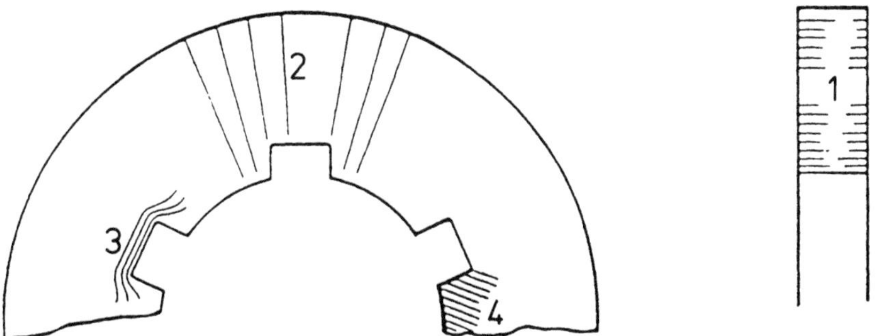

Fig. 9.22 Preferred fibre directions: 1, normal to friction surface (heat flow); 2, radially at surface (heat flow); 3, flow around notches (strength); 4, ± 45° to shear at slot (strength).

It is thus possible to identify a number of development objectives to improve the carbon–carbon composite materials.

1. an increase in thermal conductivity especially perpendicular to the friction surface;
2. an increase in the static coefficient of friction;

3. a greater strain to failure;
4. improved specific strength;
5. a reduction in costs.

One of the major reasons for adopting carbon brakes is a saving in weight. Designers are therefore reluctant to improve strength at the cost of increased weight. A critical design consideration is the material's strength in the 'fully worn' condition. An increase in specific strength is thus sought. Simultaneously, the volume of the heat sink is critical and hence a high-density composite is required which would possess the additional benefit of a higher thermal conductivity. Similarly, the use of continuous, three-dimensionally woven fibres would also increase thermal conductivity as well as specific strength and toughness. The improved toughness of a continuous fibre reinforcement network would be less sensitive to the non-uniform distribution of thermomechanical loading. An increase in the static coefficient of friction of the carbon–carbon would eliminate the unwanted high torques developed in aircraft brakes during taxiing, or racing cars when 'cold'. It is difficult to imagine, however, how this could be achieved.

Evidence suggests, then, that the optimum material configuration for brakes would be a 3-D polar weave with a high-density pitch-derived matrix. In reality, however, it is unlikely that such a material would be introduced since it would not be economically viable, costing of the order of £1000 kg^{-1} as opposed to £150 kg^{-1} for the essentially 2-D discontinuous material which, although not ideal, is adequate for the task. It is not inconceivable that such a product form will be introduced into the relatively cost-insensitive Formula 1 application, where the intrinsically more thermodynamically stable material would also alleviate oxidation problems. The brakes presently used in Formula 1 are somewhat inefficient in both materials and operation. High density 3-D material might perhaps allow later braking into corners. Furthermore, the development of a three-disc (two stators, one rotor) system along similar lines to an aircraft brake, would reduce the risk of oxidation to the friction surface, ensure a more uniformly distributed braking pressure and allow a greater friction contact surface area provided the thermal conductivity was sufficiently high to prevent overheating . The brake unit could then be packaged in a smaller configuration with a greatly reduced aerodynamic cross-section and lower inertia which would be manifest in improved acceleration out of a bend. Both discs and pads would benefit from improved oxidation protection to the non-rubbing surfaces, especially within ventilation holes.

9.2 ROCKET MOTORS

A rocket motor is essentially a venturi system, comprising a convergent portion which is embedded in the motor, a throat and an exit cone

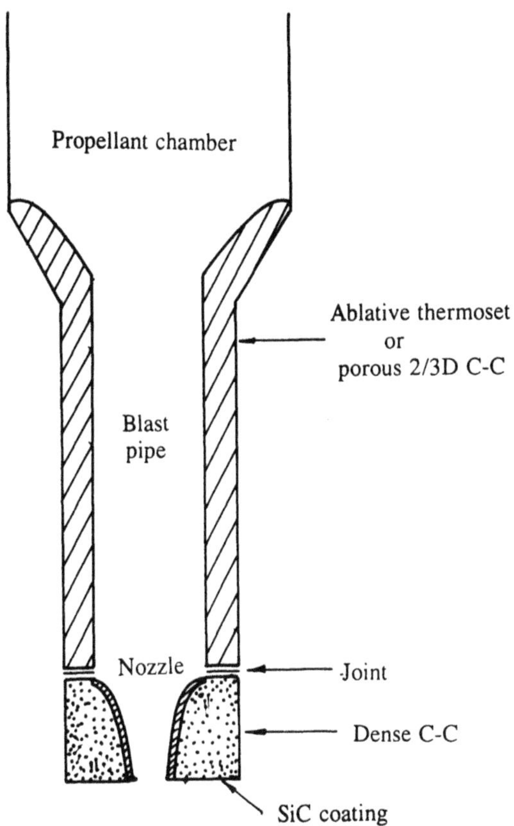

Fig. 9.23 Schematic diagram of a typical rocket motor construction.

(divergent). The exhaust gases from the propellant chamber pass out through the throat and then finally out of the exit nozzle. On average a rocket motor burns for around 30 s. The demands on the construction materials are relatively short-lived, but extremely intense. The size of components are dependent on the rocket's payload; British rockets typically possess a 75–200 mm diameter throat. The USA, of course, produce some much larger structures. A sketch of a rocket motor's construction is shown in Fig. 9.23.

Stress analysis conducted in the throat region has shown the thermomechanical loading regime to be approximately isotropic [6], thus necessitating the choice of 3-D material. A number of other concerns necessitate a preference for pitch-densified material; components are far too thick in cross-section to enable CVD to be a viable option, high-density pitch-based materials offer the best ablation resistance, mechanical stability and resistance to mechanical surface load at motor ignition. Orthogonally woven 3-D carbon–carbon has been successfully employed in rocket throat manufacture of small radius (up to a few tens of millimetres) for a number of years. Larger components require to be made from

Fig. 9.24 3-D carbon–carbon ITE for nuclear missile (courtesy Hercules Inc.).

polar woven material, either machined from cylinders or contoured shapes. Components of this type have been test fired both in the USA and France since 1982. Ongoing research and development have resulted in a successive reduction in the number of parts required such that the whole of the entry and throat can now be made as a single component whereas previously, they were separate or a number of pieces. They are now integral so that gaps, joints or other weak points, which are sites for stress corrosion, are eliminated. Furthermore, there is no longer a difference in thermal expansion. Figure 9.24 shows a three-dimensionally polar woven single piece carbon–carbon throat, known as an ITE, for integral throat exit/entry. The threaded neck section is for attaching the exit cone or nozzle.

The criteria for selecting an exit nozzle material are even more stringent than those for the throat. The nozzle is generally constructed out of a dense carbon–carbon composite which may be coated with a ceramic, generally silicon carbide, to ensure good oxidation and wear resistance. It is important to note that although the burn time of the motor is relatively short, the nozzle must possess excellent dimensional stability, or else the direction of thrust cannot be properly controlled. Temperatures often exceed 2000 °C, gas velocities are supersonic and the exit gases contain uncombusted fuel and water. Such an environment would obviously severely erode unprotected carbon–carbon composites, hence the requirement for a modicum of ceramic protection. Since rocket nozzles are not typically reused they do not necessarily require oxidation protection and may be allowed to burn partially. This burning must, of course, be taken into account at the design stage.

Dense carbon–carbon is preferred because of its superior ablation resistance. A number of companies operate the HIPIC process to manufacture the rocket motor components. These include Morton Thiokol [7], FMI, who also supply the UK's MoD with their carbon–carbon requirements, RTAC, Aerospatiale and Textron. The open literature reports the use of 2-D carbon–carbon exit cones in the USA, but such material may delaminate during manufacture and results in poor interlaminar properties and hoop strength.

The French aircraft company, Aerospatiale, have been involved in carbon–carbon composites since 1968. In order to compete successfully in the world market, the company has an industrial agreement with the carbon specialists Le Carbone-Lorraine under which they manufacture a family of products under the Aerolor banner. Aerospatiale are arguably the world leaders in the weaving of 3-D preforms. Their automated technology has been licensed to Hercules in the USA who produce rocket motor components in a joint venture with Rohr Industries under the name RTAC (Figs 9.25 and 9.26). Both Hercules and Aerospatiale produce 3-D parts up to 1200 mm in diameter for use in advanced missile systems such as Trident II.

9.3 HEATSHIELDS FOR RE-ENTRY VEHICLES

When a rocket blasts into space with velocities exceeding 17 000 mph (27 000 km h^{-1}) the heat generated at the leading edges can lead to temperatures as high as 1400 °C. Re-entry temperatures can be even higher, approaching 1700 °C, and well beyond the operation temperatures of metals [7]. Figure 9.27 shows a temperature profile across the surface of a trans-atmospheric space vehicle (such as the space shuttle) during re-entry. On a weight-for-weight basis, carbon–carbon can endure higher temperatures for longer periods of time than any other ablative material. Thermal shock resistance permits rapid transition from –160 °C in the cold of space to close to 1700 °C during re-entry without fracture. Although the space shuttle is perhaps the most well known example of a carbon–carbon re-entry heat shield, the greatest number of parts produced are used in the nose cones of ballistic missiles. All the British, French and American strategic nuclear missiles employ carbon–carbon heat shields which provide ablation and

Fig. 9.25 Automated 3-D weaving facility producing fibre preforms for carbon–carbon rocket parts (courtesy Hercules Inc.).

heat resistance within a structural member. The primary purpose of the shield is to protect the crew (in the case of a manned vehicle) or instrumentation from the searing heat of re-entry.

The loading on a nose cone during re-entry is such that 3-D carbon–carbon is the most convenient material. The surface temperature rise occurs almost immediately the vehicle enters the earth's atmosphere. The thermal conductivity of the graphitic 3-D carbon–carbon is high enough to eliminate thermomechanical surface overload and avoid surface cracking. Additionally, the specific heat is so high that the component operates as a heat sink, absorbing the heat flux without any problems. The rate of ablation/erosion, driven by the air speed, depends critically on the grain size of the material. HIPIC-densified products are characterized by their fine grain morphology and low levels of porosity and are thus exceptionally ablation resistant. A problem arises, however, in constructing large-scale components such as those used in the leading edges and nose cone of the shuttle. Despite being the superior choice, 3-D HIPIC-fabricated materials, as we have discussed in previous chapters, are limited in the size that can be produced. It is not possible to make large thick pieces by CVD, so the only route available is that using thermosetting resins. The nose cone and leading edge of the space shuttle are formed from 2-D

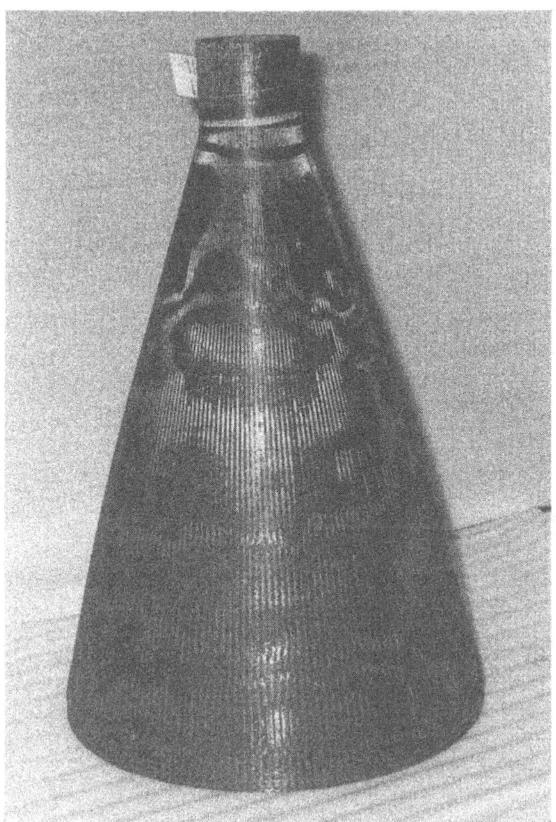

Fig. 9.26 Carbon–carbon rocket exit cone (courtesy Hercules Inc.).

carbon fabric impregnated with phenolic resins carbonized to around 1000 °C. Reimpregnation is carried out using furan or furan/pitch blends, and usually around four cycles are required. Figure 9.28 illustrates some of the re-entry loads on a typical ballistic warhead and a comparison between carbon–carbon and steel [6]. The superiority of the former is very obvious, especially in terms of ablation resistance and low density. Carbon–carbon re-entry vehicles are generally protected against oxidation, especially if, as in the case of the shuttle, the parts are designed to be reusable. Some warheads are, however, left unprotected with a controlled ablation input into the design calculations as this is believed to stabilize the missiles' flight path.

9.4 AERO-ENGINE COMPONENTS

The thermodynamic efficiency of heat engines such as gas turbines is greatly improved with an increase in operating temperature. There is therefore a

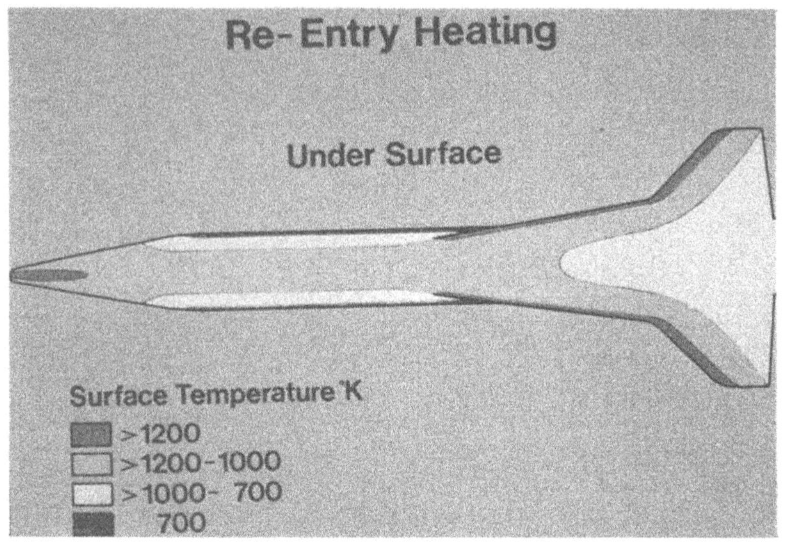

Fig. 9.27 Re-entry temperature profile across a transatmospheric space vehicle.

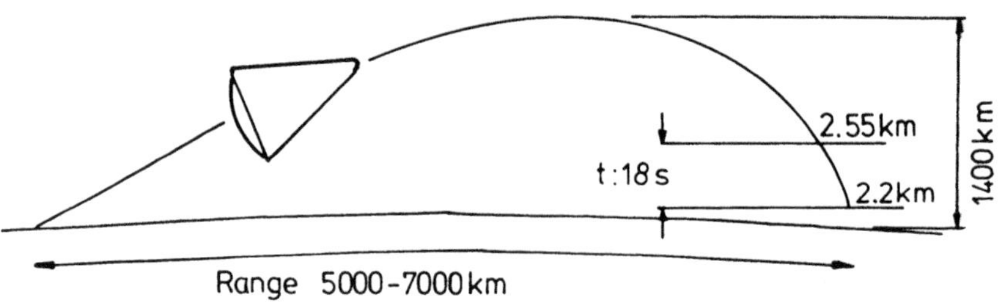

	Carbon/carbon	Steel
Wall temperature (°C)	3200	1500
Max flux (MWm^{-2})	65	80
Average flux (MWm^{-2})	15	35
Ablation ratio	1	7.5
Density(gcm^{-3})	2	7.8

Fig. 9.28 Material behaviour during re-entry.

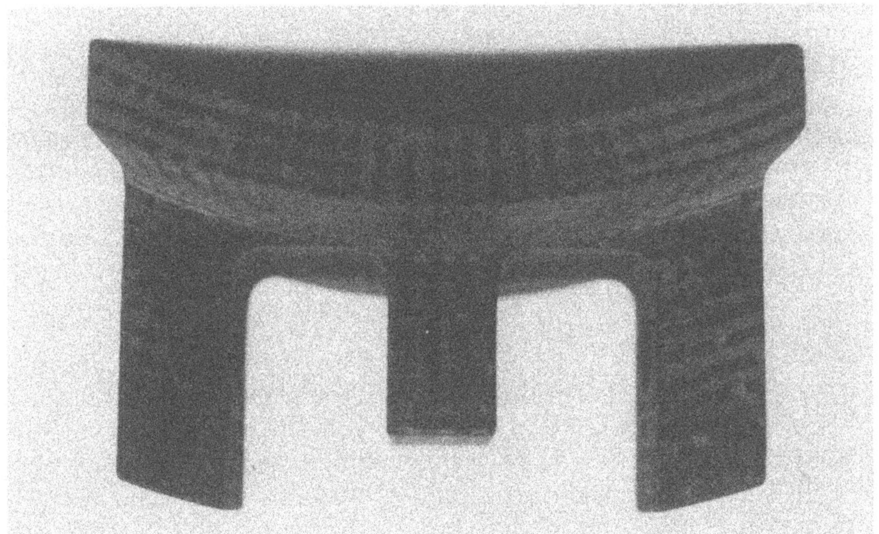

Fig. 9.29 3-D woven carbon–carbon jet engine component (courtesy Hercules Inc.).

driving force to develop carbon–carbon jet engine components. For the same power output the engines could then be made smaller, lighter and more fuel efficient. The toughness, most especially in impact, of carbon–carbon is two orders of magnitude greater than conventional ceramics. Jet engine motors can be made in single-piece format from carbon–carbon, thus saving fabrication and assembly costs. A number of countries are aiming to develop hypersonic aircraft over the next decade, capable of flying at speeds of up to mach 5. Carbon–carbon might well, therefore, be chosen as a structural material in the aircraft's basic airframe. In terms of properties, carbon–carbon is the ideal choice for hot engine components; its exploitation, though, is extremely limited due to the problems of oxidation. Until protective coatings, capable of withstanding extreme thermal cycling in aggressive oxidative atmospheres, are developed it is unlikely that the promise of carbon–carbon will ever be achieved in the theatre of jet engines. A number of materials producers are involved in joint venture research programmes with aero-engine manufacturers and a selection of prototypes have been made. To date, however, none has succeeded in fabricating a suitable flying component. In view of the high toughness requirements, it ought to be more likely that the 3-D configuration be preferred (Fig. 9.29). Having said that, production difficulties and costs may dictate a simple 2-D structure will be the first to be exploited.

9.5 INDUSTRIAL APPLICATIONS

The high price and secrecy surrounding carbon–carbon materials have in the past tended to restrict their use to aerospace and military applications. Commercial products do exist, however, such that there is an ever-increasing industrial exploitation of carbon–carbon, generally in areas where extreme heat would quickly destroy other materials. Research into the industrial applications of carbon–carbon components is especially strong in Germany [8].

9.5.1 Glass making

The machinery used to manufacture glass bottles dispenses a small amount of molten glass known as a 'gob'. The gob rolls down a channel into a mould where it is shaped by air pressure into a bottle. The equipment is designed in such a way that a damaged mould can be identified. A gob interceptor rejects the gob intended for that mould, diverting it so as not to clog up the machinery. Carbon–carbon gob interceptors, although approximately 100 times more expensive than the asbestos they replace, result in fewer replacements and less frequent shutdowns and so are exceptionally cost-effective. The two-dimensional carbon–carbon is inherently less bio-hazardous than the asbestos, and also acts as a heatshield protecting the underlying metallic components.

9.5.2 High-temperature mechanical fasteners

The tensile strength of most ceramics in, and refractory metals and alloys above 750 °C, is extremely poor. Hence mechanical fixing at high temperatures is often a cause of severe problems in engineering design. Screws, nuts and bolts, etc. made from carbon–carbon experience no loss in strength at high temperatures. The load-bearing ability of carbon–carbon is less than that of metals at low to moderate temperatures but a superior at high temperatures. Carbon–carbon nuts and bolts are shown in Fig. 9.30.

9.5.3 Hot press dies

High-quality ceramics and metals may be produced by sintering under mechanical pressure in a hot press operation [9]. The traditional material used in die manufacture is polycrystalline graphite. Such dies are required to be of considerable wall thickness due to the poor mechanical properties of the graphite. Carbon–carbon dies are specifically designed with high hoop strengths so as to reduce considerably the size of component required. A carbon–carbon die of, for example, 130 mm ID need only have a wall thickness of 15 mm. One may thus achieve shorter heating cycles and a more uniform temperature distribution in the hot pressed product [10].

Fig. 9.30 High-temperature carbon–carbon fasteners (courtesy Hercules Inc.).

9.5.4 Hot gas ducts

High-temperature strength and resistance to thermal shock and temperature gradients make carbon–carbon tubes an ideal choice as the liners in the hot gas ducting in high-temperature nuclear reactors [11].

9.5.5 Furnace heating elements and charging stages

Furnaces operating between 1000 and 3000 °C in a non-oxidizing atmosphere tend to use graphite heating elements. The brittleness and moderate strength of graphite make such elements extremely difficult to handle and damage susceptible, especially considering the complex geometries often required. Carbon–carbon elements are far less fragile and are a considerable advantage in hot isostatic pressing equipment as only a small volume is needed, thus maximizing the working zone. Aside from its mechanical property advantages, the higher electrical resistance of carbon–carbon permits higher power operation and less voluminous electrical connections.

Charging stages allow a more effective use of furnace heating volume. Such stages have historically been produced from refractory alloys, ceramics or graphite depending upon operating temperature. The use of lightweight impact-resistant carbon–carbon serves to save volume and increase the lifetime of these stages. The lower thermal mass of the carbon–carbon component increases process efficiency and greatly reduces heating and cooling cycle times.

9.5.6 Moulds for forming superplastic metals

Superplastic forming of metals is a relatively new and extremely versatile route to the fabrication of complex shaped components especially in the aerospace industry. Titanium alloys in particular require temperatures in the region of 1000 °C and pressures of several atmospheres. Carbon-carbon moulds offer significant improvements over their 'opposition', mild steel. Being two orders of magnitude lighter their reduced thermal mass affords shorter processing cycles and much easier handling. Further, the negligible thermal expansion of carbon–carbon results in greater tolerances and eliminates folding of the titanium insert.

9.6 BIOMEDICAL DEVICES

Elemental carbon is known to have the best biocompatibility of all known materials [12]. It is compatible with bones, blood and soft tissue. The high cost of carbon–carbon compared with its competitors, be they metallic or polymeric, in the biomedical field may at first glance preclude its use. Consider, however, a rigid metal plate as frequently used in the repair of fractures. Such a device when firmly attached to a long bone such as the femur, will inevitably result in an altered stress distribution. The modulus of the bone is generally an order of magnitude lower than, say, a stainless steel plate of roughly the same cross-sectional area [13]. A plate made from this alloy will transmit around 90% of the load causing the underlying bone to become porotic. It is thus paramount that such plates be removed as soon as the fracture has healed. Failure to do so would most likely result in a spontaneous refracture at some later date.

Thus, although metals possess the required strength and toughness for use during the healing phase, they are far from satisfactory. With operating theatre charges running at around £10 000 per day, surgeons and hospital administrators would much prefer to avoid the second operation. Patients too would certainly wish to forgo the further trauma associated with the removal of the prosthesis. Carbon–carbon can be engineered in such a way as to possess mechanical properties identical to those of bone (Fig. 9.31), thus eliminating the need for removal once the healing is complete. There is a considerable interest in materials such as carbon–carbon, which are considered to be 'bio-active', in many areas of implant surgery. The objective is to use a material which is able to encourage the formation of bone rather than soft tissue at the interface. Carbon–carbon has been shown experimentally to be biodegradable in such applications – the body gradually replacing it with bone [7].

For obvious reasons, licenses to allow the use of materials take a long time to obtain. As a result the biomedical uses of carbon–carbon are very

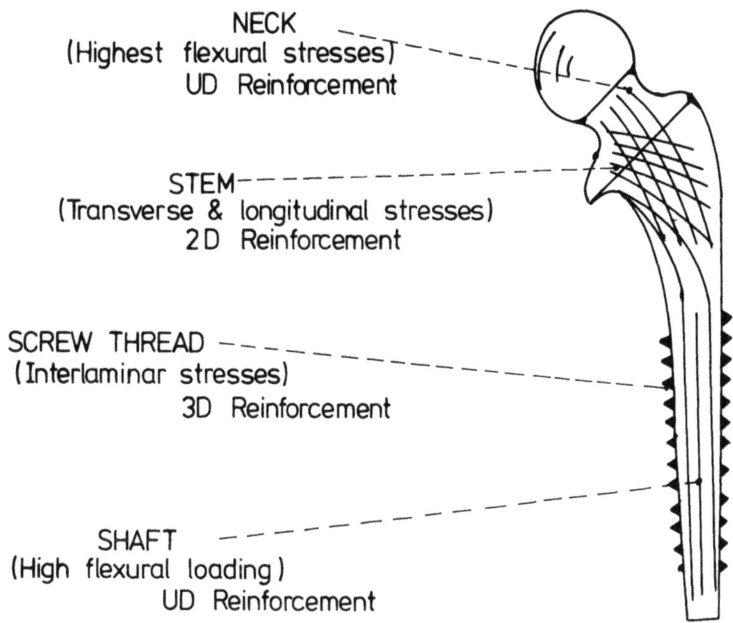

Fig. 9.31 Schematic diagram of carbon–carbon hip joint replacement designed to imitate the femur structure.

limited. There are reports in the literature, however, of its use in artificial hearts for animals [12], fixations for carbon fibre artificial ligaments [14] and bone plates in osteo-synthesis and as endoprosthesis [15].

9.7 SUMMARY

The most widespread use of carbon–carbon materials is in brakes. The rapid deceleration required for an aircraft on landing generates a considerable amount of frictional heat. Carbon–carbon composite brakes retain strength at high temperatures. Unlike steel brakes, carbon–carbon brakes maintain a more consistent performance over the life of the part, with no increase in stopping distance. The CTE of carbon–carbon composites remains fairly constant over a broad temperature range. Carbon–carbon has a high heat capacity so it can act as a lightweight heat sink. The material will slide against itself without galling, resisting wear and outlasting steel brakes 2 : 1. The lower wear rate means fewer overhauls and lower maintenance costs. Carbon–carbon brakes can endure thousands of thermal cycles with little or no fatigue. The brakes weigh far less than their metal equivalents, saving up to 800 kg on a Boeing 747. They are less prone to warping than metal and in racing cars carbon–carbon can reduce stopping distance by up to 30% compared to metal.

In a rocket throat and nozzle, hot gases rush through at high velocity, stressing and eroding the walls. A carbon–carbon composite will resist high-temperature erosion and abrasion without burning away. Rocket nozzles are typically not reused, so that they do not always require oxidation-protective coatings. They are thus allowed to burn partially. This burning must, however, be taken into account in the design of the part. In the past, the rocket motor entrance and throat were made up of a number of pieces. They are now integral such that a single-piece component is now the norm. This eases fabrication and eliminates the gap, joint or weak point, which is a site for stress corrosion. Further, there is no longer a difference in thermal expansion. Similar to a rocket nozzle, the nose cone and leading edge of the space shuttle and the heatshields of ballistic missiles must endure the searing heat of re-entry, up to 1700 °C. They must also endure solar radiation and the attack of atomic oxygen.

The USA and France have spent large sums on the development of carbon–carbon for military uses. In Germany the market focuses on the industrial applications of the material, albeit at a lesser stage of development. Carbon–carbon is widely used in glass bottle manufacture as push-outs, wear guides and transfer pads, in contact with the molten glass. It is not melted by molten glass and does not wet or require external cooling liquids. Its long life reduces maintenance and replacement costs. Other applications include hot press dies and heating elements.

The nuclear industry is interested in carbon–carbon for fusion reactors since it resists ultra high temperatures and thermal shock. If used in automotive pistons, carbon–carbon's low coefficient of friction would reduce drag and its light weight would increase engine efficiency. Higher combustion temperatures permit greater efficiency and performance while reducing engine size, weight and fuel consumption. In long-term engine use, most especially gas turbines, a ceramic coating for carbon–carbon is essential. The coating itself does not bear structural loads but it must adhere to the substrate to transfer loads.

Finally, because of its extremely good biocompatibility, controlled porosity and attractive Young's modulus which can be tailored to be close to that of bone, it is of interest as a replacement for metals used in implant surgery.

REFERENCES

1. Fisher, R. and Stimson, I. L. (1980) *Phil. Trans. R. Soc. Lond.*, **A294**, 583.
2. Awasthi, S. and Wood, J. L. (1988) *Ceram. Eng. Sci. Proc.*, **9**(7–8), 553.
3. Shigley, J. E. (1987) *Mechanical Engineering Design*, McGraw-Hill, New York.
4. *SEPCARB Brakes Assistance Manual*, Carbonne Industry France.
5. Fitzer, E., Fritz, W., Gkogkidis, A. and Mörgenthaler, K. D. (1986) *Proc. Carbon 86 Conf.*, Baden-Baden.

6. Grenie, Y. (1987) in *Looking Ahead for Materials and Processes* (eds J. de Bossu, G. Briens and P. Lissac, Elsevier, Amsterdam, p. 377.
7. Klein, A. J. (1986) *Adv. Mat. Process. Inc. Met. Prog.*, **130**(5), 64.
8. Von Gellhorn, E. (1986) in *HI-Tech – The way into the nineties* (eds, K. Brunsch, H. D. Gölden and C. M. Hekert), Elsevier, Munich, p. 67.
9. Richerson, D. W. (1982) *Modern Ceramic Engineering*, Marcel Dekker, New York.
10. Hütner, W. (1985) *Andernach*, **29**, Oct.
11. Von Gellhorn, E. and Grüber, H. (1986) *Proc. Carbon 86 Conf.*, Baden-Baden.
12. Lamicqe, P. J. (1984) *SEP Int. Carbon Conf.*, Bordeaux, France.
13. Williams, D. F. (1981) *Metals and Materials*, **7**(1), 24.
14. Burri, C. and Neugedauer, R. (1985) *Replacement of Ligaments by Carbon Fibres*, Springer-Verlag, Munich.
15. Claes, L. Fitzer, E. Hüttner, W. and Kinzl, K. (1980) *Carbon*, **18**, 383.

Technology Summary and Market Review

10

10.1 SUMMARY

Carbon–carbon composites consist of carbon fibres in a carbon matrix. The fibres may be chopped, continuous or woven and may be produced from rayon, polyacrylonitrile (PAN) or pitch (mesophase or isotopic). The carbon matrix may be deposited by chemical vapour deposition (CVD), by the carbonization of a thermosetting or thermoplastic organic material or by a combination of these. The result is a family of composites whose microstructures and properties may be multi-dimensionally tailored to a great degree for a range of applications. Unlike metals and ceramics, carbon–carbon composites retain their strength at very high temperatures. High thermal conductivity and low thermal expansion give carbon–carbon materials an excellent resistance to thermal shock. A high heat of sublimation and low CTE for carbon and graphite result in good ablation resistance. Other advantages include chemical resistance, excellent high-temperature wear characteristics, biocompatibility, shape stability and pseudo-plastic fracture behaviour.

High specific strength at high temperatures, excellent fracture toughness and thermal shock resistance have resulted in carbon–carbon composites finding use as aircraft and racing car brakes, heatshields for re-entry vehicles and rocket nozzles. All of these applications require the retention of properties to high temperatures. The excellent thermal shock resistance of carbon–carbon is, perhaps, best illustrated by rocket nozzles which can withstand a change in temperature from ambient to almost 3000 °C in the first few seconds of a missile launch. When carbon–carbon materials do break, they generally undergo a non-brittle, pseudo-plastic, type of fracture rather than catastrophic failure.

The deployment of carbon–carbons, despite their attractive properties, is severely limited, however, by two serious drawbacks. The materials are extremely expensive to produce as a result of the very slow and inefficient processes currently used in the production. Furthermore, carbon–carbon composites oxidize at temperatures as low as 400 °C unless protected in some way. Oxidation is not such a great problem in limited-life components such as nose cones and rocket nozzles of military ordnance. The brakes of aircraft and racing cars are prone to oxidation in service but this can be minimized by design, and refurbishment after predetermined increments of time. When used for furnace insulation and heating elements the atmosphere is non-oxidizing, although they are attacked by high-temperature nitrogen. Oxidation is a serious problem in potential applications such as jet engine parts and reusable transatmospheric vehicles, requiring thermal cycling in oxygen-containing atmospheres. A great deal of research has been carried out and is ongoing into the protection of carbon–carbon using various coatings and inhibitors. As yet, however, no materials system has been sufficiently successful to allow in-service deployment in jet engines. In addition to the major problems of oxidation and cost, carbon–carbons exhibit low strain to failure, poor matrix properties, poor erosion resistance, lack of resistance to attack by moisture, and severe difficulties with joining. The properties of all carbon–carbon composites are influenced by the type, volume fraction and orientation of the fibres, the matrix microstructure, the method and conditions of processing, oxidation-protective coatings and inhibitors and any other additives.

10.2 THE STATE OF THE ART

10.2.1 Carbon fibres

As is generally the case with advanced composites, the mechanical properties of carbon–carbon composites are dominated by fibre properties and orientation. Most carbon–carbon composites are reinforced with continuous carbon fibres, so that their in-plane strength and stiffness are determined by the fibre strength and modulus, weave or lay-up geometry and fibre volume fraction.

The first carbon–carbons used rayon-based fibres, although their use is now limited to between 5 and 10% of the market as they have been largely superseded by the cheaper and superior PAN-based fibres. Rayon-based firbres are lighter ($\rho = 1.45$ g cm^{-3}) than PAN fibres and have higher CTEs than either PAN or pitch fibres. The disadvantages of rayon fibres are higher processing costs than PAN, lower strength and modulus, and severe shrinkage during heat treatment. Mesophase pitch-based fibres, which are more expansive and often possess higher moduli than PAN fibres, are widely used in aircraft brakes and non-aerospace applications. They do, however, exhibit lower strengths than PAN fibres.

Chopped fibres are used extensively in brakes and industrial applications. The composites thus formed offer considerable improvements in strength and toughness over monolithic carbons for relatively little added cost. The properties of chopped-fibre composites are not enhanced by fibre loadings in excess of 25% by volume. A number of brake manufacturers claim short fibres produce more uniform frictional properties which improve performance, although no data have been presented to substantiate this.

The purity of the fibres is of prime concern when the composites are used to make missile components. In such applications the fibres require to be greater than 99% carbon and contain less than 20 ppm of sodium and other alkali metals. Alkali metal flames result in an easily detectable signature from a rocket that may be readily discerned by an enemy's sensors. Furthermore, alkali metals catalyse the oxidation of carbon–carbon, thus decreasing the life.

The fracture toughness of a carbon–carbon composite is due, in part, to the deflection of cracks running through the matrix at the fibre/matrix interface as a result of the imperfect bonding therein. The sizing compounds, weaving lubricants and any other materials added to the surface of the fibres for ease of handling are generally removed prior to fabrication in order to avoid compromising the optimum bonding level. Crenulated or ribbed carbon fibres, such as those made from rayon, aid mechanical bonding or 'keying' of the fibre to the matrix, and often result in improved out-of-plane properties such as interlaminar shear strength.

The maximum use of the mechanical properties of the fibres is achieved by keeping them straight. Although some unidirectional (UD) fibres are used in carbon–carbon composites, the overwhelming majority of techniques employ woven fabric reinforcement. Fabrics only provide a two-dimensional plane of reinforcement, and for this reason the composites are referred to as 2-D composites. Should strength be required in three dimensions, fibres may be woven together in a 3-D configuration. It is customary to describe a three-dimensional weave in terms of the number of different **directions** in which the fibres are woven. The accepted nomenclature is n-D where n represents the number of directions. Clearly, as n increases so will the complexity of the weave, the cost of production and the difficulties in densification. The addition of multiple directions of reinforcement will of course proportionally reduce the strength in the primary x–y planes of the composite.

Lower modulus fibres are much easier to weave into fabrics or preforms. High modulus fibres have a tendency to break in the loom as a consequence of their low strain to failure. The cost of weaving tends to vary with the fibre properties, complexity of weave, the amount of material to be processed, fabric width and length of production run. Weaving is a relatively expensive process and may increase the price of the fibres by up to a factor of 2 or 3. Three-dimensional weaving is very costly and can raise the cost of fibres by up to an order of magnitude. Ongoing research into automation, computer control and better weaving processes ought to reduce weaving costs. One would expect most of the reductions to arise in the three-dimensional sector since a great deal of work has already been done in two-dimensional weaving for the polymer composites industry.

10.2.2 The carbon matrix

The purpose of the carbon matrix is to hold the fibres in place and transfer the applied load to the fibres. The matrix controls the out-of-plane properties and, to a degree, the off-axis properties of the composite. Those properties depend upon the matrix precursor and the processing conditions. Possible precursors include pitches, thermosetting resins and hydrocarbon gases. Fractionated, modified pitches have carbon yields of up to 90% and are easily graphitized. Carbons thermosetting from resins are difficult to graphitize, and develop local graphitic structures. Crystalline graphite has very high strength in one plane, while strength in the other directions is low. Isotropic carbons, of course, have more isotropic properties. They generally exhibit higher strengths and lower strain to failure than graphitic matrices. Higher-density graphitic carbons have a higher resistance to ablation and oxidation than isotropic matrices.

Thermoset resins do not soften at high temperature, as does pitch. They can thus be used in low-pressure impregnation processes, especially in the form of prepregs which are easy to handle and process. Pitches are relatively easy to process in high-pressure carbonization processes, but the purity and reproducibility of these essentially 'natural' materials are extremely problematic. A great deal of research has been carried out into the use of high carbon yield polymeric precursors such as polyimides, polyphenylene and acetylene-terminated resins. Such resins have, as yet, only been used on a developmental basis. These resins, while much more expensive, allow the production of carbon–carbon with fewer processing cycles, thus reducing both the cost of fabrication and the chances of damaging a component during manufacture.

The strength of a carbon–carbon material increases with the reduction of the void content during processing. It may not be desirable to fully densify the composite, however, as it has been observed that uniform small-scale porosity can be effective as crack arrestors and in the attenuation of mechanical impulses. Pitch precursors are often purified and fractionated prior to use in order to separate high carbon yield fractions and improve product quality control. The blending of pitches with thermoset resins is often used to control grain size and degree of transformation from carbon to graphite.

10.2.3 Manufacturing processes

Chemical vapour deposition

Carbon fibre felts, 3-D preforms and porous 2-D carbon–carbon structures may be densified by chemical vapour deposition (CVD). In the CVD technique, the artefact is placed in a furnace where its pores are filled with carbon. The carbon is produced from the thermal decomposition (cracking) of a hydrocarbon gas, usually methane. Hydrogen is often added as a diluent in order to facilitate processing. As a general rule of thumb the carbon deposited at around 1100 °C has an isotropic structure and is referred to as pyrolytic carbon. Between 1000 and 1700 °C the carbon deposited has an 'intermediate' microstructure which becomes more graphitic with increasing temperature. Carbon deposited between 1700 and 2300 °C is graphitic in nature and known as pyrolytic graphite.

CVD carbon–carbon can have superior properties to composites made by any of the other routes as the matrix is built up layer by layer, thus ensuring good fibre/matrix bonding and minimizing the number of defects in the structure. CVD carbon–carbon often has a high degree of porosity, however, leading to low strength. A further advantage of the CVD process is that it can also be used to deposit non-carbon materials such as oxidation-protective coatings and inhibitors. Subtle changes in processing conditions enable the deposition of carbonaceous material of differing mechanical properties and oxidation resistance. However, CVD is an extremely slow and, therefore, expensive process requiring considerable operator skill. Processing conditions (temperature, system pressure, diluent partial pressure and gas flow rates, etc.) and even the placing of the component within the furnace all affect the manner in which the carbon is deposited. If the aforementioned conditions are not correctly achieved and controlled, the surface of the structure will overcrust, producing a weak, low-density area in the centre. In commercially operated processes it is generally necessary to compromise between an ideal system and one in which the rate of carbon deposition occurs at an economically viable rate. Consequently, parts usually require to be periodically removed from the furnace and the surface machined to open up the inner pores; the whole process commonly takes several months for completion.

Low-pressure carbonization of thermosetting resins

The low-pressure, thermoset, process commences with the lamination of a carbon fibre reinforced polymer structure using the techniques developed for the fibre-plastic composites industry. The resins are cured, usually in an autoclave, followed by a post-curing cycle to finish chemical cross-linking and expel any remaining volatiles. The composite is subsequently carbonized in an inert atmosphere at around 850–1000 °C, in a process that may take up to a week. Following pyrolysis, the part may be further heat-treated depending upon its final operating temperature or required density. The most commonly used impregnating resins, phenolics and furans, have carbon yields of only 50–60% by weight. A number of higher-yield resins have been researched, but, to date none, have been commercialized because of problems such as excessive cost and processing difficulties. The part must therefore be reimpregnated, pyrolysed and heat-treated often three, four or more times before its final density and properties are achieved.

Reimpregnation/densification may be attained using more resin, a pitch/resin blend, CVD or a combination of these. The thermoset resin process is considerably cheaper than the CVD technique but the composites formed are sometimes of poorer quality. Relatively little initial capital investment in equipment is required and the techniques perfected in the polymer composites industry may be successfully transferred and modified to allow the production of large and complex structures. The technique is, however, labour intensive, especially in the laying-up of laminates, and again requires a good degree of operator skill and experience. Chopped fibres, impregnated with a thermoset, may be compacted in a die and heated to cure the resin. The part may then be removed from the die, then carbonized, graphitized and reimpregnated as required. Carbon–carbon made in this way is used in brake manufacture and in relatively inexpensive industrial applications such as dies for hot pressing.

High-pressure impregnation/carbonization using pitches and thermoplastic resins

In the hot isostatic pressure impregnation carbonization (HIPIC) technique, a 3-D woven fibre preform is impregnated with a thermoplastic resin or pitch and loaded into a processing can with excess matrix precursor. The can is evacuated and sealed and placed in a specially designed hot isostatic press (HIP) unit. The workpiece is heated so that the precursor (usually a pitch) melts and flows. Pressure is applied, by an inert gas via the sealed can, which forces the pitch into the pores of the preform. The artefact is then carbonized by increasing the temperature and pressure. The carbonized billet may, if required, be subsequently graphitized. Impregnation, carbonization and graphitization are repeated until the desired properties are achieved. Generally fewer cycles are required to process to the equivalent density than with the low-pressure technique owing to the higher carbon yields and the higher density of pitch carbon/graphite. Typically three to five cycles are used for the production of high-density materials ($\rho > 1.95$ g cm^{-3}).

The problems incurred in the HIPIC process include fibre distortion and breakage during processing and the inability to impregnate large structures. The difficulties arise from ineffective filling of the pores in the initial preform, or partially densified substrate, with the molten pitch. Design limitations require carbonization and graphitization to be performed in separate furnaces. The size of articles which can be manufactured is also restricted to less than 2 m in diameter as the result of furnace technology limitations. HIPIC is, of course, extremely capital intensive because the high-temperature and high-pressure equipment required is very expensive, coupled with the need for materials handling equipment and safety requirements.

High-pressure carbonization without the need for metal cans may be carried out using 'exotic' thermoplastic resin precursors. These materials produce carbon–carbon composites with a unique range of microstructures but are prohibitively expensive. It is difficult to imagine, therefore, the widespread deployment of such materials except in very high added value products such as biomedical devices, space structures and fasteners. Chopped-fibre versions of such composites may be of value in brake systems as they represent a useful outlet for scrap recovery from their polymeric 'ancestors'.

10.2.4 Oxidation protection

The properties of carbon–carbon composites make them attractive for use in a great number of applications. Their usage to date has been restricted by high costs, poor matrix properties and high temperature oxidation. At the time of writing, only about 10–15% of carbon–carbon composites are protected from oxidation, but this is an area where a great number of market opportunities will arise should the technological problems be solved.

The rate at which carbon is oxidized depends upon temperature, oxygen partial pressure and the microstructure of the carbon. Oxidation of carbon starts at around 400 °C. Graphitic carbon with its denser, crystalline structure and lower proportion of reactive edge sites, does not begin to oxidize until slightly higher temperatures. Generally, a graphitic matrix microstructure will add between 50 and 100 °C to the onset of oxidation. Higher temperatures and partial pressures of oxygen result in faster oxidation rates.

Carbon–carbon composites are protected against oxidation by the use of refractory ceramic coatings and inhibitors which are generally glass-forming materials.

Oxidation-protective coatings

The coating process aims to protect carbon–carbon from oxidation by ensuring that it does not come into contact with high-temperature oxygen. The most common oxidation-resistant coating materials are silicon ceramics which are effective below 1700 °C. The coatings may be applied by a number of techniques such as pack cementation, CVD and thermal spraying. One of the most notable, coating process used to date is that developed by

Table 10.1 Thermal expansion coefficients for selected protective coating materials

Coating	*Linear CTE* (10^{-6} K^{-1})
SiO_2	0.55
SiC	4.3
Si_3N_4	3
Al_2O_3	7.5
$3Al_2O_32SiO_2$ (mullite)	5
TiC	9
TiB_2	8.5
ZrO_2 (partially stabilized)	10
Carbon–carbon	0.32–2.66

LTV missiles to protect the leading edges and nose cone of the space shuttle. The pack cementation process is executed by immersing the composite in a bed of silicon carbide, alumina and silicon powders. The bed is heated in an inert argon atmosphere to 1760 °C in order to allow a diffusion reaction between the silicon vapours and the carbon–carbon which converts the composite's outer layers to SiC. In the LTV technique, the coated component is subsequently impregnated with tetraethyl orthosilicate (TEOS) which is hydrolysed and heat-treated to produce a silica (SiO_2) residue within the coating. Any remaining surface inhomogeneities such as cracks and pores are sealed using a sodium silicate–silicon carbide mixture to produce a final coating, roughly 0.5 mm thick. A number of organizations such as GA Technologies have used CVD techniques to apply SiC and silicon nitride (Si_3N_4) coatings. These materials, along with boron nitride, are applied in layers of which a minimum thickness of 0.2–0.6 mm is required to afford adequate protection.

Silicon ceramic coatings provide protection by reacting with oxygen at high temperatures to form silica. The liquid silica flows into cracks in the coating and seals them. Unfortunately, below 1000 °C the carbon–carbon substrate is very prone to oxidation because the silica is too viscous to flow into the cracks and does not wet the carbon very well. Furthermore, thermal cycling and impurities may result in devitrification of the silica. Devitrification is a phase transformation which increases the rate of oxidation and causes the coating to lose strength and crack. The major problem with the materials used to coat carbon–carbon is that their CTEs are generally much higher than that of the substrate (Table 10.1). Under extreme thermomechanical loading it is obvious that two materials of very different CTE will eventually 'part company', resulting in spallation of the coating. Certain carbon–carbon composites will be easier to coat than others since the materials may have significantly different CTEs depending on the type of fibre and matrix. Unfortunately it is the high strength and high modulus composites, which are the most attractive for structural purposes, that are the most difficult to coat. Finally, care must be taken to ensure that processing and operating temperatures are not high enough to allow a carbothermic reduction reaction between coating and substrate to take place which would reduce the mechanical properties of the composite.

Inhibitors

Inhibitors are glass-forming materials used to protect carbon–carbon in the intermediate temperature range (450–1000 °C). Boron, silica and oxygen form borate glasses that will flow into cracks at these relatively low temperatures. Unlike silicates, borate glasses wet and adhere well to carbon fibres. Inhibitors such as boron, zirconium boride or silicon carbide powders may be added to thermoset resins prior to prepregging, or may be placed in a porous substrate by impregnation with a pre-ceramic polymer during densification or by chemical vapour infiltration (CVI). All are oxygen 'getters', forming molten borate, silicate or borosilicate glasses at high temperatures. The glasses flood into cracks and pores to prevent oxygen from penetrating the matrix. Above 1000 °C, however, oxygen will diffuse through irrespective of the method of coating. It should be noted that boron compounds can result in a decrease in composite strength by reacting with the carbon fibres to produce boron carbide at high temperatures (≈2000 °C). Additionally, boron compounds are often very brittle and may adversely affect the fracture toughness of the carbon–carbon especially when present in large amounts. A number of materials, such as lithium, zirconium and phosphorous oxides, may also be added which have the effect of lowering the viscosity of silica in the intermediate temperature range.

In addition to inhibitors, a number of other compounds have been added to carbon–carbon either to improve performance or aid fabrication. Carbon black and graphite powders are sometimes used to increase the carbon yield and reduce the pyrolysis shrinkage of thermoset resins. Milled carbon fibres have also been added to resins in order to improve the interlaminar shear strength. The milled fibres, unfortunately, tend to orientate themselves parallel to the fabric plane thus affording little, if any, improvement. Finally, some mention is given in the patent literature to the addition of ceramic powders in an attempt to improve ablation resistance.

10.3 THE CARBON–CARBON MARKET

Despite having the highest specific strength of all known materials at high temperatures, carbon–carbon composites are used in only three major applications: aircraft and racing car brakes, rocket nozzles and heatshields for re-entry vehicles. A number of minor applications also exist such as biomedical devices, space equipment and a few specialized industrial uses. Aside from their high-temperature strength, carbon–carbon composites possess a number of other desirable properties but are dogged by two glaring weaknesses: poor oxidation resistance and expensive fabrication costs. Oxidation problems severely limit the exploitation of the relatively 'cost-insensitive' jet engine market, while costs of what is essentially an inefficient labour-intensive batch process prevent deployment in a number of non-military applications. A further problem arises in that the ambient temperature properties of carbon–carbon are very mediocre. While they are stiff, strength is relatively low. They are brittle, although they fail pseudo-plastically, and are very porous with a tendency to spallation. Operators tend to circumvent such problems by periodically replacing components as is the case with brakes, or by designing for short life spans (rocket components).

Carbon–carbon is a relatively small business area accounting for around £100m. sales in 1990. The only published market survey predicts an annual growth of roughly 11% towards £200m. at the turn of the century [1]. It should be remembered, of course, that this survey was written in 1985, i.e. prior to recent sweeping political changes in the Eastern Bloc. The great majority of carbon–carbon purchases are for military use and military technology development, many of which may be curtailed or shelved in the present political climate. One would therefore expect the demand for carbon–carbon to be somewhat less than predicted. There will, however, be a significantly increased requirement for strategic materials such as carbon–carbon from emerging nations which strive to develop sophisticated weapon systems. Exploitation of such markets by Western producers would, of course, be illegal and therefore extremely hazardous! Most carbon–carbon technology is of a strategic nature, especially in the USA. Furthermore, it is not inconceivable that those nations would choose to set up their own facilities since, as Chapter 7 shows, it is not too difficult to do so provided one is prepared to bear the initial capital investment and high fabrication costs.

At the time of writing (1992), 75% of the world demand for carbon–carbon is in the USA as a result of the large aerospace and defence industry. France, again because of its aerospace and defence interests, is the leading European producer, whereas Germany leads in the limited industrial market.

The largest producer of carbon–carbon in both volume and money terms is the British company BP, by virtue of its acquisition of the Hitco corporation. Hitco manufacture both brakes and rocket parts. Second in volume, and third in sales, is the Bendix Division of the Allied-Signal Corporation who specialize in brake production. The French company

Table 10.2 Breakdown of carbon–carbon market by application

	% of total	
Application	*Volume*	*Value*
Aircraft brakes	63	31
Rocket nozzles	14	31
Nose cones and ablatives	11	37
Other	12	1

SEP splits its business between rocket parts and brakes, and ranks third in the world in volume of material produced and second in sales. The company has recently entered into a technology-licensing agreement with the chemical giant Du Pont. Other significant 'players' include Avco, Dunlop, FMI, Goodrich, Kaiser Aerotech, Schunk and Sigri. A fuller listing of the various companies involved in the carbon–carbon industry is given in Section 10.6.

The major application for carbon–carbon composites are **aircraft brakes**, accounting for between 60 and 70% of the total volume (Table 10.2). Carbon brakes are several times the price of the sintered metal brakes they replace in commercial aircraft. The cost of the discs is, however, only a fraction of the cost of a completed undercarriage system. Their advantages far outweigh the increased cost. Weight savings of up to 30% may be achieved, which accounts for up to 900 kg on a Boeing 747–400 aircraft, thus making large savings in fuel costs. Carbon–carbon brakes also possess extended service lives compared with metal brakes. Almost all military aircraft now use carbon brakes. The market for commercial airliners is, of course, much larger. All of the airbus family of aircraft (A-300, 310 and 320) fly carbon brakes as do Boeing 747–400, 757, 767, McDonnell Douglas MD-11 and, of course, Concorde. Most new commercial aircraft programmes are now specifying carbon brakes.

The military brake market is now fairly mature. Aside from the introduction of new aircraft, the bulk of the market will be for refurbishment of worn brakes. Brakes are smaller on military planes as they are much smaller aircraft. Therefore less energy has to be dissipated. Carbon–carbon brakes in commercial aircraft, on the other hand, is an emerging and expanding market expected to mature by the turn of the century. Carbon–carbon composites are also used in the brakes and clutches of Formula 1 racing cars. This is not, however, considered a prime market since the majority of teams aim to receive their materials free of charge or at least at a considerable discount as a focus of advertising and PR. It is unlikely that carbon brakes will be used in passenger cars in the foreseeable future because of costs. A more likely opportunity would arise for large trucks and, especially, high-speed trains, provided technological advancements can reduce the cost of the material to around £100 kg^{-1}. Environmental safety, improved performance and lower maintenance requirements would be the driving force of these applications.

The first applications of carbon–carbon composites were in high-temperature **rocket components**. Oxidation protection is not a critical issue in what are generally short-lived parts due to the fixed period of rocket burn. Life spans are sufficiently brief to enable 'state of the art' coatings to overcome oxidation problems, or else a 'controlled ablation' may be designed into the structure. Composites for rocket components are the most expensive, requiring a very high (graphitized) density of 1.9–2.0 g cm^{-3} to provide ablation resistance. Materials costs in excess of £1000 kg^{-1} are not uncommon.

Each of the USA's space shuttle vehicles employs roughly 750 kg of silicon carbide protected carbon–carbon in its leading edges and nose cone. Although the USA only plans to replace the *Challenger*, the EC, the former Soviet Union and Japan all have plans for similar vehicles. Doubtless, all of these vehicles would use carbon–carbon leading edges, as would any of the proposed transatmospheric airliner projects should they come to fruition.

Carbon–carbon composites could be used very effectively in high-performance jet engines to improve efficiency by reducing weight and increasing operating temperatures. The jet engine market could easily mature into a £100m. per annum business by the turn of the century provided the oxidation problem can be solved. The technological problems of providing a truly reusable high-temperature carbon–carbon system are further amplified by funding which, in the main, is provided by the military and therefore subject to the whims and discretion of politicians. Although actively researched, the only carbon–carbon to date in this application are vectoring nozzles for the afterburner of jet fighter aircraft engines produced by Pratt and Whitney, which proved unsuccessful due to material inconsistency.

Carbon–carbon has the potential for widespread use in biomedical prosthetics devices. Composites may be fabricated to possess similar stiffness to bone and can be demonstrated to be not only biocompatible but **bioactive** being porous enough for bone and tissue to grow into and anchor. The German company Schunk has carried out a great deal of work into the fabrication of carbon–carbon hip joints. Carbon–carbon prosthetics are limited by their brittleness, difficulties in effective stress analysis and cost. Further, gaining approval and qualification from the various national medical authorities is a very slow process. World sales of prosthetics account for sales of some £150m. but carbon–carbon will face stiff opposition from thermoplastic resin matrix composites which, although less biocompatible, are cheaper and easier to work with.

Table 10.3 Breakdown of production costs for a mid-price-range (£300 kg^{-1}) carbon–carbon composite

	Percentage of cost
Carbon fibres	17
Weaving	17
Matrix precursor	2
Fabrication	40
Protective coating	12
Pretax profits	12

Carbon–carbon composites are far too expensive to find use in passenger cars in the foreseeable future. They may find a limited use in racing car engines, however, as for example, pistons, provided oxidation and erosion can be overcome. Reciprocating carbon–carbon engine parts would improve efficiency by allowing higher operating temperatures coupled with dimensional stability and self-lubrication. Finally, a small market could arise for carbon–carbon structures in the space industry – lightly loaded, stiffness-critical components such as satellite and space telescope trusses. Carbon–carbon would also be a useful armour material against laser weapons as a result of its high thermal conductivity.

10.4 COMMERCIALIZING A PRODUCT

At the present time, the number of applications of carbon–carbon composites is extremely limited. To meet the requirements for current and new uses, the materials must be competitively priced, possess attractive property and performance characteristics, be able to be fabricated efficiently to high standards of quality and provide a library of well-understood design data.

The price of carbon–carbon varies considerably, depending on the end use and method of production, etc. Aircraft braking materials are the cheapest at around £120 kg^{-1} whereas high-density three-dimensional composites used in rocket nozzles may cost in excess of £1000 kg^{-1}. Fabrication costs account for a significant proportion of the price of carbon–carbon. A lowering of fabrication costs by the introduction of more efficient technology would therefore significantly reduce the price of finished articles. Table 10.3 shows a rough breakdown of a medium-priced composite costing, say, between £300 and £400 kg^{-1}. The apportioning of costs will vary considerably as a result of the wide variety of carbon–carbon parts which may be fabricated. The data do, however, serve to illustrate how finished component prices may be reduced by improvements in the various contributing factors.

Carbon–carbon raw material costs vary according to the type and geometries of fibres and matrix precursor. Carbon fibres range in price from around £10 kg^{-1} for chopped low-strength fibres to over £300 kg^{-1} for continuous high modulus (HM) types. Liquid matrix precursors cost from around £1 to £1.50 kg^{-1} for phenolic resins, and up to in excess of £50 kg^{-1} for exotic high-yield polymers such as PEEK and polyphenylene. Reductions in raw materials costs only offer a relatively small potential drop in the price of the finished composite. There currently exists an overcapacity in carbon fibre production so one would expect prices to drop significantly in the foreseeable future as the suppliers fight for market share. Even when they do, only 15–20% decreases can be expected when equated to carbon–carbon production. The most widely used liquid matrix precursors are phenolic resins which are very cheap and would not be thought to change much, if at all, in price. It is possible that the cost of the more exotic precursors such as acetylene-terminated resins will be reduced significantly over the next decade. Matrix precursors make very little direct contribution to the overall cost, but indirectly, in terms of how they influence fabrication, their effects are very noticeable. Resins possessing improved processability and/or carbon yield might well be introduced, despite what might be higher initial raw materials costs.

A reduction in weaving costs would correspondingly reduce composite prices. The cost of weaving varies according to weave geometry, number of dimensions and type of fibre. Two-dimensional fabrics add, on average, £5–£50 kg^{-1} to the cost of the fibres, whereas 3-D weaves may increase costs by up to an order of magnitude with angle interlocks falling somewhere between the two. It is doubtful whether the cost of fabrics will fall significantly since the processes are fairly advanced from the polymer composites industry. Three-dimensional weaving, on the other hand, ought to fall by a large amount as the technology develops.

Aside from brakes, all current applications may be described as **price-insensitive** in that they are in the main military applications where high performance can demand high premiums. In biomedical usage, materials costs can often be lost in surgical costs. Jet engines are another area in which high costs can be justified by improved performance and lower density. Prices ought to remain high in these theatres, although end-user protests and competition between manufacturers may lead to a reduction.

Carbon–carbon brake manufacturers must compete with lower-performance sintered metal and phenolic resin based brakes as well as each other. As a direct result, their costs have reduced steadily over the years. Prices vary according to quality and performance. Composites for fighter aircraft braking systems may cost up to £300 kg^{-1} while prices as low as £100 kg^{-1} have been quoted for inferior materials. Motor car brakes sell for less than £5 kg^{-1}. Even the cheapest carbon–carbon systems could not compete with this price unless their usage becomes mandatory due to governmental edicts on safety and environmental protection. Reduction of composite costs to around £60 kg^{-1} is, however, an attainable target, making carbon–carbon economically viable for truck and train brakes. Industrial grade composites retail at around £150 kg^{-1}. Reduction in costs would lead to corresponding increases in applications such as furnace elements, high-temperature moulds, dies and machinery.

The most important properties of carbon–carbon composites are high-temperature strength coupled with low density. Design strengths and operating temperatures vary with application, many of which are extremely sensitive both militarily and commercially. Materials requirements in terms of performance are steadily increasing with respect to both thermal and mechanical properties. Developments in carbon fibres, fabrication technology and oxidation should allow carbon–carbon to meet those demands. In missile applications and, to a lesser extent, in brakes, the ablation resistance of the material is important. Such components are generally made from high-density composites with graphitic matrices. The high density and heat of sublimation of graphite allow large amounts of energy to be absorbed when a missile re-enters the earth's atmosphere or an aircraft is brought to a halt. More important for brakes are wear and frictional properties.

Carbon–carbon brakes have a coefficient of friction of around 0.5 under low energy stop conditions. The brakes wear uniformly and are not only lighter than their metal competitors they can endure longer lifetimes before they need to be replaced. Lifetime is generally not of prime concern for missile components since they will only be used once. By contrast, jet engines require part-lives in excess of 1000 h at temperatures between 1000 and 1400 °C in an oxidizing atmosphere. Any company able to solve the oxidation problem on temperature cycling will clearly have a large potential business in this area. Furthermore, the engines required for the proposed 'hypersonic' aircraft will demand even higher temperatures, perhaps in excess of 2000 °C.

Fabrication expertise, most especially the achievement of strict quality control, is a crucial requirement for the commercialization of a carbon–carbon product range. A manufacturer must clearly understand, qualitatively at least, the complex relationships between raw materials, process variables and materials properties. A more quantitative understanding will be preferable so as to allow for efficient optimization of his operations. Carbon–carbon parts may take anything up to 9 months to produce. The high-temperature equipment required will necessitate a large initial capital investment in both hardware and highly skilled personnel. It is pertinent to consider the carbon–carbon business as a highly specialized, 'bespoke' fabrication operation, where customer contacts are therefore an important key to success. The materials supplier must understand in detail the end user's requirements and applications so that the various weaving, densification and protection processes may be tailored best to meet those requirements.

The carbon–carbon supplier must have good relations with the national government, since the majority of parts are used in military applications. Government contracts are often very lucrative and can be used to finance R&D costs and provide capital for expenditure on plant. Funding from government allows a manufacturer to develop an expertise which may, at a later date, be transferred into civilian markets. Military funding does, however, have one or two pitfalls associated with it. Difficulty may often be incurred in exporting carbon–carbon artefacts as a result of the strategic nature of the material. The administration in the USA, through its 'United States Law Preventing Export of Military Sensitive Products (ITAR) regulations', requires that all carbon–carbon researchers and manufacturers be registered with the Defence Logistics Service Centre (DLSC). The export of carbon–carbon is forbidden unless an export licence is obtained from the aforementioned organization. Indeed, no information, with respect to carbon–carbon composites, may be passed to a foreign national without government approval. The regulations have thus proved very difficult for the European companies who have entered the market via a US acquisition.

Finally, it is important that carbon–carbon producers research carefully the properties of their products and in tandem with the designers who will ultimately employ them. Design data for carbon–carbon are very limited so many designers feel very uncomfortable in specifying such unpredictable anisotropic materials. There is no typical carbon–carbon composite, rather a whole family of materials. The fabricator and designer must therefore work closely together in order to design and tailor a specific product to a specific application.

10.5 ORGANIZATION OF THE CARBON–CARBON BUSINESS

The carbon–carbon business may be divided into seven different sectors:

1. carbon fibre suppliers;
2. matrix precursor suppliers;
3. weavers;
4. prepreggers;
5. manufacturers;
6. coaters;
7. end users.

The ways in which the various sectors interact are illustrated in the flow diagram in Fig. 10.1. Carbon fibre suppliers sell products to weavers or prepreggers. Matrix precursor suppliers sell to prepreggers or directly to manufacturers. Prepreggers also supply manufacturers. Weavers may also

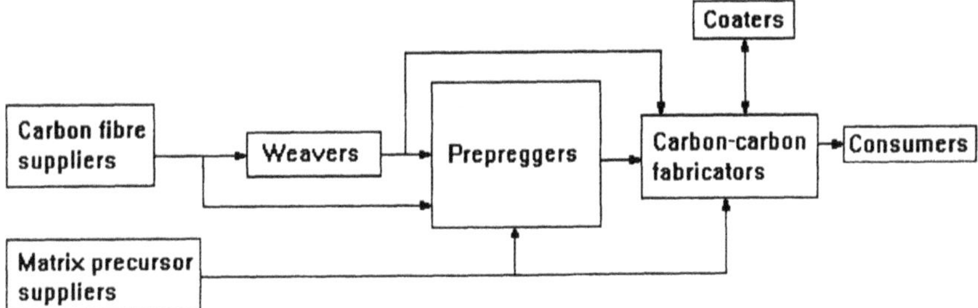

Fig. 10.1 Integration of companies involved in the carbon–carbon industry.

supply direct to the fabricators, especially 3-D preforms. Manufacturers generally contract protective coatings to outside specialist companies, although a few do coat composites in-house. Fabricators, of course, sell the completed composites to the end user. Companies active in the business often compete within a number of the market sectors. ICI (Fiberite), for example, are both a weaver and prepregger, while FMI is a fibre supplier, weaver and manufacturer. Furthermore, a number of end users are believed to be contemplating producing their own materials, unhappy at the poor quality of the raw materials with which they are supplied.

Firms generally enter the carbon–carbon market by acquiring an expertise in one of the seven business sectors. They may choose to specialize in that area, or by various processes of backward and forward integration, enter one or more of the other sectors. There are essentially three routes of entry into the business: internal development, technology licensing and acquisition or joint venture. The majority of carbon–carbon producers all have products that were originally developed internally. There is a high mobility of personnel within the business as firms 'poach' experts from their competitors in order to take a 'short cut' in development. Great care must be taken that a company has more than one expert otherwise he may be enticed elsewhere taking with him his knowledge! Generally 25% or more of the investment into a product will be required to go into R&D. Government contracts are usually used as funding and products will take at least 5 years to commercialize.

Latecomers into the market, or those with a limited technology base, may license technology from outside. Table 10.4 shows a number of firms known to be licensing aspects of their expertise to others. The US government will license its patents free of charge to firms who sell their products back to the government.

A quick entry into the business may be attained via acquisition although it is not without its problems. The acquiring company is generally required to pay a substantial premium, perhaps 20 or more times the annual earnings of the company it wishes to buy. The opportunities to acquire

Table 10.4 Licensing activity in the carbon–carbon business

Licenser	*Technology*	*Licensee(s)*
Vought	Oxidation protection	BP (Hitco)
Aerospatiale	Three-dimensional weaving	Hercules
GA Technologies	Oxidation protection	Chromalloy B.F. Goodrich Kaiser Aerotech Rohr Industries

carbon–carbon concerns are extremely limited and subject to government veto should it be deemed to 'endanger national security'. Most of the organizations involved in the business are small components of larger companies and thus difficult to capture. The purchase of smaller, privately owned, firms such as FMI would generally be easier. Notable acquisitions in recent years include Fiberite by ICI during its take-over of Beatrice Chemicals, BP's purchase of Hitco from Owens-Corning for $240m. and Amoco's acquisition of the Union Carbide carbon fibre operations. A number of firms have entered joint ventures to pool their respective resources especially in the theatre of jet engine components, most notable being the teaming of Pratt and Whitney and Hitco, and RTAC with General Electric.

10.6 MAJOR COMPANIES IN THE CARBON–CARBON MARKET

ABS (Aircraft Braking Systems) grew from Goodyear Aerospace and is one of the largest carbon–carbon brake suppliers after Hitco and Bendix. They are based in Akron, Ohio and have material on some A320 and Fokker 100 programmes.

Aerolor from France is a joint venture formed between Aerospatiale and Le Carbone-Lorraine in 1975. The company's products are densified by a variety of methods and their technology has been licensed to a number of other companies such as Hercules. Aerospatiale are particularly expert in 3-D weaving technology.

Amoco, the American oil company, has acquired the carbon fibre business formally owned by Union Carbide. They produce fibres from rayon, PAN and pitch precursors and are the dominant fibre suppliers to the carbon–carbon industry in the USA. Sold under the Thornel brand name, Amoco fibres range in price from £20 to £500 kg^{-1}.

Ashland Carbon Fibres based in Ashland, Kentucky produce pitch-based carbon fibres and fractionated pitches for use as matrix precursors. Ashland fibres are sold under the trade name of Carboflex in the form of mat, chopped and milled grades. Aerocarb pitch, costing around £6–10 kg^{-1}, has a theoretical yield of close on 90% by weight. Ashland are engaged on a programme of research aiming to improve the quality and reproducibility of their pitches, and methods of blending them with thermosetting resins.

AVCO's speciality materials division from Lowell, Massachusetts produces PAN-based carbon fibres, called Avcarb, which are woven into satin fabrics. Carbon–carbon composites are densified using the thermoset resin and CVD processes. The company has filed a number of patents around the addition of ceramic powder to improve ablation resistance.

BASF from Germany acquired the Celion line of carbon fibres from Celeanese in the USA in 1985. These fibres are used both in polymer and carbon matrix composites. BASF also own the prepregger Narmco, situated in Anneheim in California. Narmco produce phenolic and bismaelamide precursor prepregs in addition to epoxy prepregs and other advanced materials.

The Bendix Aircraft and Strut Division is based in South Bend, Indiana and forms part of the Allied-Signal Corporation. The company produces carbon–carbon brakes from phenolic resin impregnated, chopped pitch based fibres.

BF Goodrich (Super Temp Division) produce carbon–carbon composites for use in brakes, by CVD at their plant in Alto Pueblo in Colorado. The brakes are used on the space shuttle and Boeing 747–400 aircraft. In an attempt to diversify beyond brakes they have licensed oxidation protection know-how from GA Technologies.

Borden chemicals from Columbus, Ohio produce the phenolic resin SC-1008. Originally developed by Monsanto, SC-1008 is the most commonly used thermoset resin matrix precursor, selling for between £1 and £1.50 kg^{-1}. The resin is supplied in the form of a solution containing 60% solids.

Carbon–carbon Advanced Technologies (CCAT) is owned by Alco Standard and located in Fort Worth, Texas. The company was founded by former employees of Vought Missiles and produces 2-D composites from phenolic prepreg precursors. CCAT is active in the jet engine market, producing oxidation-protected composites by impregnating partially densified substrates with pre-ceramic polymers which decompose to silicon carbide on pyrolysis.

Chromalloy Research hail from Orangeburg in New York State and are one of the largest coaters of carbon–carbon components using technology licensed from GA Technologies.

Dunlop in the UK produce carbon–carbon brakes by the isothermal CVD method which are used on Concorde and by Boeing.

Fiberite have production and research facilities in Winona, Minnesota and Tempe, Arizona and are owned by the UK's ICI. They supply a variety of phenolic matrix precursor prepregs along with a chopped fibre brake moulding compound. In addition, Fiberite weave their own fabrics at their plant in Greenville, Texas.

Fibre Materials Inc. (FMI) are based in Biddeford, Maine, and are primarily a manufacturer of carbon–carbon substrates for use in rocket nozzles and nose cones. They also produce carbon fibres on a limited scale and have developed proprietary 3-D weaving technology. FMI were the first people commercially to exploit the HIPIC process.

The Garrett Turbine Engine Company is a division of Allied-Signal and based in Los Angeles. Garrett are evaluating carbon–carbon jet engine components supplied by LTV's Vought Missiles Division.

General atomic (GA) Technologies are located in San Diego, California and are a subsidiary of the Chevron Oil Corporation. GA undertake protective CVD coating of carbon–carbon and licenses its technology to a number of other organizations. BF Goodrich, Kaiser Aerotech and Rohr industries have all licensed complete oxidation protection systems, while Chromalloy have acquired only coating technology. The system is based on boron inhibitors, a boron-containing 'paint' and CVD-deposited silicon carbide.

General Electric from Philadelphia is yet another engine manufacturer engaged on carbon–carbon research. GE have teamed with Rohr Industries and SAIC in order to further their research.

Hercules have an outstanding reputation as a supplier of advanced composites and carbon fibres. The company are based in Salt Lake City, Utah and produce a number of carbon fibres and fabrics under their Magnamite trade name. Most notable for carbon–carbon are their range of intermediate modulus (IM 6, 7 and 8) and high modulus (HMU and UHMS) fibres made from PAN precursors. IM-6 in particular has a very high tensile strength and good dimensional stability at extreme temperatures. HMU fibres have been found to be especially suited for use in carbon–carbon. The propulsion systems division of Hercules assemble rocket motors. Historically they have bought in carbon–carbon components made elsewhere, but recently in-house production facilities have been constructed. Their plant contains HIPIC and CVD equipment along with 3-D weaving apparatus using a process licensed from Aerospatiale. In 1989 Hercules and Rohr Industries formed a joint venture company, Refractory Technology Aerospace Components (RTAC). RTAC develops and manufactures carbon–carbon rocket motor, turbine engine, missile and aerospace components.

Hitco from Gardena in California are arguably the world's largest producer of carbon–carbon. Purchased from Owens Corning by BP, Hitco concentrate mainly on brake materials which they sell almost exclusively to Aircraft Braking Systems. The majority of Hitco's production is 2-D, employing all the densification techniques except HIPIC, although this is being extensively researched by the parent company (BP). Hitco are involved with Pratt and Whitney in developing oxidation-resistant jet engine parts.

Kaiser Aerotech from San Leandro in California produce a 2-D carbon–carbon composite called K-Karb which is used in industrial applications such as glass making.

Kobe Steel Company of Japan manufacture very high quality, high-density carbon–carbon by the HIPIC route.

Mitsubishi Kasei Corporation, based in Tokyo, produce a range of ultra-high modulus coal-tar pitch-based fibres. These fibres possess very high thermal conductivity and are used in aircraft, motor racing, passenger train and main battle tank (MBT) brakes. Mitsubishi carbon–carbon is also used in fusion reactors, space vehicles and an number of other applications. They possess expertise in all aspects of production (CVD, thermoset resin and HIPIC).

Rhône-Poulenc from France market Kerimid 601, a polyimide precursor resin with a carbon yield of 80% which is cited in a number of patents.

Rohr Industries are based in Chula Vista, California. They are teamed with general Electric for the manufacture of jet engine components. They have worked closely with SAIC in the past and are presently involved with Hercules in their joint venture company RTAC.

Schunk from Germany produce brakes, industrial and biomedical components at their plant in Giessen.

Science Applications Incorporated (SAIC) are based in San Diego in the USA, producing rocket parts and inhibited composites. SAIC are not a production company but, rather, a development house concerned with R&D for other companies and government agencies. SAIC generally employ low-pressure resin techniques and the HIPIC process.

Sigri hail from Munich and supply industrial products such as heat exchanger tubes, metal-forming moulds and furnace elements etc.

Société Européene de Propulsion (SEP) are located in France and are one of the world's leaders in carbon–carbon technology. SEP's carbon–carbon business has evolved from its work on the rocket nozzles and warheads for France's independent nuclear deterrent. The materials are marketed under the trade name SEPCARB for use in aircraft and racing car brakes and missile parts. The company is also reported to be actively researching bioimplants although, as yet, boasts no commercial products. CVD, HIPIC and low-pressure resin technologies are all employed in their various densification processes. A range of carbon–ceramic and ceramic–ceramic composites are produced, the most notable of which are C–SiC and SiC–SiC (SEPCARBINOX) which are densified by CVD.

Textron, based in Massachusetts, is a leading supplier of 3-D carbon–carbon materials for rocket nozzle integral throat entrances used in tactical, strategic and space propulsion systems.

Tonen from Tokyo in Japan, produce a range of ultra-high modulus pitch-based fibres which are used in brake manufacture.

United Technologies Corporation (UTC), in the guise of their Pratt and Whitney Division, are actively researching jet engine components. A well-documented programme involved fabrication of carbon–carbon 'afterburner' flaps for their F-100 jet engine. The parts were rejected not on cost grounds, but rather due to poor quality and reproducibility. UTC's research laboratories have demonstrated a degree of success in silicon ceramic coatings of oxidation protection using pack cementation and CVD techniques.

Vought Missiles are a division of the LTV Corporation from Dallas, Texas. They developed the 2-D carbon–carbon composite used in the leading edges and nose-caps of the space shuttle. Vought no longer produce carbon–carbon on a large scale but are engaged in extensive research programmes funded by the US government and the Garrett Turbine Company.

10.7 CONCLUSION

Carbon–carbon composites are a unique family of composite materials, retaining their mechanical properties to extremely high temperatures. This, along with a high degree of toughness and inertness, makes them ideal candidates for application in a large theatre of advanced engineering. There are, however, two major drawbacks limiting the widespread development of carbon–carbon; firstly, the fabrication processes used are extremely inefficient and secondly, they tend to oxidize quickly in air at temperatures as low as 400 °C. The cost of carbon–carbon varies from around £100 kg^{-1} to well in excess of £1000 kg^{-1} with an average of the order of £300–400 kg^{-1}. As a result, markets are limited mainly to military applications where high premiums are paid for improved performance. Surveys of the carbon–carbon business prospects have generally predicted an increasing usage of the materials well into the next century. With recent changes in the political climate, however, it remains to be seen how much of this military spending will come to pass. A number of companies are already 'feeling the pinch', finding that defence contracts are not the 'blank cheque' they once believed them to be!

It is possible to be extremely simplistic in describing the future market prospects of carbon–carbon, splitting the business into two distinct sectors. In the 'commercial' sector, usage, such as brakes, heat exchangers and furnace elements, is constrained simply by cost. There are a whole host of applications ideally suited to the properties of carbon–carbon provided the price is lowered as a result of more efficient fabrication. In the 'military' sector it is performance rather than price which is of concern. Efforts to

Table 10.5 Predicted growth for the carbon–carbon market (sales in £m.)

Business area	*1985*	*1990*	*1995*	*2000*
Brakes				
Aircraft	20	45	60	75
Land transport	—	—	5	10
Rocket components				
Exit cones	20	25	25	30
Nose cones/ heatshields	20	25	25	25
Jet engine parts	—			
Space	—	0.5	1	2
Biomedical	—	—	0.5	1
Total	60	95	117.5	143

develop oxidation-resistant carbon–carbon parts for heat engine components and the heatshields of hypersonic aircraft are intensive in the drive towards a truly reusable system. The long-term reality, however, is that carbon–carbon composites will remain highly specialized performance materials at least until the turn of the century. It is very doubtful that there will be sufficient reduction in price or advances in technology in the short to medium term to bring carbon–carbon into widespread use. Table 10.5 shows projected estimates for carbon–carbon sales up to the year 2000.

REFERENCES

1. '*Advanced Materials Technologies Report 8, Carbon–Carbon Composites*, C. H. Kline & Co. (1987).

Index